Television Engineers' Pocket Book

Television Engineers' Pocket Book

Edited by

P. J. McGOLDRICK, C.Eng., M.I.E.E., M.S.M.P.T.E.

*Senior Lecturer, Dept. of Electrical and Electronic Engineering,
Plymouth Polytechnic*

Specialist Contributors

J. P. HAWKER
GORDON J. KING
J. A. REDDIHOUGH
J. I. SIM
G. R. WILDING
'TELEGENIC'

LONDON
NEWNES–BUTTERWORTHS

THE BUTTERWORTH GROUP

ENGLAND   Butterworth & Co (Publishers) Ltd
London: 88 Kingsway, WC2B 6AB

AUSTRALIA   Butterworths Pty Ltd
Sydney: 586 Pacific Highway, NSW 2067
Melbourne: 343 Little Collins Street, 3000
Brisbane: 240 Queen Street, 4000

CANADA   Butterworth & Co (Canada) Ltd
Scarborough: 2265 Midland Avenue, Ontario M1P 4S1

NEW ZEALAND   Butterworths of New Zealand Ltd
Wellington: 26–28 Waring Taylor Street, 1

SOUTH AFRICA   Butterworth & Co (South Africa) (Pty) Ltd
Durban: 152–154 Gale Street

First published in 1954 by George Newnes Ltd

Sixth edition published in 1973 by Newnes-Butterworths,
an imprint of the Butterworth Group

Second impression, 1975

© Butterworth & Co (Publishers) Ltd, 1973

ISBN 0 408 00102 X

Filmset by Ramsay Typesetting Ltd,
London and Crawley

Printed and bound in England by
Hazell Watson & Viney Ltd, Aylesbury, Bucks.

PREFACE TO SIXTH EDITION

Since the last edition of this popular pocket manual was published in 1968, colour television has really sprung into life in the U.K. With the major television networks radiating the majority of their material in colour, the demands on the knowledge and ability of all those concerned with television have dramatically increased.

Even so, the installation and servicing of monochrome receivers will also be of concern for many years, and the U.K. faces a period of at least fifteen years where the v.h.f. 405-line service, and therefore the servicing of 405-line receivers, will continue. This cannot, yet, be a reference manual of colour receivers, and I hope that I have presented an edition which will be of use to all, whatever their receiver, whatever their problem.

Apart from up-dating and revising all sections of the book, a considerable amount of absolutely new material has been introduced on the subjects of circuits, colour, integrated circuits and servicing. This latter material, it is felt, is particularly important because an attempt is made to swing away from the fundamentally crude system of servicing by intuition that has existed in television for too long. With colour servicing there is no place for it whatever.

Television-receiver technology has changed rapidly in the last few years, and I hope that the relevant material of those changes has been introduced without upsetting the readability of any part of the book.

Grateful acknowledgement is made to J. P. Hawker and J. A. Reddihough, who edited the first five editions of the book between 1954 and 1968, and established it as an invaluable pocket reference book. Other commitments have prevented them from further involvement, and I am grateful for the opportunity to take over the responsibility of editing the sixth edition.

P. J. McG.

CONTENTS

1

STANDARDS AND WAVEFORMS

The basic principle of transmitting pictures over wire or radio circuits by breaking down a series of images into a large number of tiny points, or picture elements; transmitting a signal corresponding to the light value of each element; and then later reassembling an identical series of light points in the same sequence at the receiving end is now so widely understood as to require little detailed explanation. For practical television purposes, it is, however, important to appreciate that, although the number of elements into which each picture is broken down provides the final measure of the definition of the picture that can be achieved on any given transmission system, a number of other factors are of importance in determining the fidelity of the reproduced picture. These factors include the bandwidth of the transmission channel, the amount of flicker that can be tolerated, the spot size of the cathode-ray-tube beam at the receiver, contrast, and the linearity of the deflection systems.

It should be noted that, although it is usual to refer to horizontal scanning, in fact the lines are not precisely horizontal, but incline downwards as they proceed from left to right, owing to the influence of the field deflection, which produces the second dimension of the picture.

Picture-repetition frequency

The rate at which the process of scanning must be repeated is determined by the following factors: the process must appear continuous to the eye, and must be repeated sufficiently often to give the impression of continuous movement of objects to the viewer, and to avoid noticeable flicker. This depends for its success on the 'persistence of vision' of the human eye, that is to say the fact that the retina retains an impression of an image for an appreciable fraction of a second after the object itself has disappeared. A series of still images presented at a rate of about 14 per second will provide an illusion of continuous movement, but would be accompanied by considerable flicker. If the rate is increased to 25 per second, the flicker will be considerably reduced, but will still be noticeable, particularly where the picture is bright: a repetition rate of 50 per second will eliminate flicker for all practical purposes.

If, however, a television system were to adopt a repetition rate of 50 complete pictures a second, the video frequencies involved in the transmission of a high-definition system would rise to an extremely high figure, necessitating an excessive transmission bandwidth. This difficulty has led to the adoption of what is termed 'interlaced' scanning. In this system, instead of transmitting each line of the picture in sequence, alternate lines of the picture are first scanned, i.e., the lines 1, 2, 3, 4, of *Figure 1.1(b)*. The remaining *even* lines (A, B, C, D, etc.) are

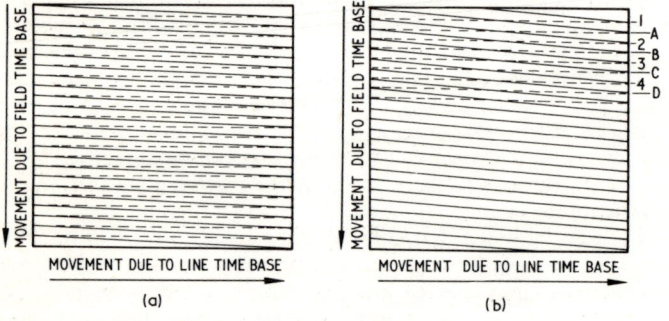

Figure 1.1   Scanning. (a) Sequential; (b) interlaced

then scanned. By this means, although only 25 *complete* pictures are scanned each second in the British system, the screen is evenly illuminated 50 times per second. Since, with high definition, it is difficult for the eye to perceive the scanning of the individual lines, the effect is to raise the repetition rate to 50 per second while keeping the amount of detail and video frequencies to those for sequential scanning at 25 fields a second. With the British 405-line system, these are divided into two fields of $202\frac{1}{2}$ lines each, the half-line playing an important role in automatically controlling the interlace of the receiver scanning system. It is important to note that where, for any reason, the control of the receiver interlace is faulty, the scan will tend to trace out two almost identical paths for both fields, producing a picture in which the lines are clearly visible and reducing picture quality.

Note that in some older texts and television receiver circuits the field may be referred to as the *frame*. This term is deprecated because it causes confusion with *picture* (*i.e.* two fields).

Picture transmission

Before discussing the make-up of the British television waveform, it is necessary to consider briefly why the transmission of pictures should be more complex than the transmission of sound.

The transmission from a sound broadcasting station must be capable of reproducing in the receiver the audio frequencies (pitch and harmonic content) of the programme at each moment and also the intensity (amplitude) of the sound; this is done basically by modulating a steady radio-frequency carrier with electrical frequencies corresponding to the audio frequencies, and thus in effect producing a slight variation of transmitter frequency in the form of sidebands, while at the same time varying the amplitude (voltage output) of the transmitter to correspond with the intensity of the sound.

In picture transmission it is necessary to indicate the relative brightness of each picture element in turn by means of scanning, and when this differs, a high video frequency will automatically occur. Thus, as in sound broadcasting, a radio-frequency carrier is modulated by varying the amplitude of the transmitted wave, in this case to correspond with the intensity of illumination over the range black to white; and this in turn produces sidebands, or frequency variations, whenever a change in illumination takes place. Thus, so far, vision transmission is not basically different from sound transmission, except in the width of the sidebands, which will be much greater for vision than for sound. However, for successful picture transmission, it is also necessary to provide two additional items of information that are not required in sound transmission; these are the line- and field-synchronisation (sync.) signals, which are required to ensure that the raster on the television-receiver screen is traced out exactly in step with the scanning of the transmitted image in the camera. It is also desirable that the cathode-ray-tube trace be suppressed during the flyback periods. We thus have in effect a number of different types of information that must be radiated by means of a single radio carrier in such a way that each item can readily be separated and made to perform its particular function at the receiving end.

Number of lines

The British 405-line system dates from the commencement of regular
public television broadcasting by the B.B.C. in 1936. At that time the
standard television cathode-ray-tube screen size was about 9 in, for
which 405 lines provided good definition. With increase in the size of
picture tube screens to the present standards between 19 and 25 in,
the lines have become more noticeable, so that with the commencement
of u.h.f. television broadcasting by the B.B.C. in 1964 it was decided to
increase the number of lines to 625, which had meanwhile become the
European standard (the U.S.A. and Japanese standard is 525 lines).
Whichever standard is adopted is something of a compromise decision
since although increasing the number of lines improves picture quality
it also needs a greater bandwidth for transmission, and the frequencies
available for television transmission are limited.

British 405-line television system

The methods adopted in the 405-line television service to convey these
items of information may be summarised as follows:

Picture brightness or light value of each picture element is transmitted by amplitude modulation of the carrier, adopting a figure of
30 per cent of peak output to correspond with 'black' and the full peak
output to correspond with 'peak white'. This is termed 'positive'
modulation, to distinguish it from the alternative system in which peak
output corresponds to the sync. level (see *Figure 1.2*).

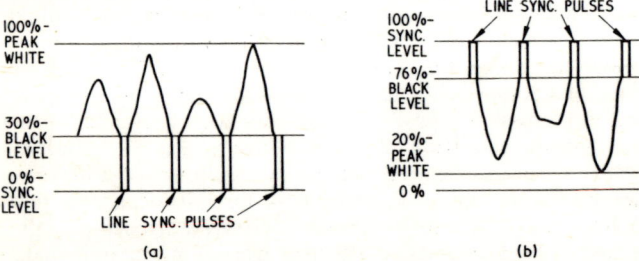

*Figure 1.2   Positive modulation (a) with negative-going sync. pulses is used
in the British 405-line system. Negative modulation (b) with positive-going sync.
pulses is used on all 525- and 625-line systems*

The video frequencies produced by the variation of picture brightness levels form sidebands varying from 0 MHz when transmitting an even tone picture (i.e., all black, all white, all grey, etc.) up to approximately 3 MHz for a fine network of black and white lines. These video frequencies are separated from the carrier frequency in the receiver by a detector or demodulator as in sound broadcasting.

Synchronisation is effected by using the 'blacker than black' portion of the carrier output, the majority of the output from the transmitter being suppressed (3 per cent of peak output) for greater or lesser periods to provide field and line pulses. In the receiver these pulses are separated from the video signals by means of an amplitude limiter (usually termed the sync. separator), and are then further separated into field- and line-triggering pulses by means of circuits capable of distinguishing between pulses of different durations.

A section of this basic waveform, showing one complete line period (98·8 μs), is shown in *Figure 1.3* from which it will be seen that this period is made up of 80·3 μs of picture information; a 'front porch' pre-line sync. signal suppression period of 1·7 μs, to allow time for the

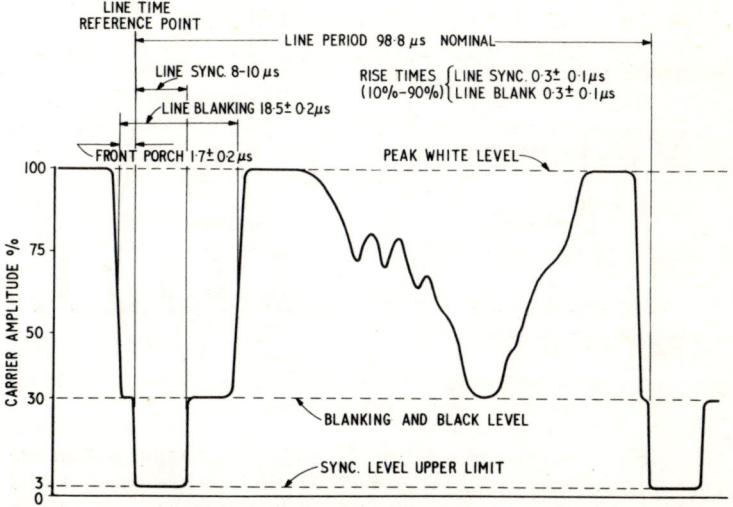

Figure 1.3   Detailed transmission waveform showing the line sync. period, British 405-line system

carrier to drop to the black level in cases where the line ends on a point of high picture amplitude (i.e., white or near white); the 9·0 μs line sync. pulse period, during which the carrier amplitude drops to 3 per cent (as percentage of peak vision carrier); and a 'back porch' or post-line sync. signal suppression period of 7·8 μs. Thus during each 'line' there is a period of 18·5 μs when picture information is not transmitted, and this provides the flyback time for the cathode-ray-tube trace. It will be noted that the line pulse is not shown with vertical edges, and that a slight margin of time is quoted for each of the figures so far given. This is because a slight amount of time is required for the transmitter to change from the black to the sync. level. Nevertheless, the

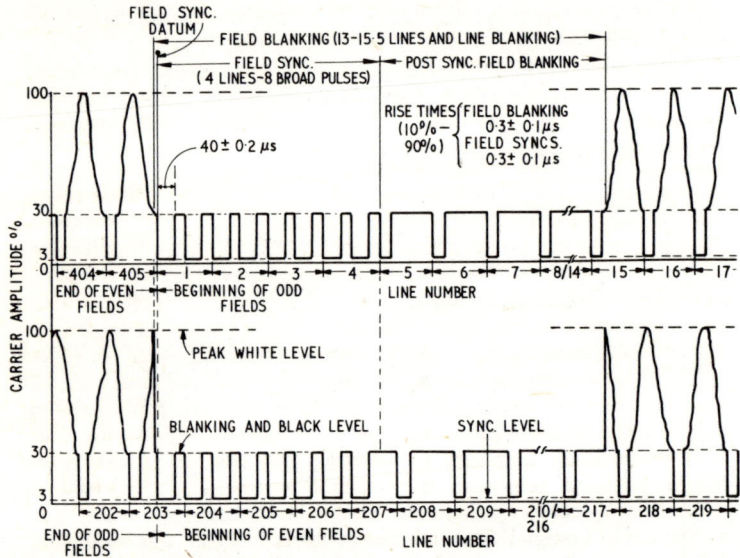

Figure 1.4 Detailed transmission waveform showing the field sync. period, British 405-line system

leading and trailing edges of the sync. pulses are extremely steep and it is most important that the leading edge of this pulse, which is used to synchronise the line-scan generator, retains this shape in the receiver.

So far we have considered only the line-sync. pulses, but the waveform for an interlaced system must also contain pulses that will trigger off the field-scan oscillator twice in each complete picture: at the end of the 405th line and halfway through the 203rd line, as well as making provision for the suppression of the flyback trace during the return of the spot from the bottom of the picture to the top. The field pulses take the form of eight broad pulses of 40 μs duration occupying the space of 4 line periods as shown in *Figure 1.4*. This is followed by the suppression of picture information for a further period of 11 lines, to allow time for the field flyback trace to occur; this is known as the post-field sync. suppression period. During this time it is necessary for line pulses to be inserted to maintain the line-scan oscillator at the correct frequency. It will thus be appreciated that, although we consider the Band I/III system as having 405 lines, 30 of these lines are lost in so far as the transmission of picture information is concerned, so that in practice an enlarged viewing screen would show a picture made up of 374 complete and 2 half-lines.

Vestigial sideband transmission

The requirements of a national television broadcasting service make it necessary to use the available frequency bands as effectively as possible, and to this end vestigial sideband transmission has been adopted. In sound broadcasting the sidebands transmitted extend on each side of the carrier frequency up to a frequency equal to the highest modulating frequency. In the 405-line system frequencies up to 3 MHz are transmitted, so that double-sideband transmission would require a total vision bandwidth of 6 MHz, to which must be added the sound channel bandwidth and the guard bands between channels; on 625 lines the corresponding frequencies would be 5·5 and 11 MHz. Since, however, the information in the upper and lower sidebands is the same, it is theoretically necessary to transmit only one set of sidebands. Single sideband television transmission would, however, lead to difficulties in receiver design. An acceptable compromise is provided by the use of vestigial sideband transmission, in which one set of sidebands is partially suppressed. As an example, *Figure 1.5* (left) shows the Channel 3 sound and vision bandwidths, with sound carrier frequency of 53·25 MHz and vision carrier frequency of 56·75 MHz. As can be seen, double sideband transmission of the vision signal is retained up to 0·75 MHz on both sides of the carrier frequency, vision frequencies above 0·75 MHz being transmitted on the lower sideband only. In this way the total bandwidth required for the vision and sound signals and guard bands is reduced to 5 MHz.

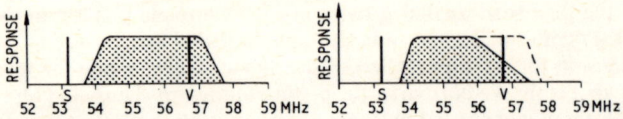

Figure 1.5 (left) R.F. spectrum diagram of 405-line transmission showing vestigial sideband transmission of the vision channel to conserve bandwidth. (right) Correct response curve for reception of vestigial sideband signals

Response (right) at the vision carrier frequency (V) should be half that over the single sideband section of the vision channel. At v.h.f. the local oscillator frequency is above the signal frequency, so the response on the right will be reversed at i.f. since the sidebands are then transposed.

Table 1.1 BRITISH TELEVISION SYSTEM STANDARDS

Standard	405-line	625-line
Channel bandwidth	5 MHz	8 MHz
Upper sideband (vision signal)	0·75 MHz	5·5 MHz
Lower sideband (vision signal)	3 MHz	1·25 MHz
Vision modulation	a.m. positive	a.m. negative
Sound modulation	a.m.	f.m.*
Sound carrier relative to vision carrier	−3·5 MHz	+6 MHz†
Aspect ratio	4:3	4:3
Blanking and black level	30% peak	76% peak
White level	100% peak	20% peak
Sync. level	3% peak	100% peak
Video bandwidth	3 MHz	5·5 MHz
Field frequency	50 Hz	50 Hz
Line frequency	10 125 Hz	15 625 Hz
Field sync. signal	8 broad pulses in 4 line periods	5 equalising then 5 broad pulses, followed by 5 equalising pulses in 7·5 line periods
Field sync. and flyback intervals	2 × 14/15 line periods	2 × 25 line periods
Line period (approx.)	99 μs	64 μs
Line sync. pulses (approx.)	9 μs	4·7 μs
Line blanking (approx.)	18·5 μs	12 μs
Field sync. pulses (broad)	40 μs	27·3 μs
Field sync. pulses (equalising)	—	2·3 μs

*Maximum deviation ±50 kHz; pre-emphasis 50 μs
†The exact spacing is +5·9996 MHz ±500 Hz

Vestigial sideband transmission means that the receiver is supplied with twice the power of vision signals up to 0·75 MHz than it is with vision frequencies above 0·75 MHz. Thus the receiver must be aligned so that its response curve is roughly as shown in *Figure 1.5* (right), with the response at the vision carrier frequency half that at the higher vision frequencies (6 dB down).

On 625-line transmissions, the vestigial sideband is 1·25 MHz so reducing the total bandwidth requirements to only 8 MHz.

British 625-line television system

A comparison of the standards used for the 405- and 625-line systems is given in Table 1.1, from which it will be seen that there are a number of differences other than in the number of lines per field. The main differences are the use of negative modulation (see *Figure 1.2(b)*) for the vision signals, with 20 per cent of peak output representing peak white, 76 per cent output the black level and 100 per cent output the sync. level; the use of the upper sideband for the transmission of the higher vision frequencies instead of the lower sideband; the placing

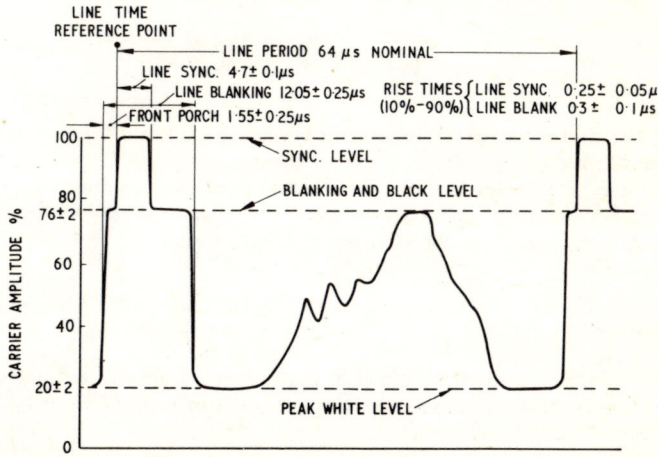

Figure 1.6 Detailed transmission waveform showing the line sync. period, British 625-line system

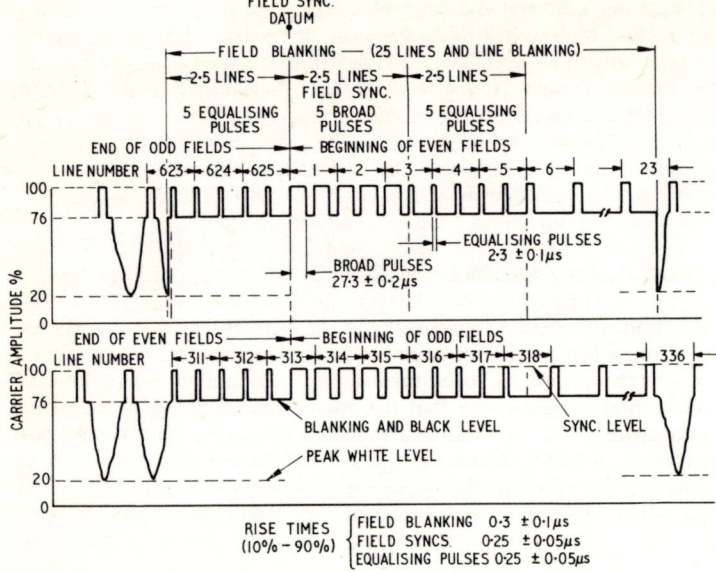

Figure 1.7   Detailed transmission waveform showing the field sync. period, British 625-line system

of the sound carrier above the vision carrier in the 625-line system; and the use of f.m. for the sound channel.

Figure 1.6 shows part of the 625-line waveform including line sync. signals, with front and back porches, and *Figure 1.7* shows the field sync. signals used in the 625-line system.

Bandwidth

The video-frequency bandwidth required to transmit a television picture faithfully with equal vertical and horizontal resolution is given by the formula:

$$\text{Bandwidth} = \frac{N^2 \times A \times P(1 - S_F)}{2 \times (1 - S_L)} \text{ Hz}$$

where N is number of lines, A is aspect ratio, P number of pictures per second, S_F fraction of field period occupied by field suppression, S_L fraction of line period occupied by line suppression. For 405-line standards $N = 405$, $A = \frac{4}{3}$, $P = 25$, $S_F = 0\cdot069$, $S_L = 0\cdot172$ giving a theoretical bandwidth of $3\cdot06$ MHz.

The maximum theoretical definition is reduced in practice by deficiencies in the transmitter and receiver, and the amount of detail which can be seen is further dependent on the brightness and contrast in the picture, and the distance from which it is viewed.

Methods of minimising the ill effects of ambient light, using dark cathode-ray-tube screens, neutral or coloured plastic filters and black net over the face of the tube have all been used more or less successfully, and are effective because the light from the tube is attenuated only once when coming through the filters, whereas stray light shining on the tube is reflected back through the filter and attenuated twice.

It is interesting to note that the eye is capable of resolving detail about twice as fine as is observable in a high-quality television picture at a viewing distance of four times the picture height. The actual ratio varies considerably according to the brightness and contrast of the detail, but under no circumstances is the eye called upon to work at its maximum effort when looking at television, so that eye-strain should not result from this cause.

Colour bars

A common test waveform used by both the I.B.A. and B.B.C. on u.h.f. is a set of vertical colour bars. These involve only the primary colour hues and their complements at a particular luminance and saturation. They are used to check the correct operation of studio, transmission and receiver decoding equipment. *Figure 1.8* shows one of the common transmission versions of these bars which are generally known as *95% bars*. From the line sync. time the luminance levels descend across the line period, the colours being represented (at 95 per cent display saturation) as: yellow, cyan, green, magenta, red and blue. The period before chroma information begins and immediately after it ends will give additional displays of white and black respectively.

At the time of publication the I.B.A. were not using 95 per cent bars but another version known as E.B.U. bars (European Broadcasting Union) where the luminance signal for the chroma is reduced by 25 per cent—i.e. all the chroma maximum and minimum voltages in *Figure*

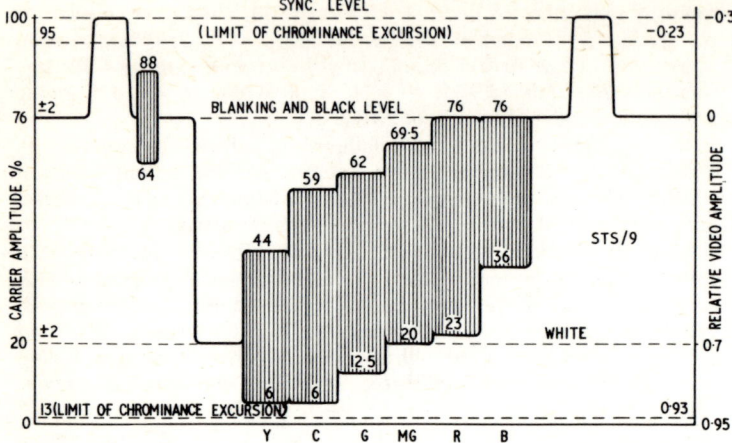

Figure 1.8   95 per cent transmission colour bars
Note: The two limits of chrominance excursion would correspond to the transmission of full saturation colours.

1.8 are moved towards sync. tips by a relative 0·175 video amplitude—without affecting the saturation. The signal is a less stringent test of the transmission system.

BRITISH TELEVISION NETWORK

Television transmission in the United Kingdom is on the very- and ultra-high-frequency bands shown in Table 1.2. Band I is used for B.B.C.-1 transmissions, Band II is used for the v.h.f./f.m. sound broadcasting networks, Band III for I.B.A. television transmissions and also for some B.B.C.-1 transmissions where programme service

Table 1.2   V.H.F./U.H.F. BROADCAST AND TELEVISION BANDS

Band I – – –	41–68 MHz	}	
Band II – – –	87·5–108 MHz	}	v.h.f.
Band III – – –	174–216 MHz	}	
Band IV – – –	470–585 MHz	}	
Band V – – –	610–960 MHz	}	u.h.f.

extensions (e.g. B.B.C.-Wales) took place after exhaustion of Band I allocations. Bands IV and V are used for B.B.C.-2 television transmissions and the 625-line duplication of B.B.C.-1 and I.B.A. programmes. All programmes are originated in the studio centres using 625 lines, the v.h.f. transmitters putting out 405-line pictures converted from the original 625 signals. The majority of programmes are originated in colour but the converted, 405-line pictures on v.h.f. are in monochrome.

Later extensions in u.h.f. services will exhaust the present Band IV and V channel allocations. It is probable that a higher group of frequencies will be allocated in Band VI (approx. 1 000 MHz–11 000 MHz: i.e. 1–11 GHz) for normal transmissions and satellite broadcasting.

Table 1.3 lists the channels in Bands I and III, with the sound and vision carrier frequencies, and Table 1.4 the channels and carrier frequencies of Bands IV and V.

Table 1.3   BAND I/BAND III 405-LINE CHANNELS

Channel	Sound, MHz	Vision, MHz	Channel	Sound, MHz	Vision, MHz
1	41·5	45	7	181·25	184·75
2	48·25	51·75	8	186·25	189·75
3	53·25	56·75	9	191·25	194·75
4	58·25	61·75	10	196·25	199·75
5	63·25	66·75	11	201·25	204·75
Band III			12	206·25	209·75
6	176·25	179·75	13	211·25	214·75
			14	216·25	219·75

The allocation of channels to stations is done on a geographical basis to avoid interference between stations operating on the same channels.

Tables 1.5, 1.6 and 1.7 list the stations at present in operation or due to come into operation in Bands I, III, IV and V. Note that a large number of stations are low-power relay stations designed to *fill-in* areas poorly served by the main stations. On Band I interference from continental stations can be severe during warmer weather and a large number of the relays on this band are required to overcome these effects.

Table 1.4 BAND IV/V 625-LINE CHANNELS

Channel	Channel limits MHz	Carrier frequencies		Channel	Channel limits MHz	Carrier frequencies	
		Vision MHz	Sound MHz			Vision MHz	Sound MHz
21	470–478	471·25	477·25	45	662–670	663·25	669·25
22	478–486	479·25	485·25	46	670–678	671·25	677·25
23	486–494	487·25	493·25	47	678–686	679·25	685·25
24	494–502	495·25	501·25	48	686–694	687·25	693·25
25	502–510	503·25	509·25	49	694–702	695·25	701·25
26	510–518	511·25	517·25	50	702–710	703·25	709·25
27	518–526	519·25	525·25	51	710–718	711·25	717·25
28	526–534	527·25	533·25	52	718–726	719·25	725·25
29	534–542	535·25	541·25	53	726–734	727·25	733·25
30	542–550	543·25	549·25	54	734–742	735·25	741·25
31	550–558	551·25	557·25	55	742–750	743·25	749·25
32	558–566	559·25	565·25	56	750–758	751·25	757·25
33	566–574	567·25	573·25	57	758–766	759·25	765·25
34	574–582	575·25	581·25	58	766–774	767·25	773·25
35*	582–590	583·25	589·25	59	774–782	775·25	781·25
36*	590–598	591·25	597·25	60	782–790	783·25	789·25
37*	598–606	599·25	605·25	61	790–798	791·25	797·25
38†	606–614	607·25	613·25	62	798–806	799·25	805·25
39	614–622	615·25	621·25	63	806–814	807·25	813·25
40	622–630	623·25	629·25	64	814–822	815·25	821·25
41	630–638	631·25	637·25	65	822–830	823·25	829·25
42	638–646	639·25	645·25	66	830–838	831·25	837·25
43	646–654	647·25	653·25	67	838–846	839·25	845·25
44	654–662	655·25	661·25	68‡	846–854	847·25	853·25

* Not used in U.K.   †Not used in U.K., to protect radio astronomy interests.
‡No Channel above 854 MHz assigned at present.

V.H.F. relays are indicated by the letter R after the station name; u.h.f. relays are shown in Table 1.7 inset after the main station of the area.

In the case of the u.h.f. stations listed in Table 1.7, it will be noticed that four channels are allocated to each transmitter. As it is proposed eventually to provide a service on four u.h.f. channels in each area, u.h.f. aerials are designed as broadband devices. To simplify the provision of aerials for Bands IV/V, there are five colour-coded aerial groupings. Aerials in group A cover channels 21–34 and are colour coded red; aerials in group B cover channels 39–51 and are colour coded yellow; aerials in group C cover channels 50–66 and are colour coded green; aerials in group D cover channels 49–68 and are colour coded blue; aerials in group E cover channels 39–68 and are colour coded brown.

Table 1.7 includes all stations either in service already or being planned at the time of publication. Those not in operation should be before 1974, although some of the information must be regarded as provisional. There will eventually be some 60 main and 500 relay stations—the relays gap-filling the coverage of the main transmitters.

For convenience, station names are arranged in regional groups with the main station listed first and relay station names inset alphabetically after. The coverage indicated at the end of each group and in the map of *Figure 1.9* must only be taken as guidance to the *total* area served by the group; there will always be some pockets of poor reception within the area, and other pockets outside the area where reception is acceptable. In drawing up the coverage details, only the principal towns or communities have been included. Where a situation may be served by more than one station, only the principal service station has that situation in its coverage list.

The regional grouping of stations coincides with the B.B.C. Regional Boundaries, while the I.T.V. programme company is listed for the main station and does not necessarily apply to *all* the relay stations in the group: Weymouth relay, for example, will carry the programmes of Westward Television.

The aerial group given for each station refers to the list given at the top of this page and the u.h.f. channels given refer to those of Table 1.4. It will be noticed that all main stations use horizontal polarisation for transmission and that all relay stations use vertical polarisation.

Further details on opening dates or specific transmitter coverage can be obtained from the Engineering Information departments of the I.B.A. and the B.B.C. (addresses given in Chapter 12).

Table 1.5 B.B.C. BAND I/III TELEVISION TRANSMITTERS

Station	Channel	Aerial polarisation	Maximum vision E.R.P.
Abergavenny (B.B.C.-Wales) R . .	3	H	30 W*
Aldeburgh R	5	V	25 W*
Ammanford (B.B.C.-Wales) R . .	12	H	20 W*
Ashkirk R	1	V	18 kW*
Ayr R	2	H	50 W*
Ballachulish R	2	V	100 W*
Ballater R	1	V	10 W*
Ballycastle R	4	H	50 W*
Barnstaple R	3	H	200 W*
Bath R	6	H	250 W*
Bedford R	10	H	500 W*
Belmont R	13	V	20 kW
Betws-y-Coed (B.B.C.-Wales) R	4	H	35 W*
Bexhill R	3	H	150 W*
Blaen-plwyf (B.B.C.-Wales) . . .	3	H	3 kW*
Bodmin	5	H	10 W*
Bressay R	3	V	6 kW*
Brighton	2	V	400 W*
Brougher Mountain R	5	V	7 kW*
Bude R	4	V	100 W*
Cambridge R	2	H	100 W*
Campbeltown R	5	V	500 W*
Canterbury R	5	V	30 W*
Cardigan (B.B.C.-Wales) R . . .	2	H	45 W*
Carmarthen (B.B.C.-Wales) R . .	1	V	20 W*
Churchdown Hill R	1	H	250 W*
Crystal Palace	1	V	200 kW*
Divis	1	H	35 kW*
Dolgellau (B.B.C.-Wales) R . .	5	V	25 W*
Douglas	5	V	3 kW*
Dundee Law R	2	V	10 W*
Eastbourne R	5	V	50 W*
Ffestiniog (B.B.C.-Wales) R . .	5	H	50 W*
Folkestone R	4	H	40 W*
Forfar R	5	V	5 kW*
Fort William R	5	H	1·5 kW*
Girvan R	4	V	20 W*
Grantown R	1	H	400 W*
Hastings R	4	H	15 W*

Table 1.5 *Continued*

Station	Channel	Aerial polarisation	Maximum vision E.R.P.
Haverfordwest (B.B.C.–Wales) *R* .	4	H	10 kW*
Hereford *R*.	2	H	50 W*
Holme Moss	2	V	100 kW
Holyhead (B.B.C.–Wales) *R* . . .	4	H	10 W*
Hungerford *R*	4	H	25 W*
Kendal *R*	1	H	25 W*
Kilkeel *R*	3	H	25 W*
Kilvey Hill (B.B.C.–Wales) *R* . .	2	H	500 W*
Kingussie *R*	5	H	35 W*
Kinlochleven *R*	1	V	5 W*
Kirk o'Shotts	3	V	100 kW
Larne *R*	3	H	50 W*
Les Platons *R*	4	H	1 kW
Llanddona (B.B.C.–Wales) . . .	1	V	6 kW*
Llandrindod Wells (B.B.C.–Wales) *R*	1	H	1·5 kW
Llanelli (B.B.C.–Wales) *R* . . .	3	V	15 W*
Llangollen (B.B.C.–Wales) *R* . .	1	H	35 W*
Llanidloes (B.B.C.–Wales) *R* . .	13	H	20 W*
Lochgilphead *R*	1	V	20 W*
Londonderry *R*	2	H	1·5 kW*
Machynlleth (B.B.C.–Wales) *R* . .	5	H	50 W*
Maddybenny More (Portrush) *R* .	5	H	20 W*
Manningtree *R*	4	H	5 kW*
Marlborough *R*	7	H	25 W*
Meldrum	4	H	17 kW*
Melvaig *R*	4	V	25 kW*
Millburn Muir *R*	1	V	10 W*
Moel-y-Parc (B.B.C.–Wales). . .	6	V	20 kW*
Morecambe Bay *R*	3	H	5 kW*
Newhaven *R*	8	V	50 W*
Newry *R*	4	V	30 W*
Northampton *R*	3	V	90 W*
North Hessary Tor	2	V	15 kW*
Oban *R*.	4	V	3 kW*
Okehampton *R*	4	V	40 W*
Orkney	5	V	15 kW*
Oxford *R*	2	H	650 W*
Penifiler *R*	1	H	25 W*
Perth *R*.	4	V	25 W*

Table 1.5 *Continued*

Station	*Channel*	*Aerial polarisation*	*Maximum vision E.R.P.*
Peterborough	5	H	1 kW
Pitlochry *R*.	1	H	200 W*
Pontop Pike	5	H	17 kW
Port Ellen *R*	2	V	50 W*
Redruth *R*	1	H	10 kW*
Richmond (Yorkshire) *R*. . . .	3	V	
Rosemarkie	2	H	20 kW*
Rosneath	2	V	20 W*
Rowridge	3	V	100 kW*
Rye *R*	3	H	50 W*
Sandale (North)	4	H	30 kW*
Sandale (Scotland)	6	H	70 kW*
Scarborough *R*	1	H	500 W*
Scilly Isles *R*	3	H	20 W*
Sheffield *R*	1	H	50 W
Sidmouth *R*	4	H	30 W*
Skegness *R*.	1	H	60 W
Skriaig *R*	3	H	12 kW*
Sutton Coldfield	4	V	100 kW
Swindon *R*.	3	H	200 W*
Swingate	2	V	1·5 kW*
Tacolneston	3	H	45 kW*
Thrumster	1	V	7 kW*
Toward *R*	5	V	250 W*
Ventnor *R*	5	H	10 W*
Weardale *R*	1	H	150 W*
Wensleydale *R*.	1	V	20 W*
Wenvoe (B.B.C.1)	5	V	100 kW
Wenvoe (B.B.C.–Wales) . . .	13	V	200 kW*
Weymouth *R*	1	H	50 W*
Whitby *R*	4	V	40 W*
Winter Hill	12	V	125 kW*

*Directional aerial. *R* Relay station.
E.R.P. Effective radiated power. Aerial polarisation: V vertical, H horizontal.

Table 1.6 I.B.A. BAND III TRANSMITTERS

Station	Channel	Aerial polarisation	E.R.P. (kW)
Abergavenny, Mon. *R*	11	H	0·1
Angus, Dundee and Perth *R*	11	V	50*
Arfon. Caern. *R*	10	H	10*
Aviemore, Inverness *R*	10	H	1
Bala, N. Wales *R*	7	V	0·1
Ballycastle, N. Ireland *R*	13	H	0·1
Bath, Somerset *R*	8	H	0·5
Belmont, East Lincs.	7	V	20*
Black Hill, Central Scotland	10	V	475*
Black Mountain, N. Ireland	9	H	100*
Brecon, Brec. *R*	8	H	0·1
Burnhope, N.E. England	8	H	100*
Caldbeck, The Borders	11	H	100*
Caradon Hill, Cornwall	12	V	200*
Chillerton Down, I. of W.	11	V	100*
Croydon, London	9	V	350*
Dover, Kent	10	V	100*
Durris, Aberdeen	9	H	400*
Emley Moor, Yorkshire	10	V	200*
Ffestiniog, N. Wales *R*	13	V	0·1
Fremont Point, Channel Islands	9	H	10*
Huntshaw Cross, Devon *R*	11	H	0·5
Lethanhill, Ayrs. *R*	12	V	2
Lichfield, Midlands	8	V	400*
Llandovery, N. Wales *R*	11	H	0·1
Llandrindod Wells, N. Wales *R*	9	H	3
Membury, Berks. *R*	12	H	30*
Mendelsham, East Anglia	11	H	200*
Moel-y-Parc, Flints.	11	V	25*
Mounteagle, Inverness	12	H	50*
Newhaven, Sussex *R*	6	V	1
Presely, Pembrokeshire	8	H	100*
Richmond Hill, I. of M. *R*	8	H	10*
Ridge Hill, Hereford *R*	6	V	10
Rosneath, Dunb. *R*	13	V	0·1
Rothesay, Bute *R*	8	V	1
Rumster Forest, Caithness and Orkney *R*	8	V	30*
Sandy Heath, Beds. *R*.	6	H	30*
Scarborough, Yorks *R*	6	H	1*

Table 1.6 *Continued*

Station	Channel	Aerial polarisation	E.R.P. (kW)
St. Hilary, S. Wales (West)	10	V	200
St. Hilary, S. Wales (Welsh) . . .	7	V	100
Selkirk, Berwick *R*	13	V	25*
Sheffield, Yorks. *R*	6	H	0·1
Stockland Hill, Devon	9	V	100*
Strabane, West Ulster *R*.	8	V	100
Whitehaven, Cumb. *R*	7	V	0·1
Winter Hill, Lancs.	9	V	100

*Directional aerials
R Relay station

Table 1.7   BAND IV/V TELEVISION TRANSMITTING STATIONS
(625-LINE COLOUR SERVICES)

STATION NAME	CHANNELS				Aerial Group	Polarisation	Peak Sync. Power—E.R.P. (kW)	I.T.V. Programme Company of Main Station
	B.B.C.-1	B.B.C.-2	I.B.A.	Fourth				

London and South East

CRYSTAL PALACE	26	33	23	30	A	H	1000	Thames/L.W.T.
Guildford	40	46	43	50	B	V	10	
Hemel Hempstead	51	44	41	47	B	V	10	
Hertford	58	64	61	54	C	V	2	
High Wycombe	55	62	59	65	C	V	0·5	
Reigate	57	63	60	53	C	V	10	
Tunbridge Wells	51	44	41	47	B	V	10	
Wooburn	49	52	56	68	D	V	0·1	
Woolwich	57	63	60	67	D	V	0·63	

This Group covers Greater London and an approximate area bordered by High Wycombe, Berkhamstead, St. Albans, Hertford, Bishop's Stortford, Brentwood, Gravesend, Wrotham, Sevenoaks, Tonbridge, Tunbridge Wells, East Grinstead, Crawley, Guildford, Hindhead, Aldershot, Camberley, Maidenhead, Marlow.

Table 1.7 *Continued.*

STATION NAME	CHANNELS				Aerial Group	Polarisation	Peak Sync. Power—E.R.P. (kW)	I.T.V. Programme Company of Main Station
	B.B.C.-1	B.B.C.-2	I.B.A.	Fourth				
DOVER	50	56	66	53	D	H	100	Southern

Dover, Folkestone, Deal, Ramsgate, Margate, Canterbury, Rye, Ashford.

HEATHFIELD	49	52	64	67	D	H	100	Southern
Hastings	22	25	28	32	A	V	1	
Newhaven	39	45	43	41	B	V	2	

Lewes, Eastbourne, Bexhill, Hastings, Newhaven, Uckfield, Hurst Green, Battle, Tenterden.

MIDHURST	61	55	58	68	D	H	100	Southern

Horsham, Petworth, Alton, Haslemere, Petersfield.

NORTH KENT (BLUEBELL HILL)	40	46	43	65	E	H	30	Thames/L.W.T.

Faversham Whitstable, Sheppey, Maidstone, Sittingbourne, Rochester, Chatham.

OXFORD	57	63	60	53	C	H	500	A.T.V.

Oxford, Swindon, Thame, Chinnor, Risborough, Aylesbury, Buckingham, Bicester, Chipping Norton, Witney, Wantage, Abingdon, Wallingford, Woodstock.

South

HANNINGTON	39	45	42	66	E	H	250	Southern

Basingstoke, Andover, Hungerford, Newbury, Reading, Camberley, Winchester.

ROWRIDGE	31	24	27	21	A	H	500	Southern
Brighton	57	63	60	53	C	V	10	
Salisbury	57	63	60	53	C	V	10	
Ventnor	39	45	49	42	B	V	2	
Weymouth	40	46	43	50	B	V	2	

Table 1.7 *Continued.*

STATION NAME	CHANNELS				Aerial Group	Polarisation	Peak Sync. Power—E.R.P. (kW)	I.T.V. Programme Company of Main Station
	B.B.C.-1	B.B.C.-2	I.B.A.	Fourth				

Isle of Wight, Southampton, Bournemouth, Portsmouth, Dorchester, Romsey, Chichester, Bognor Regis, Swanage, Worthing, Brighton, Hove, Salisbury, Weymouth.

West

STATION NAME	B.B.C.-1	B.B.C.-2	I.B.A.	Fourth	Aerial Group	Polarisation	Peak Sync. Power	I.T.V. Company
MENDIP	58	64	61	54	C	H	500	Harlech
Bath	22	28	25	32	A	V	0·25	
Bristol (Ilchester Crescent)	40	46	43	50	B	V	0·5	
Marlborough	22	28	25	32	A	V	0·5	

Bristol, Weston Super-Mare, Axbridge, Wells, Glastonbury, Taunton, Bridgewater, Minehead, Chippenham, Devizes, Frome, Trowbridge, Bath, Marlborough.

South West

STATION NAME	B.B.C.-1	B.B.C.-2	I.B.A.	Fourth	Aerial Group	Polarisation	Peak Sync. Power	I.T.V. Company
BEACON HILL	57	63	60	53	C	H	100	Westward

Torbay, Brixham, Totnes, Newton Abbot, Bovey Tracy.

STATION NAME	B.B.C.-1	B.B.C.-2	I.B.A.	Fourth	Aerial Group	Polarisation	Peak Sync. Power	I.T.V. Company
CARADON HILL	22	28	25	32	A	H	500	Westward
Plymouth	58	64	61	54	C	V	2	

Plymouth, Saltash, Liskeard, Tavistock, Launceston, Bodmin, Wadebridge, Padstow, Camelford, Bude, Modbury, St. Austell.

STATION NAME	B.B.C.-1	B.B.C.-2	I.B.A.	Fourth	Aerial Group	Polarisation	Peak Sync. Power	I.T.V. Company
HUNTSHAW CROSS	55	62	59	65	C	H	100	Westward

Barnstaple, Westward Ho!, Chulmleigh, Holsworthy, Hatherleigh.

STATION NAME	B.B.C.-1	B.B.C.-2	I.B.A.	Fourth	Aerial Group	Polarisation	Peak Sync. Power	I.T.V. Company
REDRUTH	51	44	41	47	B	H	100	Westward

Cornwall West from a NW/SE line through St. Austell.

Table 1.7   *Continued.*

STATION NAME	B.B.C.-1	B.B.C.-2	I.B.A.	Fourth	Aerial Group	Polarisation	Peak Sync. Power—E.R.P. (kW)	I.T.V. Programme Company of Main Station
	CHANNELS							

STOCKLAND HILL | 33   26   23   29 | A | H | 250 | Westward
Exeter, Tiverton, Crediton, Sidmouth, Axminster, Chard, Bridport, Portland Bill, Beaminster, Honiton.

Midlands

RIDGE HILL | 22   28   25   32 | A | H | 100 | A.T.V.
Hereford, Ross-on-Wye, Monmouth, Gloucester, Cheltenham.

SHROPSHIRE
  (SALOP) | 33   26   23   29 | A | H | 100 | A.T.V.
Ludlow, Oswestry, Shrewsbury, Whitchurch, Wrexham, Montgomery.

SUTTON

	B.B.C.-1	B.B.C.-2	I.B.A.	Fourth	Aerial Group	Polarisation	Peak Sync. Power	I.T.V.
COLDFIELD	46	40	43	50	B	H	1000	A.T.V.
Brierley Hill	57	63	60	53	C	V	10	
Bromsgrove	31	27	24	21	A	V	10	
Kidderminster	58	64	61	54	C	V	2	
Lark Stoke	33	26	23	29	A	V	10	
Leek	22	28	25	32	A	V	1	
Malvern	56	62	66	68	D	V	10	
Nottingham	21	27	24	31	A	V	2	
Stoke-on-Trent (Fenton)	31	27	24	21	A	V	10	

Birmingham, Wolverhampton, Coventry, Nuneaton, Rugby, Leamington, Warwick, Burton-on-Trent, Derby, Stafford, Evesham, Worcester, plus relay towns.

WALTHAM | 58   64   61   54 | C | H | 250 | A.T.V.
Nottingham, Leicester, Loughborough, Grantham, Oakham, Stamford.

East Anglia

	B.B.C.-1	B.B.C.-2	I.B.A.	Fourth	Aerial Group	Polarisation	Peak Sync. Power	I.T.V.
SANDY HEATH	31	27	24	21	A	H	1000	Anglia
Northampton	51	44	41	47	B	V	1	

Table 1.7 *Continued.*

STATION NAME	CHANNELS				Aerial Group	Polarisation	Peak Sync. Power—E.R.P. (kW)	I.T.V. Programme Company of Main Station
	B.B.C.-1	*B.B.C.-2*	*I.B.A.*	*Fourth*				

Peterborough, Kettering, Wellingborough, Northampton, Buckingham, Biggleswade, Dunstable, Luton, Hitchin, Cambridge, Ely, Newmarket, Huntingdon, Saffron Walden.

SUDBURY	51	44	41	47	B	H	250	Anglia

Ipswich, Colchester, Chelmsford, Bury St. Edmunds, Stowmarket, Felixstowe, Harwich, Walton-on-the-Naze, Clacton-on-Sea, Southend, Braintree.

TACOLNESTON	62	55	59	65	C	H	250	Anglia
Aldeburgh	33	26	23	30	A	V	10	
West Runton	33	26	23	29	A	V	2	

Norwich, Gt. Yarmouth, Lowestoft, Southwold, Thetford, East Dereham, Fakenham, Cromer, Aldeburgh, Saxmundham.

North

BELMONT	22	28	25	32	A	H	500	Anglia

Grimsby, Lincoln, Mansfield, Kingston upon Hull, Scunthorpe, Skegness, Boston, Kings Lynn, Spalding, Sleaford, Bridlington.

EMLEY MOOR	44	51	47	41	B	H	1000	Yorkshire
Chesterfield	33	26	23	29	A	V	2	
Cop Hill	22	28	25	32	A	V	1	
Halifax	21	27	24	31	A	V	2	
Hebden Bridge	22	28	25	32	A	V	0·25	
Idle	21	27	24	31	A	V	1	
Keighley	58	64	61	54	C	V	10	
Oxenhope	22	28	25	32	A	V	1	
Sheffield	31	27	24	21	A	V	5	
Skipton	39	45	49	42	B	V	10	
Wharfedale	22	28	25	32	A	V	2	

Huddersfield, Leeds, Barnsley, Doncaster, Wakefield, Rotterham, York, Market Weighton, Selby, Gainsborough, Worksop, Alfreton, Bradford, plus relay towns.

Table 1.7 *Continued.*

STATION NAME	CHANNELS				Aerial Group	Polarisation	Peak Sync. Power—E.R.P. (kW)	I.T.V. Programme Company of Main Station
	B.B.C.-1	B.B.C.-2	I.B.A.	Fourth				

North West

STATION NAME	B.B.C.-1	B.B.C.-2	I.B.A.	Fourth	Aerial Group	Polarisation	Peak Sync. Power—E.R.P. (kW)	I.T.V. Programme Company of Main Station
WINTER HILL	55	62	59	65	C	H	500	Granada
Bacup	40	46	43	53	E	V	0·25	
Birch Vale	40	46	43	53	E	V	0·25	
Buxton	21	27	24	31	A	V	1	
Congleton	51	44	41	47	B	V	0·2	
Darwen	39	45	49	42	B	V	0·5	
Glossop	22	28	25	32	A	V	1	
Haslingden	33	26	23	29	A	V	10	
Kendal	58	64	61	54	C	V	2	
Ladder Hill	33	26	23	29	A	V	1	
Lancaster	31	27	24	21	A	V	10	
Littleborough	21	27	24	31	A	V	0·5	
Pendle Forest	22	28	25	32	A	V	2	
Saddleworth	52	45	49	42	E	V	2	
Sedbergh	40	46	43	50	B	V	0·5	
Todmorden	39	45	49	42	B	V	0·5	
Whitworth	22	28	25	32	A	V	0·25	
Windermere	51	44	41	47	B	V	0·5	

Liverpool, Manchester, Birkenhead, Bolton, Crewe, Macclesfield, Oldham, Rochdale, Southport, Preston, Blackburn, Blackpool, Fleetwood, Barrow, Warrington, Ormskirk, Chester, Flint, West Kirby, plus relay towns.

North East

STATION NAME	B.B.C.-1	B.B.C.-2	I.B.A.	Fourth	Aerial Group	Polarisation	Peak Sync. Power—E.R.P. (kW)	I.T.V. Programme Company of Main Station
BILSDALE (WEST MOOR)	33	26	29	23	A	H	500	Yorkshire
Swaledale	40	46	43	50	B	V	1	
Whitby	55	62	59	65	C	V	0·25	

Harrogate, Richmond, Darlington, Stockton-on-Tees, Redcar, West Hartlepool, Whitby, Northallerton, Middleham, Pickering, Hawes, Askrigg.

Table 1.7 *Continued.*

STATION NAME	CHANNELS				Aerial Group	Polarisation	Peak Sync. Power—E.R.P. (kW)	I.T.V. Programme Company of Main Station
	B.B.C.-1	B.B.C.-2	I.B.A.	Fourth				
CALDBECK	30	34	28	32	A	H	500	Border
Whitehaven	40	46	43	50	B	V	2	

Workington, Cockermouth, Carlisle, Penrith, Appleby, Longton, Annan, Dumfries, Moniaive, Whithorn, Whitehaven.

| CHATTON | | 39 | 45 | 49 | 42 | B | H | 100 | Tyne Tees |

Berwick, Alnwick, Coldstream, Balford, Lindisfarne, Bamburgh, Wooler.

PONTOP PIKE	58	64	61	54	C	H	500	Tyne Tees
Alston	52	45	49	42	D	V	0·4	
Fenham	31	27	24	21	A	V	2	
Morpeth	22	28	25	32	A	V	0·5	
Newton	33	26	23	29	A	V	2	
Weardale	51	44	41	47	B	V	1	

Newcastle upon Tyne, Whitley Bay, Blyth, Hexham, Durham, Consett, Middlesborough, Bishop Auckland, plus relay towns.

Wales

| BLAENPLWYF | | 31 | 27 | 24 | 21 | A | H | 100 | Harlech |

Aberystwyth, Aberayron, Aberdovey.

| CARMEL | | 57 | 63 | 60 | 53 | C | H | 100 | Harlech |

Carmarthen, Ammanford, Llandovery.

LLANDDONA	57	63	60	53	C	H	100	Harlech
Bethesda	57	63	60	53	C	V	0·025	
Betws-y-Coed	21	27	24	31	A	V	2	
Conway	40	46	43	50	B	V	2	

Anglesey, Holyhead, Llandudno, Bangor, Caernarvon, Pwlheli, Conway, Bethesda, Betws-y-Coed, Llanrwst.

Table 1.7   *Continued.*

STATION NAME	B.B.C.-1	B.B.C.-2	I.B.A.	Fourth	Aerial Group	Polarisation	Peak Sync. Power—E.R.P. (kW)	I.T.V. Programme Company of Main Station
	CHANNELS							
MOEL-Y-PARC	52	45	49	42	E	H	100	Harlech

Denbigh, Ruthin, St. Asaph, Rhyl, Prestatyn, Colwyn Bay, Mold.

STATION NAME								
PRESELY	46	40	43	50	B	H	100	Harlech

Cardigan, Newport, Fishguard, Haverfordwest, Milford Haven, Tenby, Pembroke, Lampeter.

STATION NAME								
WENVOE	44	51	41	47	B	H	500	Harlech
Aberdare	21	27	24	31	A	V	0·5	
Abergavenny	39	45	49	42	B	V	1	
Abertillery	22	28	25	32	A	V	1·4	
Abertridwr	57	63	60	53	C	V	0·05	
Bargoed	21	27	24	31	A	V	1·5	
Blaenavon	57	63	60	53	C	V	0·75	
Blaina	40	46	43	50	B	V	0·1	
Croeserw	58	64	61	54	C	V	0·12	
Cwmavon	21	27	24	31	A	V	0·07	
Ebbw Vale	55	62	59	65	C	V	0·5	
Gilfach Goch	21	27	24	31	A	V	0·05	
Kilvey Hill	33	26	23	29	A	V	10	
Llangeinot	55	62	59	65	C	V	1	
Llanhilleth	39	45	49	42	B	V	0·03	
Maesteg	22	28	25	32	A	V	0·5	
Merthyr Tydfil	22	28	25	32	A	V	0·5	
Mynydd Bach	58	64	61	54	C	V	0·25	
Mynydd Machen	33	26	23	29	A	V	2	
Pontypool	21	27	24	31	A	V	1	
Pontypridd	22	28	25	32	A	V	2	
Porth	40	46	43	50	B	V	0·25	
Rhondda	33	26	23	29	A	V	4	
Rhymney	57	63	60	53	C	V	0·75	

Cardiff, Barry, Newport, Caerphilly, Swansea, Llanelly, Neath, plus relay towns.

Table 1.7 *Continued.*

STATION NAME	CHANNELS				Aerial Group	Polarisation	Peak Sync. Power—E.R.P. (kW)	I.T.V. Programme Company of Main Station
	B.B.C.-1	B.B.C.-2	I.B.A.	Fourth				

Scotland

ANGUS	57	63	60	53	C	H	100	Grampian
Perth	39	45	49	42	B	V	1	

Forfar, Arbroath, Dundee, Alyth, Blairgowrie, St. Andrews, Perth, Dunbar.

BLACK HILL	40	46	43	50	B	H	500	Scottish

Glasgow, Clydebank, Paisley, Hamilton, Coatbridge, Motherwell, Wishaw, Airdrie, Armadale, Lanark, Falkirk, Stirling, Alloa, Dunfermline.

CRAIGKELLY	31	27	24	21	A	H	100	Scottish

Edinburgh, Kirkcaldy, Dalkeith.

DARVEL	33	26	23	29	A	H	100	Scottish
Lethanhill	57	63	60	53	C	V	0·25	

Kilmarnock, Ayr, Ardrossan, Brodick, Cumnock, Maybole, Girvan.

DURRIS	22	28	25	32	A	H	500	Grampian
Garthy Moor	58	64	61	54	C	V	10	
Peterhead	55	62	59	65	C	V	0·1	
Rosehearty	51	44	41	47	B	V	2	

Aberdeen, Montrose, Brechin, Banchory, Aboyne, Stonehaven, Inverbervie, Inverurie, Peterhead.

KNOCK MOOR	33	26	23	29	A	H	100	Grampian

Turriff, Banff, Fraserburgh.

ROSEMARKIE	39	45	49	42	B	H	100	Grampian

Burghead, Lossiemouth, Nairn, Inverness, Dingwall, Cromarty, Invergordon, Dornoch, Brora, Helmsdale, Buckie.

ROSNEATH	58	64	61	54	C	H	50	Scottish

Dumbarton, Greenock, Dunoon.

Table 1.7 *Continued.*

STATION NAME	B.B.C.-1	B.B.C.-2	I.B.A.	Fourth	Aerial Group	Polarisation	Peak Sync. Power—E.R.P. (kW)	I.T.V. Programme Company of Main Station
			CHANNELS					
RUMSTER FOREST	31	27	24	21	A	H	100	Grampian
Wick, Thurso, Lybster.								
SELKIRK	55	62	59	65	C	H	50	Border
Eyemouth	33	26	23	29	A	V	2	
Selkirk, Kelso, Coldstream, Eyemouth.								

Northern Ireland

STATION NAME	B.B.C.-1	B.B.C.-2	I.B.A.	Fourth	Aerial Group	Polarisation	Peak Sync. Power	I.T.V. Company
BROUGHER MTN.	22	28	25	32	A	H	100	Ulster
Strabane	57	63	60	53	C	V	2	
Omagh. Enniskillen, Strabane.								
DIVIS	31	27	24	21	A	H	500	Ulster
Carnmoney Hill	40	46	43	50	B	V	0·1	
Kilkeel	39	45	49	42	B	V	2	
Killowen Mtn.	31	27	24	21	A	V	0·15	
Larne	39	45	49	42	B	V	2	
Newry	58	64	60	54	C	V	0·5	

Belfast, Bangor, Newtonards, Lurgan, Lisburn, Banbridge, Downpatrick, Armagh, Cookstown, Antrim, Ballymena, Maghera, Newcastle, Larne, Newry, Kilkeel.

LIMAVADY	55	62	59	65	C	H	100	Ulster
Londonderry	51	44	41	47	B	V	8	

Limavady, Dungiven, Coleraine, Bushmills, Londonderry, Ballymoney.

IRISH TELEVISION NETWORK

The Irish national broadcasting organisation, Radio Telefís Éireann, has a television network of seven main high-power transmitters and thirteen satellite (relay) transposers, listed in Table 1.8. The standard

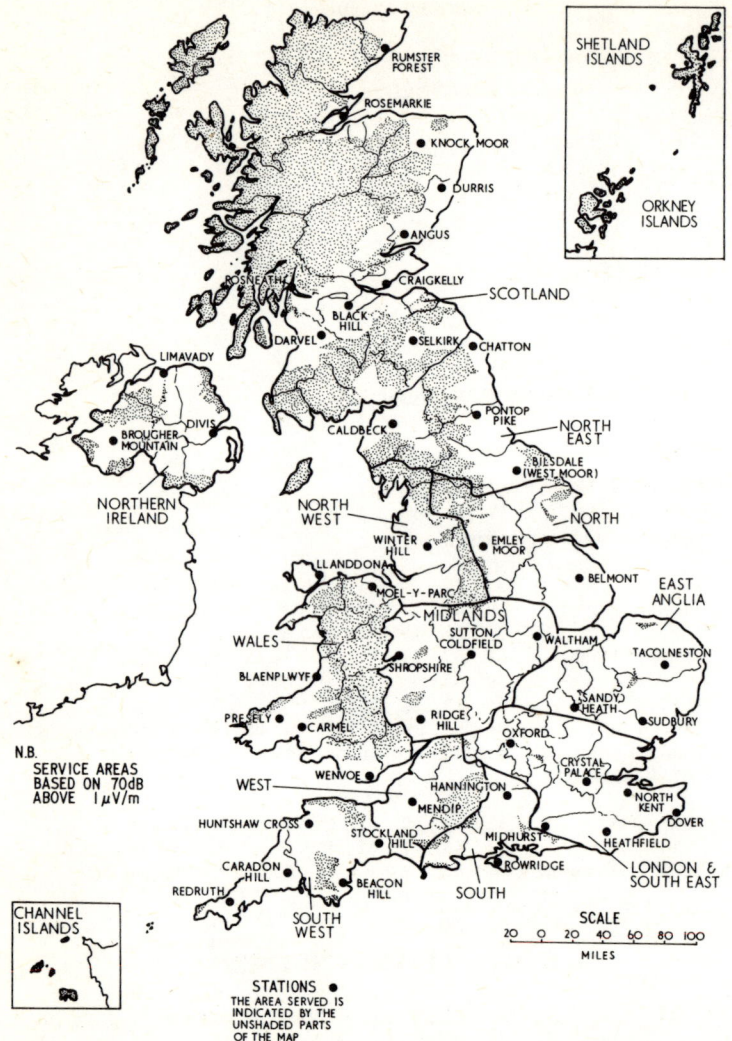

SHETLAND ISLANDS

ORKNEY ISLANDS

RUMSTER FOREST
ROSEMARKIE
KNOCK MOOR
DURRIS
ANGUS
CRAIGKELLY
SCOTLAND
ROSNEATH
BLACK HILL
DARVEL
SELKIRK
CHATTON
LIMAVADY
PONTOP PIKE
NORTH EAST
DIVIS
BROUGHER MOUNTAIN
CALDBECK
BILSDALE (WEST MOOR)
NORTHERN IRELAND
NORTH WEST
NORTH
WINTER HILL
EMLEY MOOR
LLANDDONA
BELMONT
EAST ANGLIA
MOEL-Y-PARC
MIDLANDS
WALES
SUTTON COLDFIELD
WALTHAM
TACOLNESTON
BLAENPLWYF
SHROPSHIRE
SANDY HEATH
PRESELY
CARMEL
RIDGE HILL
OXFORD
SUDBURY
CRYSTAL PALACE
WEST
WENVOE
HANNINGTON
NORTH KENT
DOVER
MENDIP
HUNTSHAW CROSS
STOCKLAND HILL
MIDHURST
HEATHFIELD
CARADON HILL
ROWRIDGE
LONDON & SOUTH EAST
REDRUTH
BEACON HILL
SOUTH
SOUTH WEST

N.B.
SERVICE AREAS BASED ON 70dB ABOVE 1 μV/m

SCALE
20 0 20 40 60 80 100
MILES

CHANNEL ISLANDS

STATIONS ●
THE AREA SERVED IS INDICATED BY THE UNSHADED PARTS OF THE MAP

Figure 1.9   U.H.F. television main transmitting stations

is a 625-line one partly allied to the general European 625-line standard, but programmes are also broadcast on the 405-line standard to serve owners of 405-line receivers in northern parts of the country. The 625-line standard has three channels (A–C) on Band I and six channels (D–I) on Band III. The 405-line standard has five channels (B1–B5) on Band I and nine channels (B6–B14) on Band III. The 625-line channels are indicated in Table 1.9; the 405-line channels are those used in the U.K. (Table 1.3).

Table 1.8   RADIO TELEFÍS ÉIREANN TELEVISION TRANSMITTERS

Station	Band	Channel	Aerial polarisation	Maximum vision E.R.P.
405-lines				
Kippure	III	B7	H	100 kW
Truskmore	III	B11	V	100 kW
Dublin *S*.	I	B3	V	250 W
Fanad *S*	III	B7	H	1 kW
Letterkenny *S* . . .	III	B6	V	1 kW
Monaghan *S* . . .	III	B10	H	1·6 kW
Moville *S* . . .	III	B12	H	2 kW
625-lines				
Kippure	III	H	H	100 kW
Maghera	I	B	H	100 kW
Mt. Leinster	III	F	V	100 kW
Mullaghanish . . .	III	D	V	100 kW
Truskmore	III	I	V	100 kW
Achill *S*	III	F	V	1 kW
Cahirciveen *S* . . .	III	F	H	1 kW
Cappoquin *S* . . .	III	H	H	250 W
Castlebar	III	H	V	1 kW
Castletownbere *S* . .	III	H	H	250 W
Clifden *S*	III	D	H	1 kW
Cork *S*	III	H	V	250 W
Crosshaven *S* . . .	III	G	H	120 W
Dublin *S*.	I	C	H	250 W
Suir Valley *S* . . .	III	I	V	400 W

S Satellite station.

Table 1.9   RADIO TELEFÍS ÉIREANN 625-LINE SYSTEM

Channel		Vision (MHz)	Sound (MHz)
Band I:	A	45·75	51·75
	B	53·75	59·75
	C	61·75	67·75
Band III:	D	175·25	181·25
	E	183·25	189·25
	F	191·25	197·25
	G	199·25	205·25
	H	207·25	213·25
	I	215·25	221·25

2

BASIC RECEIVER UNIT CIRCUITRY

A television receiver comprises essentially: (*a*) a means of receiving the picture information and using this to modulate the electron beam of a cathode-ray tube (often referred to as the picture tube); (*b*) the local timebase oscillators and associated amplifiers to provide saw-tooth current signals for deflecting the beam of the picture tube so that it traces out the raster, together with arrangements for keeping these oscillators accurately in step with those used at the studio in the original scanning of the picture; (*c*) a sound-channel receiver; and (*d*) the necessary power supplies, including a source of extra high tension (e.h.t.) required for the final anode of the picture tube.

Although there has been some trend towards standardisation of the major features of design in modern receivers, there is still considerable variation in circuit details. It should be appreciated that the arrangements used are by no means the only possible ones; they have evolved primarily as a result of economic competition between manufacturers to provide low-cost but relatively dependable receivers capable of providing a medium-sized picture bright enough for daylight viewing. For this reason commercial receivers rarely do full justice to the quality of the transmitted signals.

For a number of years, a considerable problem existed in the U.K. for receiver operation on both the 405- and 625-line standards (the dual-standard receiver). This involved not only switching the line time-base frequency to suit the different standards: use of u.h.f. for the 625-line transmissions and v.h.f. for 405-line necessitated a switching between the tuners used; the signal bandwidths are different between standards and the intermediate frequencies (i.f.s) used are changed; negative modulation is used on 625-line vision and positive modulation on the 405-line standards; the sound systems employed are different as well—frequency modulation (f.m.) on 625-lines and amplitude modulation (a.m.) on 405-line transmissions.

All operational u.h.f. stations carry all the programme services. In any one area of the U.K. therefore there is now no necessity for dual-

standard operation. As a large percentage of the total population is
served by such transmissions this could be said to be a general story.
In some of the less densely populated areas, however, viewers must
still receive the original 405-line transmissions.

Both the broadcasting authorities in the U.K.—the B.B.C. and the
I.B.A. (latterly the I.T.A.)—are committed to maintaining the original
405-line, v.h.f. services for some years to come (probably to about
1985), so that viewers in remote areas are not deprived of a television
service or forced to buy a more modern receiver. The 405-line transmitters use the 625-line programme service as their source and use an
electronic standards-converter to derive monochrome, 405-line pictures
for the v.h.f. service.

This section deals with the basic circuitry used in modern receivers—
both 625-line single-standard receivers and late model 405/625-line
dual-standard receivers—and encompasses both semiconductor and
valve circuitry. The next section offers the same details for the timebase and power supply parts of the receiver.

Receiver arrangements

Up to about 1963, U.K. receivers were designed for single-standard
405-line operation. The block outline of such a receiver is shown in
Figure 2.1. During the period of dual-standard receiver operation
(from 1963 to about 1969) the receiver became rather more complex
because of the switching involved.

Some of this added complexity of a typical dual-standard model is
indicated in *Figure 2.2.* In this diagram the switches indicate changes
in the main signal path for 405- and 625-line reception, but the diagram
does not attempt to show the detailed changes, for example in the line
timebase, required to adjust the circuits and waveforms for 10 125 Hz
and 15 625 Hz operation; these will be discussed later in this section.

A dual-standard model can be considered as comprised of the
following main sections:

Band IV–V tuner.   This provides variable or pre-set tuning to
stations that are within the frequency band 470–960 MHz, incorporating an r.f. amplifier and frequency changer to convert the incoming
frequencies to the standard 625-line i.f.s of 39·5 MHz (vision carrier)
and 33·5 MHz (sound carrier). The tuner may be built as a separate
sub-unit, although integrated u.h.f./v.h.f. tuners, using transistors,

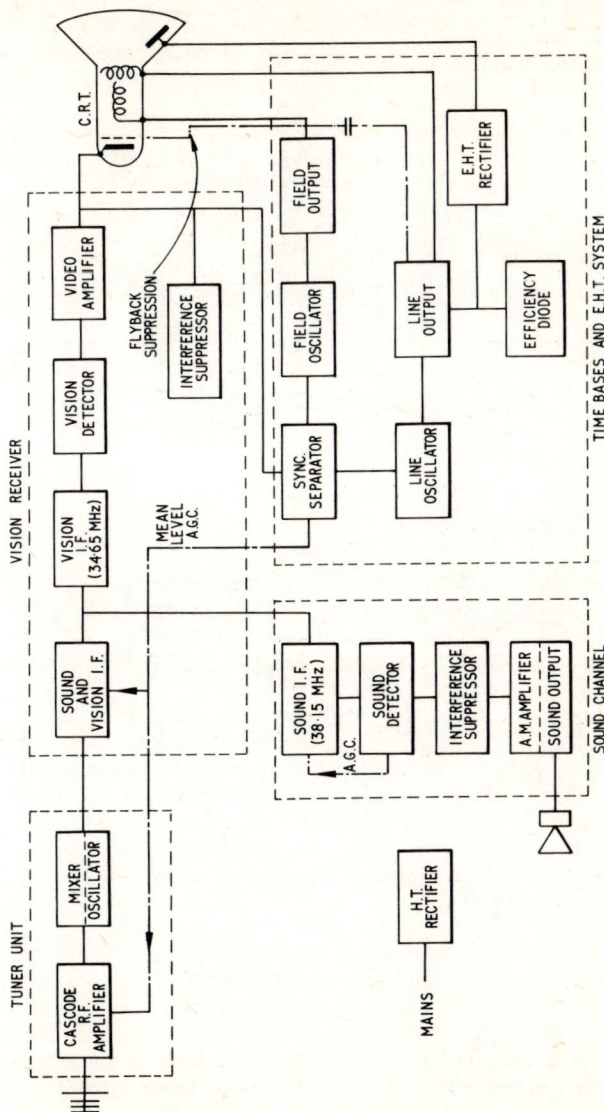

Figure 2.1 Block diagram indicating main units of a typical band 1/111 television receiver (405-line, single standard)

36

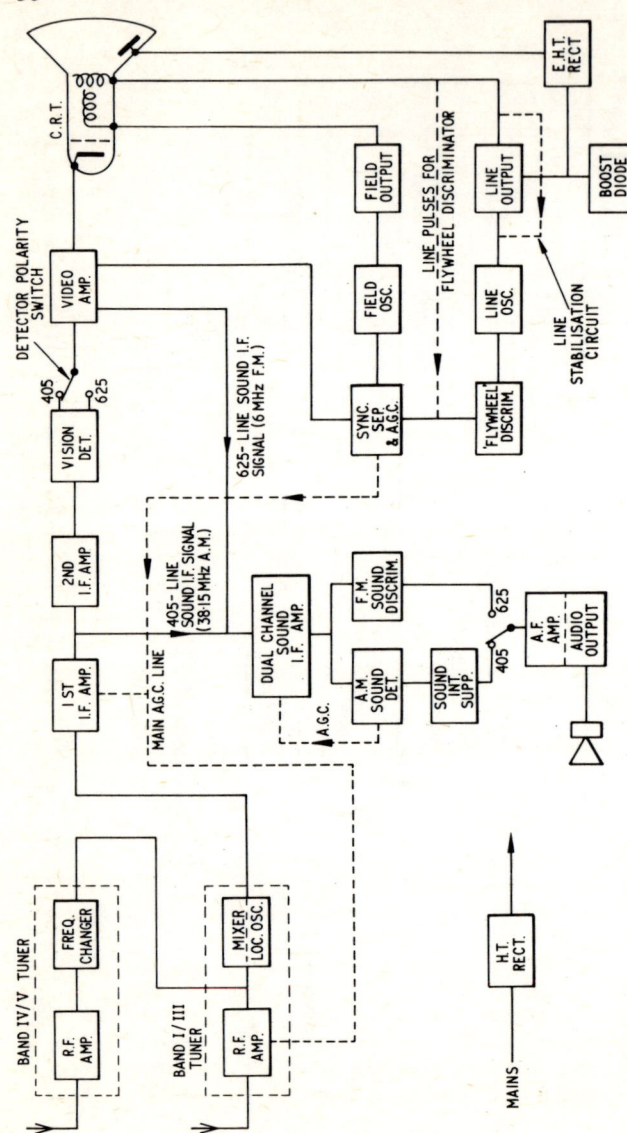

Figure 2.2 Block diagram of a typical dual-standard receiver, incorporating flywheel line sync.

were widely used. For optimum sensitivity the u.h.f. tuner employs low-noise transistors, generally in the common-base mode. These are often employed in a u.h.f. tuner even where valves are used for the remaining sections of the receiver. Where valves were used, these usually take the form of grounded-grid triodes (one as r.f. amplifier, and a second as additive self-oscillating mixer). Since the overall gain of a tuner of this type is appreciably below what can be obtained in a v.h.f. (Bands I and III) tuner, the output is often taken to the mixer input circuit of the v.h.f. tuner so that this stage functions as an additional i.f. stage for u.h.f. reception. Both valves of a u.h.f. tuner are normally operated at fixed gain, with no attempt to connect the a.g.c. line to the tuner; however, this does not apply to transistorised u.h.f. tuners.

Band I/III tuner. This tuner is also generally in the form of a separate sub-unit. It comprises an r.f. amplifier which may be either a double-triode cascode stage or a neutralised beam-triode amplifier. The frequency-changer is almost invariably a triode-pentode, with the pentode section used as a mixer, and the triode as a local oscillator.

In the U.K. the intermediate frequencies for Band I/III reception have, for a number of years, been standardised at 34·65 MHz vision carrier and 38·15 MHz sound carrier.

It should be noted that for British u.h.f. signals, the vision carrier is *lower* in frequency than the sound carrier, whereas it has already been noted that the reverse condition applies with v.h.f. transmission.

Vision receiver. The vision i.f. stages usually comprised two high gain i.f. amplifiers in valved models, of which the first was connected to the a.g.c. line. The first amplifier is normally common to both vision and sound signals on 405-line reception, after which the sound i.f. signal is 'taken off' and diverted to the sound i.f. stages; on 625-lines, with its f.m. sound system, a technique called 'intercarrier' becomes desirable, and this involves passing both sound and vision i.f. signals through all the vision i.f. stages and through the video detector, although the sound signals have to be kept to a value well below that of the vision signals. The techniques of intercarrier sound are explained later.

The i.f. stages not only have to provide sufficient gain but also form the different bandpass characteristics for the two systems, a process which may involve rather complex switching of response-shaping components and traps. The output from the final vision i.f. stage is fed to a video detector, usually a semiconductor diode. High amplitude peaks of ignition or other impulsive interference may then be sliced off by means of a limiter circuit. To cater for both positive and negative modulation, the polarity of the detector output has to be reversed when changing from 405- to 625-line reception. The signals are then amplified in the video amplifier stage(s) and passed to the picture tube and also to the sync. separator.

Sound-channel receiver.   Since the sound transmissions are spaced exactly 3·5 MHz (405-line) or 6 MHz (625-line) away from the vision carrier, both sound and vision transmissions may be thought of as one broadband signal and amplified together in the same r.f. amplifier and converted, by a single local oscillator, to intermediate frequencies maintaining the same separation. But since both signals are passing simultaneously through the stages concerned, cross-modulation and intermodulation can occur if the signal levels are too high.

With an a.m. sound system, as in the U.K. 405-line system, the sound i.f. signals are separated from the vision i.f. signals at a fairly early stage; this may be immediately after the frequency-changer in the v.h.f. tuner or, more commonly, after one combined vision/sound i.f. stage. After separation, the sound i.f. signal at 38·15 MHz is then amplified in a further high-gain i.f. stage (or in some designs in a two-stage amplifier). This amplifier is designed using dual-frequency techniques similar to those developed for v.h.f./f.m., functioning also at 6 MHz as the intercarrier i.f. amplifier/limiter, for the 625-line f.m. sound signal generated in the video detector (see later).

Two different detectors (often using altogether three semiconductor diodes) are required to form an a.m. envelope detector and an f.m. discriminator, usually of the two-diode ratio-detector form, although some chassis used the EH90 locked-oscillator type of f.m. demodulator. For a.m. sound an additional diode may be used as an impulse interference limiter. The a.f. amplifier is most commonly a two-stage unit, using a single triode pentode, often with a degree of negative feedback over one or both stages. In transistorised receivers a number of stages are required.

Field timebase.   This timebase is required only to provide a synchronised 50 Hz waveform for the field (vertical) deflection of the picture-tube beam; this remains true on both 405 and 625 systems, and represents one of the few sections of a dual-standard receiver free of system switching. The field generator, commonly a form of multivibrator, drives a pentode or complementary transistor output stage. The field generator is triggered by the field sync. pulses either directly from the sync. separator or after pulse shaping, the sync. separator simply being an amplitude limiter stage which removes the picture information and passes on the sync. signals in the transmitted waveform.

Line timebase. The line timebase supplies the 10 125 Hz (405-line) and 15 625 Hz (625-line) deflection waveforms, and forms the heart of the e.h.t. system which in all modern receivers utilises the step-up of the comparatively high voltage produced across the primary winding of the line-output transformer by virtue of the rapidly changing current during the return of the scan stroke to the beginning of a new line ('line flyback period'). This provides a peak pulse voltage of the order of a few kilovolts which can then be stepped up to 15 to 20 kV by means of an overwind on the primary of the line-output transformer, and subsequently rectified. Alternatively the overwind may provide only one third of the required e.h.t. to be subsequently raised to the final value using a semiconductor multiplier. To provide the damping needed to reduce 'ringing' and also to improve the efficiency of the system, an 'efficiency diode' is incorporated. As this diode provides a boost voltage which is added to the h.t. for the line-output valve (and can be used for other stages), it is sometimes referred to as a boost diode.

A line generator may either be controlled directly from the sync. pulses, or alternatively these pulses may be used indirectly to control a free-running line generator in what is termed a 'flywheel' system. Flywheel systems are particularly desirable on systems having negative modulation (as in the British 625-line system), since, in these systems, noise pulses tend to appear in the sync. pulse region and can cause incorrect triggering of a directly controlled line generator. However, the relative freedom of the u.h.f. bands from impulse interference means that some older models used direct control for both 405- and 625-line reception.

Power supply circuits. In British practice, the h.t. supply is almost invariably obtained direct from the mains supply by half-wave rectification; the rectifier may be a valve, a metal (selenium) rectifier, or a silicon power diode. Since silicon diodes offer the greatest efficiency and are little affected by ageing, they provide a higher h.t. rail than can be obtained from the other rectifiers and are now universally used.

Within the line timebase, a higher 'boost voltage' source is available from the efficiency diode (see later), and this may be used for such applications as supplying a scanning generator, focusing, etc.

The e.h.t. required for the final anode of the picture tube is, as we have seen, produced in the line output stage.

Since one side of the mains supply is usually connected directly to chassis, without any form of isolating transformer, care must always be taken to avoid the risk of shock during servicing operations and to leave the receiver so that there is no risk to the users. The original reason for using this *live chassis* technique was to enable receivers to be operated on d.c. mains supplies. There is no longer any need for this provision and some receivers derive a higher h.t. voltage using a mains autotransformer technique. However, the chassis is still connected to one side of the mains on the vast majority of receivers and it is not safe to assume that a receiver marked 'a.c. operation only' does not use the live chassis technique.

Single-standard u.h.f. receivers

After the period of relative complication in dual-standard receivers, the present u.h.f.-only receivers are very much simpler: no standards switching is required. (*Figure 2.3*).

Parts of the receiver noted above for 625-line operation are the only ones contained in the new receiver and some of the compromises necessarily adopted for dual-standard operation (particularly in the i.f. strip) are no longer present. The essential circuit design changes lie in a general use of semiconductors throughout, including the use of a number of integrated circuits (i.c.s). Because there is now no v.h.f. tuner through which to gain additional amplification for the output of the u.h.f. tuner, it is normal for the combined vision/sound i.f. strip to have an additional stage of gain.

Tuners

The basic purpose of the tuner unit at either v.h.f. or u.h.f. is to change the vision and sound channel frequencies down to a fixed frequency— the intermediate frequency (i.f.)—whatever the channel. The i.f. signals formed can then be treated without any complicated tuning systems (that would be needed with the t.r.f. principle) when the received channel was changed.

To perform this change of frequencies requires a *mixer* which is supplied with an internal oscillator voltage at a frequency *higher* than the received channel. The oscillator frequency is varied according to

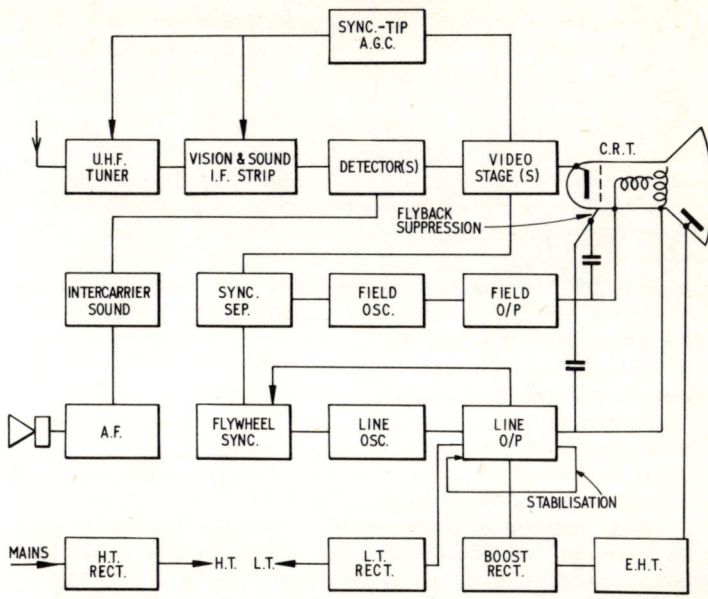

Figure 2.3   Block diagram of a typical single-standard 625-line receiver

the channel so that the i.f. is always constant—the difference between the two frequencies. The process of mixing is an inherently noisy one and to preserve a reasonable signal-to-noise ratio on the signal it is necessary to have an r.f. amplifier preceding the mixer stage. By making this stage of amplification relatively narrow-banded, some additional selectivity can be obtained, reducing the possibility of adjacent channel interference and cross-modulation. The additional stage inserted between the aerial and the mixer also helps to isolate the aerial from the oscillator voltage—so minimising the risk of radiating the voltage which could cause interference with other receivers or services.

An r.f. amplifier is thus found in both u.h.f. and v.h.f. tuners although the different frequency ranges dictate rather different circuit techniques.

On Bands I and III, conventional inductors and capacitors can be used for the tuned circuits required, although on Band III the inductors consist of relatively short lengths of wire, and the stray inductances and capacitances represented by the leads, etc., become increasingly important. On Bands IV or V such inductors would be little more than a half loop of wire, and would involve major constraints on the design and performance of the tuner. However, at these frequencies it is possible to use distributed rather than lumped constants, by using *transmission line* techniques.

The fundamental form of a transmission line circuit (Lecher line) comprises two parallel conductors. An open-ended line of this type when equivalent to an electrical half-wavelength has the characteristics of a parallel-tuned circuit; or a series-tuned circuit if the far-end is short-circuited. In practice, the twin Lecher line is seldom used, but is replaced by a single length of rod in conjunction with the *trough line* formed by the sectionalised chassis: the rod conductor and the surrounding chassis act rather like a short length of coaxial transmission line. In order to tune the resonant circuit provided by a trough line arrangement, the line is capacitively loaded at the 'far end', and trimmers are attached for tracking purposes at the nodal points for the two extremes of tuning range: the active device (valve or transistor) represents a capacitance across the 'near end' of the line.

While most valve-type u.h.f. tuners used lines representing electrical half-wavelengths (although appreciably shorter than a physical half-wavelength, because of the capacitive loading), it becomes possible with transistors to use quarter-wave lines. This is because valve holders are situated to one end of the line, whereas the smaller physical size of transistors allows them to be fitted inside the actual body of the tuner. Quarter-wave lines allow smaller tuners to be designed, with lower-capacitance tuning capacitors and fewer trimmers. However, a problem arises in the use of quarter-wave lines in integrated u.h.f./v.h.f. tuners since the capacitors may not allow tuning throughout the range, and for this reason half-wave lines may be used.

To couple two trough line circuits together, as in interstage band-pass tuning, it is necessary only to drill a hole in the adjoining screens, or alternatively to have a small loop of wire pass from one section to the other.

Most u.h.f. tuners have four tuned trough-line circuits: aerial input, r.f. amplifier load, mixer input and local oscillator. All four circuits are tuned simultaneously by means of a ganged capacitor. The tuner body is normally in the form of a drawn steel box with inter-screens forming the trough line compartments. To prevent oscillator radiation, which could otherwise cause interference with nearby receivers, the entire unit is often provided with a tightly fitted lid, with a copper gasket to ensure good electrical contact. Alternatively, with transistor units, the open side of the tuner box may be covered by copper foil, held in place with a plastics lid and it is important that this is always tightly replaced.

In general, u.h.f. tuner adjustment and alignment calls for extreme care and specialised alignment equipment: for this reason most makers recommend that little or no local servicing of such units should be attempted.

A typical valve u.h.f. tuner would have a noise factor of 10–12 dB, and a power gain of 12–15 dB.

The corresponding figures for a transistor unit would be 6–10 dB, and power gain 15–18 dB.

Normally with valve tuners, no attempt is made to apply a.g.c. to the r.f. amplifier; but with transistor units it becomes practicable to apply forward a.g.c. (see later in this Section) to this stage.

On v.h.f. tuners, turret, permeability and switch-tuning have been widely used; early u.h.f. tuners used continuous tuning by variable capacitance over the whole range of Bands IV and V. Because this is unsatisfactory to the normal viewer for reception of more than one station on u.h.f.—a certain amount of tuning skill being needed— push-button tuners have become popular. In the standard form these buttons move the sections of the tuning capacitor to preset positions. It is now common to use variable capacitance electronic tuning employing reverse biased silicon diodes (e.g. varactors) as the variable tuning elements. Contained near the tuner are a number of preset variable resistances which control the d.c. bias supply to the diodes. As the bias is varied the tuning changes. In this case the push-buttons on the control panel of the receiver change over which preset resistor is in circuit so giving a number of preset channel tunings. It is not practicable to employ a single resistor which is mechanically moved to different positions as the channel is changed, because the resetting accuracy required would demand an extremely expensive component.

It is also now common to incorporate a form of automatic frequency control (a.f.c.) in the tuner which holds the tuning steady by means of a d.c. signal fed from later in the receiver. This is particularly necessary in a colour receiver where a slight change in tuning can cause an annoying change in the saturation of the colour picture.

V.H.F. tuners

It is fairly standard practice for the first valve in a v.h.f. tuner to act as a cascode r.f. amplifier. This is, in effect, a grounded-cathode amplifier with its anode load being formed by a grounded-grid amplifier (*Figure 2.4*). The principle is that a grounded-cathode stage (V1), if required to give a large gain, is liable to go into oscillation because of signal fed back through the anode/grid stray capacitance of the valve itself. The stage has however the advantage of high input impedance.

In the grounded-grid stage (V2), the oscillation problem is overcome because any feedback passing to the grid is at a.c. earth (through the decoupling capacitor). The stage gives the same gain as the grounded-cathode because the signal is still being applied between the grid and cathode. But the grounded-grid stage has a very low input impedance when giving any reasonable gain because the anode current is also the

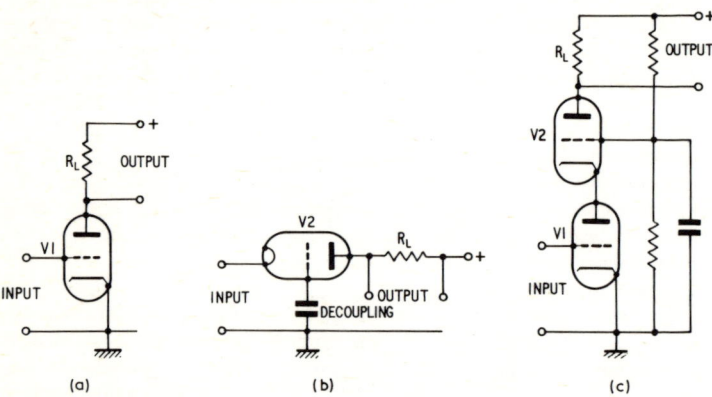

Figure 2.4   Derivation of basic cascode amplifier (c) by combination of (a) grounded-cathode and (b) grounded-grid stages

input, cathode current; as the anode current increases with gain, the cathode current also increases so the input impedance reduces. This would be no problem if the source of signal is low-impedance, and the stage has been popular in v.h.f./f.m. radio receivers.

The ideal way to arrange an input impedance to match the grounded-grid stage is to drive it with a valve that must have the same impedance—that is the same current and voltage. The cascode does just this with the lower stage providing the correct match to the upper valve. The lower stage need not have much gain so the dangers of self-oscillation are reduced, and yet the grounded-cathode gives a high input impedance

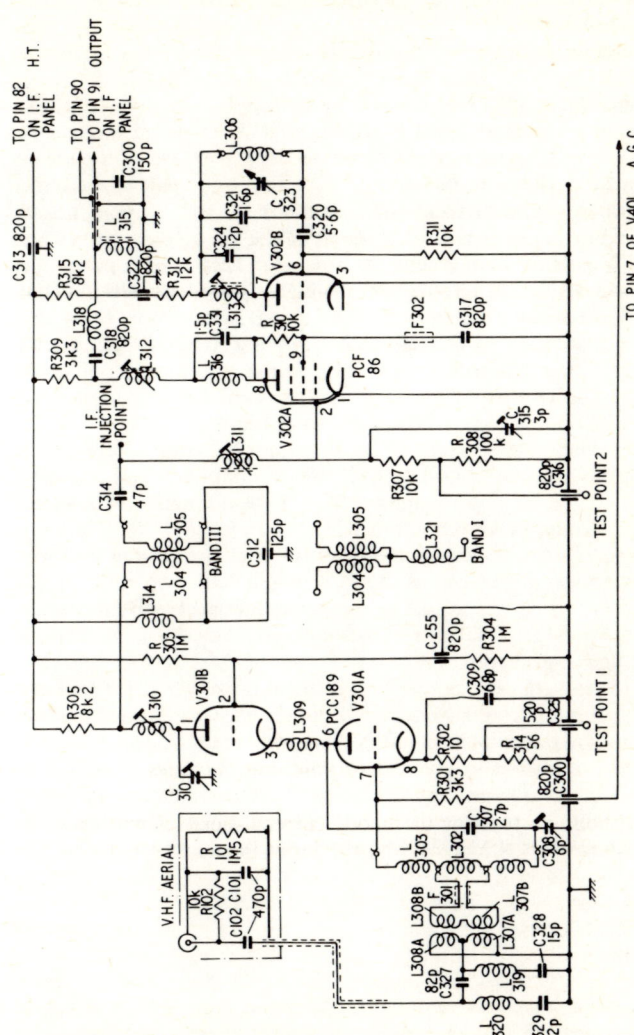

Figure 2.5   Representative v.h.f. tuner using cascode r.f. amplifier (Stella ST1093A series)

which can be easily matched from the aerial through an aerial transformer.

A representative example of a cascode r.f. amplifier tuner is shown in *Figure 2.5*.

The cascode amplifier is formed by a double-triode valve V301. The input is broadband tuned to the channel selected and the output is coupled through a tuner transformer to the signal grid of the mixer stage formed by the pentode part of V302. The triode part of the same valve operates as a local oscillator, and internal valve coupling of the local oscillator signal produces a signal at the i.f. to pass out to the i.f. strip. Important components to note are: L309 which peaks the gain of the cascode r.f. amplifier on the higher frequencies (Band III), the aerial isolation panel which gives signal conduction from the centre and outer of the coaxial socket into the tuner proper whilst preventing either external point from being at chassis potential. Automatic gain control of the r.f. amplifier is effected by varying the d.c. potential on the grid of the lower part of the stage.

Channel changing is arranged by switching between different sets of coils in the aerial tuning (L301/2/3), the r.f. amplifier load coupling (L304/5) and the oscillator tuning (L306). Fine tuning in the selected channel is effected by variation of C323.

An alternative to the cascode amplifier is sometimes used in the form of a beam-triode valve (such as the PC97 or PC900). The effect of the beam grid in this case is to reduce the valve internal feedback capacitance from anode to signal grid so that there is less chance of oscillation in the grounded-grid configuration.

The field strength of a received v.h.f. signal is usually larger than on u.h.f., and the v.h.f. aerial also has a larger *aperture* than the u.h.f. one; more of the available signal is therefore received. The gain of a v.h.f. tuner can also be greater than a u.h.f. one (because of the fall in gain that occurs with increasing frequency), and to match signal levels in a dual-standard receiver the u.h.f. tuner output is often fed back into the v.h.f. tuner to use the frequency-changer as an additional i.f. amplifier.

U.H.F. tuners

Valves at u.h.f. have less gain and higher noise factors than at v.h.f., and u.h.f. tuners using valves differ somewhat in their configuration. In general a triode valve (such as the PC88) is used as a grounded-grid

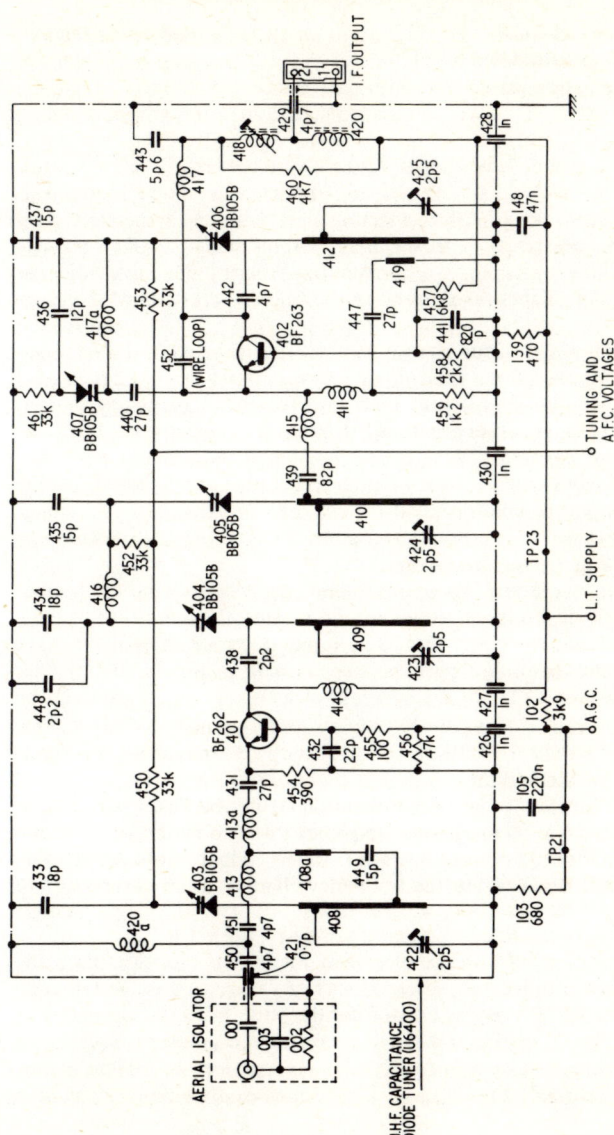

Figure 2.6 Philips u.h.f. varactor diode tuner with a.f.c. and a.g.c. facilities

r.f. amplifier and another triode (such as the PC86) is used in a grounded-grid format as both local oscillator and mixer. A very similar technique is used when transistors are employed in the u.h.f. tuner. A typical example is given in *Figure 2.6* which also incorporates varactor tuning and a.f.c.

At u.h.f., useful advantages result from the small size of semiconductor devices. Lead lengths are important at these frequencies because a short length of connecting wire with its associated stray capacitance can form a self-resonant circuit which could seriously interfere with performance. A conventional thermionic valve, together with its holder, places restrictions on the shape and layout of the tuner unit.

The input and output characteristics of a transistor differ somewhat from those of the thermionic valve it replaces, so that careful coupling into the source and load impedances is necessary. Tuned couplings between stages are usually formed by quarter-wave resonant lines, lumped constants being unmanageable at these frequencies, and it is a relatively straightforward matter to couple the transistor into the lines at the points which present the required impedances. Two stages are usually involved, a common-base tuned r.f. amplifier followed by a self-oscillating frequency mixer.

Minimum noise and maximum gain requirements tend to conflict in the design of the tuned aerial circuit because the condition of optimum power match to the transistor input electrode does not always coincide with best noise performance. It is usual to design for best noise figure and to suffer a slight loss of gain. Although gain can be made up in later stages, no amount of attention will correct a poor first-stage noise performance. The noise contribution from the mixer stage is usually small.

Under the system for national u.h.f. coverage, there are certain areas having two signals spaced by 10 channel widths. With a channel width of 8 MHz and a standard i.f. in the 35–40 MHz range, this situation demands good image rejection in the tuner. By careful design the required rejection of > 50 dB can be realised using a single quarter-wave tuned line in the aerial circuit.

The varactor diodes in the circuit of *Figure 2.6* simply replace the mechanically coupled tuning capacitors of earlier tuners, across each tuned line. The a.f.c. feed is taken from the output of a discriminator at the end of the i.f. chain, and the d.c. varies the oscillator frequency to keep the tuner adjusted correctly. To prevent *lock-out* when a new channel is selected, the a.f.c. feed is broken and switched in with a relatively long time constant when the channel selector button is released.

Future developments in u.h.f. tuners will be the use of field-effect transistors giving better protection from crossmodulation because of their inherently better linearity and possible higher sensitivity and better noise performances.

Integrated tuners

Some dual-standard receivers make use of a single tuner for both u.h.f. and v.h.f. reception. These are always transistorised. As already noted, v.h.f. tuners use a separate oscillator stage whilst u.h.f. tuners employ a self-oscillating mixer. The reason is one of stability in that a v.h.f. tuner must tune right down to the bottom end of Band I (41 MHz) and there would be difficulties in building a self-oscillating mixer when the intermediate and signal frequencies are not well separated from one another. This problem does not arise at u.h.f.

The lower signal level and the lower device gain at u.h.f. mean that in any integrated tuner there must be an additional stage of gain provided so as to give more or less matched signals into the i.f. strip.

A block diagram of such a tuner is given in *Figure 2.7*. The arrangement is much as discussed; a grounded-base AF186 transistor operates as r.f. amplifier on all channels, the requisite input tuning being between the appropriate aerial socket and the transistor. On v.h.f. an AF178 is used as the local oscillator with another AF178 as the mixer stage. This mixer doubles up as the first i.f. amplifier when the tuner is switched to u.h.f., and the second AF186 transistor operates as self-oscillating mixer on u.h.f.

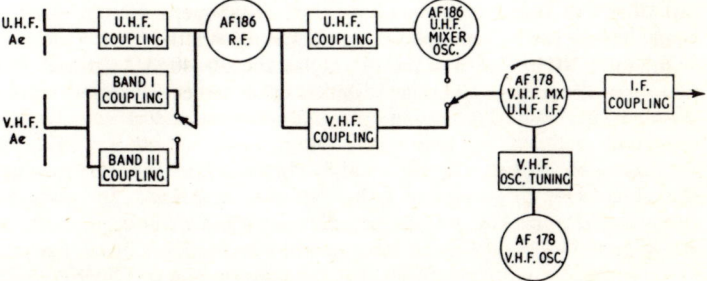

Figure 2.7   Block diagram of an integrated v.h.f./u.h.f. tuner unit

Wired relay operation

Even with an ever-increasing number of receivers operating on 625-lines/u.h.f. only, there are also, and will continue to be, areas where the geographical situation prevents direct reception of quality v.h.f. or u.h.f. signals. In a number of such areas, national or local organisations provide wired relay distribution of the broadcast signals by changing their carrier frequencies, at a common receiving point, down to suitable v.h.f. channels.

For reception of 625-line wired signals on a dual-standard receiver, it is common for the receiver unit (i.f., timebase, detector, sound system, etc.) to be left in the 625-line condition but with the v.h.f. tuner switched into circuit.

On receivers designed for single-standard u.h.f. operation, two alternatives are possible: a v.h.f. tuner may be fitted in place of the u.h.f. tuner or, more commonly now on new installations, the receiver is left in its original condition and a u.h.f. *transposer* is fitted before the aerial socket. This transposer changes the frequencies of the wired 625-line signals up to suitable channels in the u.h.f. band. It has the advantage that manufacturers no longer have to make v.h.f. tuners for the home market but it has only been possible to achieve since stable enough circuitry has been available in the transposer.

Intercarrier sound

Where both vision and sound signals employ amplitude modulation (as with 405-line/v.h.f. transmissions), it is necessary to separate these signals before the video detector; and in practice this separation is carried out early in the vision i.f. strip or immediately after the tuner. But, where frequency modulation is used in the sound channel (as with U.K. u.h.f. transmissions), the two signals can be passed through the entire vision i.f. strip and mixed together in the video detector to produce an f.m. heterodyne signal on a frequency equal to the frequency difference between the vision and sound carriers (that is 6 MHz in the case of the British 625/u.h.f. standard). Arrangements which utilise this heterodyne principle are called *intercarrier* systems. In countries using the 625-line C.C.I.R. or the 525-line F.C.C. standard, both of which have f.m. sound on all frequencies, intercarrier is often, but not always, used for v.h.f. reception. The intercarrier is in fact 5·9996 MHz in the U.K. This 400 Hz offset has been found quantitatively to reduce interference products between PAL colour signals and the sound.

With intercarrier sound, the video detector has two distinct functions: it acts as an a.m. envelope detector producing the usual video/sync. signals; and it also acts as a non-linear device to produce heterodynes between the vision and sound carriers. The resulting 6 MHz beat signal then carries the sound f.m. information but is largely unaffected by the amplitude modulation of the main vision signals. The 6 MHz signal may be taken off immediately after the video detector, or it may be further amplified in the video amplifier, in each case afterwards being taken to a 6 MHz sound i.f. amplifier and sound f.m. discriminator. In dual-standard receivers this 6 MHz amplifier is normally the same stage(s) used for the v.h.f. sound signals at 38·15 MHz, by means of dual frequency arrangements similar to those used in a.m./f.m. sound radio receivers.

With intercarrier sound, the sound receiver thus comprises a double-conversion superhet, with the second local oscillator signal taking the form of the i.f. vision carrier. The advantage of using the intercarrier principle for u.h.f. reception is that the 6 MHz difference between vision and carrier frequencies is determined accurately at the transmitter and is not affected by the degree of stability achieved in the u.h.f. oscillator within the u.h.f. tuner. The sound reception is thus considerably more tolerant of oscillator drift than would be the case for conventional sound i.f. techniques; intercarrier also reduces the effects of microphony or hum.

The importance of this stability is dual: many people have difficulty in tuning an f.m. sound transmission because tuning is made for minimum *distortion* rather than maximum level (as is the case for a.m. reception). Using intercarrier sound techniques the 6 MHz signal cannot be mistuned so distortion will be a minimum. If the front-end tuning of the receiver is maladjusted or drifts in frequency, the intercarrier level drops, but until the level passed to the f.m. detector goes below the limiting level no untoward effects will be noticed. The resultant picture impairment of the front-end mistuning would be more noticeable than the effects on the sound channel.

However, to achieve satisfactory intercarrier reception, the relative strengths of the sound and vision signals at various stages of the receiver are of considerable importance. Vision modulation should not fall below 10 per cent of peak vision carrier otherwise there would be insufficient vision carrier to demodulate the 6 MHz intercarrier. The presence of amplitude modulation on this small region of carrier would also cause 'buzz' on the intercarrier sound, and care must be taken at the transmitter to prevent overmodulation—particularly on fully saturated colours such as yellow.

To avoid the production of 'buzz' on the sound due to the wide amplitude variation between peak video and sync. tip levels, the sound carrier should be 24 dB or more down on the vision carrier at the video detector; or in other words we need only a weak sound signal to be presented to the video detector, and the sound signal must be attenuated by the response characteristics of the i.f. stages.

In practice not all these factors are achieved, partly because of variations between the ratio of the carrier powers of the vision and sound signals, and for a variety of reasons it is difficult to avoid some degree of vision-on-sound ('buzz') with intercarrier sound. Many different causes of such buzz have been identified, and it has also been found that it is more difficult to avoid buzz in the case of receivers which use the video amplifier stage as a 6 MHz sound i.f. amplifier.

In these cases, the detected video and the 6 MHz intercarrier interfere with one another, because of the amplifier nonlinearity, producing intermodulation distortion frequencies close to 6 MHz. These frequencies would be detected in the f.m. discriminator to give an audio output. As a lot of the intermodulation is due to the frequently recurring information in the video such as field, the resultant 'buzz' from the loudspeaker will also tend to be continuous with increases according to the type of picture information being transmitted.

It is also very difficult for the television transmitter to produce a modulated vision signal at u.h.f. in which the carrier does not have some unwanted phase modulation superimposed on it (known as incidental phase modulation). This is very akin to frequency modulation and in the detection of the intercarrier this modulation is superimposed on the wanted f.m. signal. Again this will be detected to give some 'buzz'.

As mentioned earlier, the fact that incorrect sound to vision carrier power ratios at the video detector causes buzz is a further constraint on the i.f. response curve, since this has to include a fairly well-defined 'shelf' which allows the sound signals to pass through the vision i.f. stages but restricts them to a sufficiently low level.

From a servicing viewpoint, an important parameter is the ability of the f.m. sound detector to reject a.m. signals, and most detectors (usually ratio detectors) include a balancing adjustment which is preset to provide optimum a.m. rejection. In some designs, the screen voltage on the final 6 MHz amplifier stage is reduced for f.m. reception to provide a degree of pre-detector limiting, or an envelope-feedback (a.g.c.) loop is provided. Alternatively, a diode limiter is used across the load of one of the intercarrier i.f. transistors.

Troublesome intercarrier buzz in a receiver can usually be found to be due to the ratio detector balance being maladjusted, the 6 MHz take-off point in the video stages being mistuned (allowing video information to be passed), or a fault in the 6 MHz i.f. channel giving low gain—in which case full limiting action may not be given. Out of balance diodes can also cause poor a.m. rejection and these should be checked at an early stage in any servicing.

Intermediate frequency amplifiers (i.f. strips)

The purpose of the vision i.f. strip is to amplify the i.f. signal up to a suitable level for detection whilst preserving the bandwidth of the signal and shaping the response as necessary to give the correct ratios between particular portions of the signal.

A variety of techniques are used in the vision i.f. amplifier strip to attain (as nearly as possible) the differences required in the overall frequency response to 625-line and 405-line signals. In order to understand the purposes of the filter techniques used in the actual circuits, it is essential first to consider the basic aims. The response curves must differ not only in overall bandwidth, so as to accommodate the much broader 625-line signal, but also in the location of the various carriers, adjacent channel rejection notches, and in the creation on 625 lines of a sound shelf for intercarrier as opposed to the full sound notch needed with 405-a.m. sound signals.

It should be noted that in the i.f. stages the relative positions of the vision and sound carriers are reversed on u.h.f. and v.h.f. stations: whereas, on v.h.f., the sound carrier is 3·5 MHz *above* the vision carrier; on u.h.f., the sound carrier is 6 MHz *below* the vision carrier.

Figure 2.8 shows these differing requirements. Ideally, if the transmitted signal were either double-sideband or single-sideband, no shaping would be required because the correct level of signal would be received at all frequencies—all that would be necessary would be sufficient bandwidth with enough gain. However, with vestigial sideband transmissions there is effectively double the required signal power being transmitted at the low video frequencies around the vision carrier compared to the higher video frequencies.

This could be negated by using an i.f. filter with a sharp cut off at the vision carrier position (i.e. rejecting all frequencies above 39·5 MHz

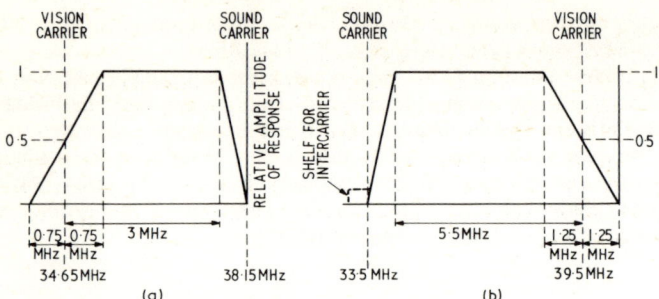

Figure 2.8   Idealised vision i.f. responses for (a) *405-line system;*
(b) *625-line system*

on 625 and below 34·65 MHz on 405). This would demand an expensive filter with a number of sections and one that would have to be set up very accurately in frequency so that no low-frequency signals, or the carrier, were lost; it is not practicable.

Instead a gentle 'roll-off' is given to the filter and half the required low-frequency signals are derived from one sideband and half from the other. This requires a filter that crosses the vision carrier i.f. at the half amplitude point—the 6 dB point on the slope of the curve shaped to give the correct vestigial sideband response (i.e. vision carrier $\pm 1·25$ MHz for 625, and $\pm 0·75$ MHz for 405). The curve should remain substantially flat out to a distance from the carrier frequency equal to the maximum video frequency, although this may not always be achieved in practice, particularly on 625-line signals: this is thus ideally vision carrier $-5·5$ MHz for 625, vision carrier $+3$ MHz for 405.

The presence, either switched-in or -out of circuit, of so many frequency-conscious reactive components in a dual-standard i.f. amplifier tends to impair the phase (envelope delay) characteristics of the amplifier, and means that the vision sidebands on different frequencies may undergo different phase shifts (i.e. signals at different frequencies take slightly different times to pass through the amplifier). Such i.f. amplifiers thus depart considerably from the ideal phase-linear characteristic. This is of especial importance in the case of colour receivers where the phase characteristics may have marked effect on the quality of colour reproduction. Good phase-linear characteristics require accurate neutralisation of feedback reactances over the full bandwidth. The i.f. response characteristics are likewise of particular importance in colour, since in this case (see Chapter 5) the colour information is carried on a sub-carrier at the upper-end of the video frequency (roughly 3·0 to 5·5 MHz for 625). In monochrome receivers (particularly older ones) it is not unusual to find that the response at such frequencies is appreciably below that at the lower frequencies, since the slight degradation in fine detail response is acceptable: this will be apparent from the frequency gratings on the Test Card signal.

There are two basic methods of providing the required i.f. gain and selectivity. On 405-line and dual-standard receivers it is common to use a number of tuned amplifier stages passing a relatively narrow band of frequencies. To achieve the overall bandwidth requirements the amplifiers are *stagger-tuned;* i.e. each stage is tuned to a slightly different centre frequency. On later single-standard 625-line receivers it is becoming common practice to use a wideband amplifier strip or integrated circuit with all the selectivity achieved in a separate filter unit at the input of the strip. This unit may be constructed using discrete components or it may be formed using a ceramic filter or a surface-wave filter.

Providing all the filtering at the input point has distinct advantages in allowing the use of simplified amplifier stages or integrated circuit(s) and also ensures that, should there be any amplifier nonlinearity, it will not cause crossmodulation interference because of adjacent channels being passed into the amplifier strip.

Fortunately the stability of i.f. transformers and circuits in modern receivers is of a high order, and realignment is seldom required.

The actual shaping of the vision i.f. response may be achieved by various combinations of bandpass i.f. transformers and resonant circuits, by series or parallel tuned traps and bridged-T notch filters.

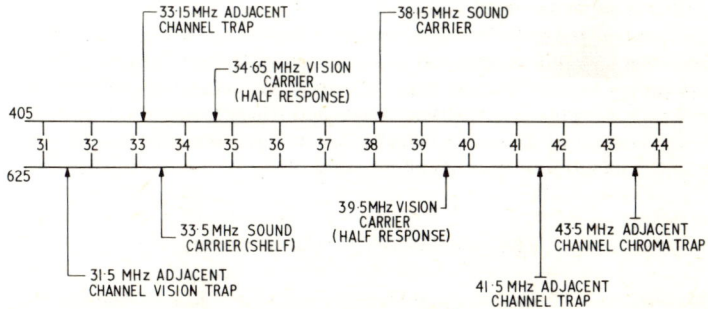

Figure 2.9   Important alignment frequencies for 405- and 625-line i.f. amplifiers

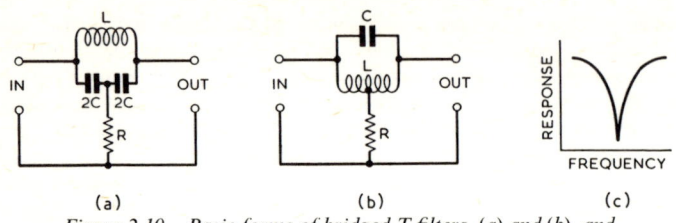

Figure 2.10   *Basic forms of bridged-T filters, (a) and (b), and
their response, (c)*

This form of band-stop filter can provide a much sharper rejection 'notch'
(point of maximum attenuation) than a simple parallel-tuned trap, and has
less effect on adjacent frequencies. The notch is most pronounced when R is
equal to one-quarter of the resonant impedance of LC.

In some dual-standard designs the response is shaped primarily to meet
the broader 625 requirement with the narrower 405 response obtained
by additional filters at the output of the v.h.f. tuner; in other cases
different bandpass characteristics are obtained stage by stage by changing the value of the coupling reactances. Because of the need to insert
notches (heavy attenuation at specific frequencies *(Figure 2.9)*), the
design and switching of dual-standard i.f. circuits are necessarily
complex. A representative example is described in Chapter 8 (Alignment).

A parallel-tuned trap provides high impedance at the resonant
frequency; a series-tuned trap very low impedance at resonance.
Another circuit technique is to include a parallel-tuned circuit in the
cathode circuit of a valve or emitter circuit of a transistor where it
provides rejection (that is attenuation) at the resonant frequency by
introducing heavy negative feedback.

A commonly used filter is a bridged-T network, which is capable of
providing a much deeper and sharper attenuation at resonance than
a parallel-tuned rejection trap, and has less effect on off-resonant
frequencies. *Figure 2.10* shows basic forms of this type of filter; the
notch is most pronounced when the resistor R is equal in value to one-
quarter of the resonant impedance of $L \times C$.

TRANSISTOR I.F. AMPLIFIERS

With the sort of tuner arrangements already mentioned in the previous
section, an i.f. gain of about 80 dB is needed to produce a sensitivity

of better than 10 µV at the aerial, which is a desirable figure for fringe-area reception. Three transistor stages are needed to give this, and, for reasons of economy, one or more of the stages will also be used to amplify the sound carrier as well as the picture information, the a.m. sound commonly being extracted at an intermediate coupling circuit, while the f.m. sound on 625-line transmission is deliberately passed right through the amplifier so that it can be extracted as intercarrier sound at the video detector or video amplifier.

The vision selectivity characteristics can be broadly divided into two parts, one concerned with the response shape over the pass-band, the other with the rejection of unwanted signals outside the pass-band. This has always been so, and the only way in which the selectivity requirements for a transistor or hybrid receiver are likely to differ from those of an all-valve receiver arises from the rather inferior signal-handling performance of present transistors. This broadly means that unwanted signals should be effectively rejected at as early a point in the chain as possible, because once the wanted carrier becomes cross-modulated by an unwanted signal, due to nonlinearities, no amount of rejection after that point will help the situation.

As with valves, the weakest stages in this respect are the gain-controlled stages, but the conventional type of reverse a.g.c. transistor stage as commonly used in broadcast receivers is noticeably inferior to the valve and not good enough for television i.f. application where signal levels are high and unwanted signals abound. The critical condition occurs when the transistor is biased well back to near-zero collector current, which corresponds to high signal input, the transistor then tending to work under Class C conditions with resultant distortion.

Forward a.g.c.

For this reason a special range of transistor types is available which have been designed to exhibit a collapse of gain with *increasing* collector current above a certain critical level. Thus, under high-signal conditions, a more linear portion of the device characteristics is traversed.

These devices are known as forward a.g.c. transistors and some are capable of giving a control range of 60 dB over a collector current range of 4–10 mA. This is sufficient to require only a single controlled stage in a vision i.f. strip, invariably the first stage, but some early hybrid

receivers using narrow-range devices had two controlled i.f. stages
and these can usually be identified by tracing the source of their base
bias potentials.

For some servicing jobs it is advisable temporarily to remove the
a.g.c. voltage from a stage, for example when using wobbulator display
of the receiver response, or when localising certain faults. In a valve
receiver this is fairly simple, because the standing bias is usually self-
developed across a cathode resistor and remains even when the grid
is returned back to earth; but a transistor standing bias has to be
separately applied, it cannot be established solely by the passage of
collector current through an emitter resistor. So, to remove a.g.c.
volts without upsetting standing bias, one must go right back to the
point where the control voltage is derived and either short-circuit or
disconnect the source from the system without interfering with the
quiescent d.c. conditions of any a.g.c. amplifiers.

A transistor television receiver is almost certain to have d.c. amplifiers in the a.g.c. circuit because of the relatively high control power
required. Forward a.g.c. stages are particularly bad in this respect
because they require to be pushed into heavy conduction and into a
region where even the d.c. gain of the transistor is collapsing. In some
cases two d.c. amplifiers are used, one of them being a high dissipation
type.

Change of overall response shape with a.g.c. action may be rather
more troublesome than in a valve receiver, particularly in a forward-
controlled stage, because of the wide variation of transistor parameters
with current. It is, however, easy to confuse this effect with that associated with stray feedback in a high gain receiver, and before con-
demning a control transistor it is as well to check for the presence of
overall feedback, perhaps by exploring with a small earthed screen
under maximum-gain conditions. If some compromise has to be
adopted on response shape, remember that very few receivers will
be called upon to operate at maximum gain and it is best to align at an
intermediate gain condition.

Feedback capacitance and distortion

The most promising line of development in television i.f. transistors
is reduction of internal feedback capacitance. In the early days of the
radio valve the capacitance between grid and anode provided a limit
to the stable gain which could be obtained, and in the same way so
does the capacitance between the collector and base of a transistor.

External neutralising can be used to raise this threshold, and the limitation is then set by the degree of neutralising error which can arise. In mass-produced equipment such as television receivers this error arises due to production tolerances, because it would be impracticable to neutralise each chassis individually. The best that can be done is to fix the neutralising components to suit an average device and accept the possibility of an error equivalent to an adverse combination of device parameter spreads and fixed capacitor tolerance. As the maximum stable gain which can be used is then closely related to the error, the source and load resistances must be set, at the design stage, at the right level to establish this gain.

The practical limitation to gain is not so much the possibility of complete instability but the distortion of response shape which occurs before the instability condition is reached. This may manifest itself as change of response when a.g.c. begins to operate.

The rate of device improvement in this respect has been impressive, and for many years now transistors for radio receivers with feedback capacitance sufficiently low not to require neutralising have been available. With television the problem has to some extent come back again because of the higher operating frequencies, but further development has now produced screened silicon planar devices with a capacitance of less than 0.2 pF (e.g. the BF167 and BF196). This is again approaching the level at which it is no longer justifiable to attempt neutralising.

While neutralising is being used it is important for the service engineer to be aware of the critical nature of the process. The essence of the arrangement is to produce a delicate balance between the internal feedback path and the external anti-phase circuit. When the feedback capacitance is as low as a fraction of 1 pF, and perhaps comparable with stray capacitance levels, one can appreciate the ease with which this balance can be offset by careless servicing. Printed circuitry helps considerably because it tends to fix the position of components. If any components associated with neutralising are found to be faulty, replacements should be as near identical to the original as possible and mounted with care to reproduce the same stray capacitances.

Rejector circuits

The shaping of the response curve to produce the required rejection of unwanted frequencies outside the pass-band is carried out in a similar way to that customary in valve receivers, by high Q factor suck-out tuned circuits, or, in some instances, by bridged-T traps, and the alignment procedure is similar. The relatively low level of transistor input and output impedances may modify the LC ratios a little, but generally the trap circuits can be readily identified by anybody with valve receiver experience.

It should be remembered that bridged-T traps may be sensitive to a change of transistors. If preceding or following transistors have to be replaced, or indeed any other associated component, it may be that the balanced conditions will be upset sufficiently to cause trouble. In this case, slight adjustment of the balancing resistor will restore maximum rejection.

Final i.f. stage

A rather special type of transistor is needed for the final vision i.f. stage, because it has to handle a considerable amount of power at high frequency. The required drive may be as high as 3–4 V to cater for the limit conditions, and this voltage must be developed linearly across the detector diode load. After allowing for detector inefficiency and power loss in the coupling transformer, the collector dissipation will be about 100 mW. For a germanium transistor this may demand the use of a small heat sink clip.

Apart from dissipation problems, further difficulties are introduced in dual-standard receivers because the two carrier frequencies are situated on opposite sides of the pass-band and it is difficult to equalise the load conditions at the collector circuit. For this reason, manufacturers guidance on re-alignment of the final i.f. transformer should be particularly carefully followed as there will be a risk of producing a serious mismatch at one or other of the carrier frequencies, even though the overall transfer response shape is apparently satisfactory. The symptoms would be compression of peak whites on 405-line transmissions or clipping of sync. pulses on 625-line transmissions.

Because of the special characteristics of the final i.f. transistor, together with its high operating current, the working mutual conductance is very high, in the order of 200 mA/V, with instantaneous values somewhat higher. This is many times larger than in a receiving valve and the emitter decoupling arrangements are consequently more critical than cathode circuit equivalents in a valve i.f. stage.

The most important item is the decoupling capacitor and it can be very misleading to check it in the time-honoured fashion of temporarily

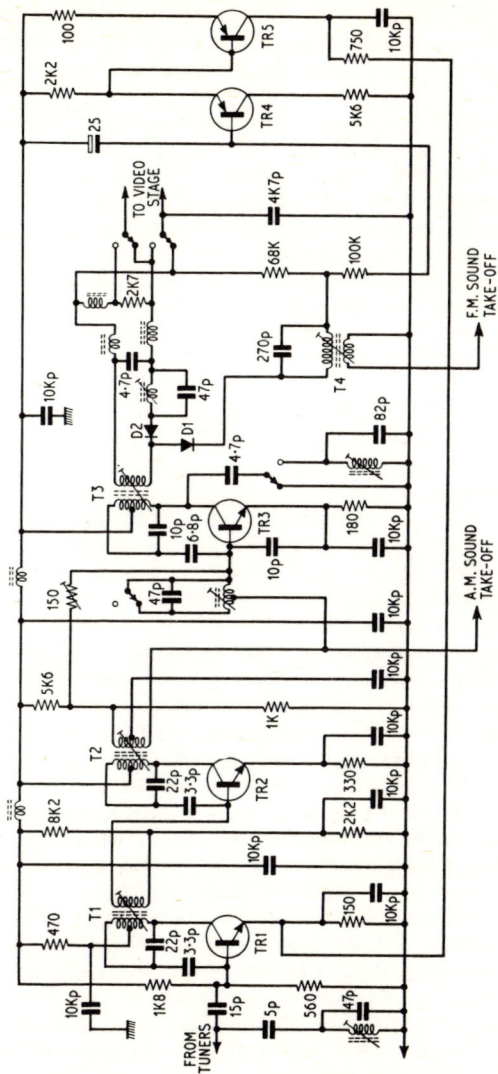

Figure 2.11   Vision i.f. and a.g.c. stages of the Pye 31F hybrid receiver

shunting another across it, particularly as there is a good chance that it has had its leads carefully cropped to produce self-resonance at the operating frequency in order to achieve minimum impedance. Exact substitution is the only safe approach.

Many of the features discussed above can be identified in *Figure 2.11*, which shows the circuit diagram of the vision i.f. and a.g.c. stages in an early Pye group hybrid receiver. TR1 is gain-controlled, the control power being provided by TR4 and TR5, while TR2 and TR3 are straight i.f. amplifiers. In each stage neutralising is applied by a small-value capacitor from the anti-phase end of the tapped collector winding on the interstage transformer. D2 is the vision detector diode, a separate diode (D1) being used for producing the intercarrier sound and a.g.c. signals. The required pass-band is obtained by stagger-tuning the single-peak coupling transformers.

625-LINE SINGLE-STANDARD I.F. AMPLIFIERS

In the dual-standard receiver, the v.h.f. tuner has a fairly high gain, and its mixer is usually employed as an i.f. amplifier for the u.h.f. tuner output, as we have already noted. A single-standard receiver will be equipped with a fairly low gain u.h.f. tuner only. The control range of the a.g.c. and the requirements of i.f. gain are therefore altered.

A maximum receiver sensitivity of 17 μV for 2 V across the detector load with a tuner noise figure of 8 dB would give a barely entertainment value picture with a signal-to-noise ratio of 15 dB. This is the worst condition that must be catered for and it demands a total gain of about 100 dB. Assuming a u.h.f. tuner gain of about 20 dB this then requires an i.f. amplifier gain of about 80 dB.

Even with forward a.g.c. the stringency of cross-modulation protection needed would not be met in a standard transistor i.f., and rejection of unwanted components must occur before the amplifier. The strip can then be quite broadband in its tuning and if the basic v.s.b. tuning is also incorporated before the amplifier, it can be broadband with no tuning whatever. One example is given in *Figure 2.12* which has the majority of the rejection and selectivity provided by the input filters and the first i.f. amplifier transistor circuit (2VT1). The gain stages consist of 2VT2 and 2VT3 arranged in cascode with 2VT4 driving the video detector. The Q of the tuned circuits in the load of both gain stages is quite low so that they are very broadband tuned around the i.f. centre frequency.

63

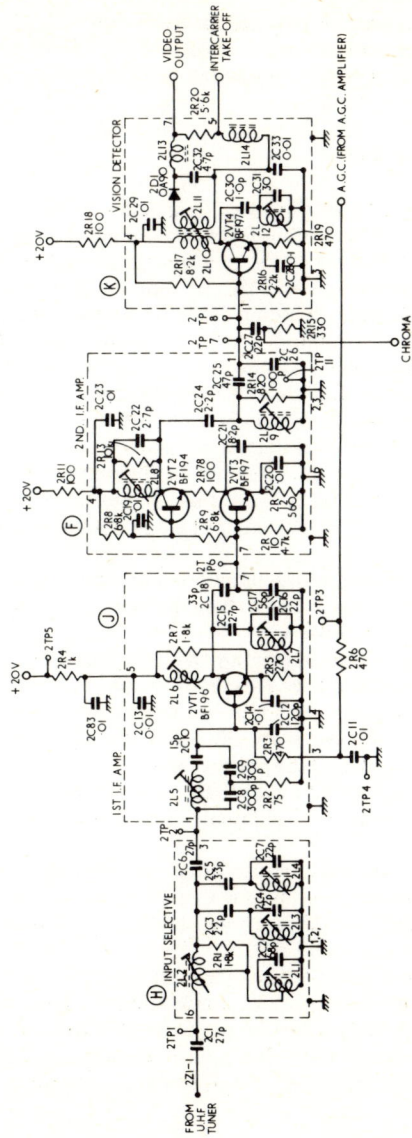

Figure 2.12 Single-standard u.h.f. receiver i.f. strip (Bush CTV 1825 series)

In future years ceramic and surface-wave filters will replace the conventional tuned circuits used for selectivity, and the broadband amplifiers will be replaced by integrated-circuit wideband amplifiers.

Video detector and amplifier

The vision i.f. stages provide the bulk of the gain needed to build up the aerial input signal voltage from a few microvolts to a value which will provide an output from the video detector of 2 volts or more (peak white on 405, sync. peaks on 625). In modern receivers the video detector is always a semiconductor diode, although in earlier models thermionic diodes were common. A relatively low value of diode load is used in order that the performance does not fall off sensibly up to the highest video frequencies in spite of the effect of stray capacitances. Without compensation, the response will be 3 dB down at the frequency for which $2\pi fCR = 1$. On 405 a load resistor of 5 k would postulate stray capacitance of only 9 pF. Generally compensation is achieved by connecting an inductor in series with the resistive load. For the larger bandwidth of 625-line signals, a smaller load of about 3·9 k is generally used.

In order to compensate for the change in modulation on 405 (positive) and 625 (negative), it is necessary to arrange in dual-standard receivers for the polarity of the output to be switched when changing between systems. By means of this switching the video and sync. signals as presented to the video amplifier remain of the same form regardless of system, and the biasing of the video amplifier can then be altered to bring the signals to the same level. In some chassis, polarity change is effected by using a video phase splitter stage.

The video detector output circuit has also to include some form of low-pass filter in order to eliminate the intermediate-frequency component.

After the video detector there is often, on 405 lines, a 3·5 MHz rejection trap to eliminate patterning caused by the vision and sound carriers beating together. This trap is often in the cathode or emitter circuit of the video amplifier.

To drive a modern cathode-ray tube fully, and to allow for certain limit conditions, a peak-to-peak output of about 120 V is demanded from the video stage. In solid-state receivers to realise the bandwidth requirements there is an upper limit to the collector load resistance so, to develop the volts, high collector currents are involved.

A video driver transistor is needed and is usually operated as an impedance converter in the common-emitter configuration to match the low detector load to the high Miller effect input capacitance of the amplifier.

On 625, the 6 MHz intercarrier signal is recovered and fed to the 6 MHz sound i.f. stage(s), either at the output of the video detector or after passing through the video amplifier; subsequently any remaining 6 MHz component in the vision signal may be filtered out by means of a 6 MHz trap. Generally, where the video amplifier is transistorised, the intercarrier feed is taken from the detector or from a driver stage between the detector and the main video amplifier. Two representative designs are shown in *Figure 2.13*.

To improve the high-frequency response of the video amplifier, various forms of 'top-lift' circuits are used, often based on peaking coils. For example an inductor may be included in the resistive load of the video amplifier and chosen to resonate with the stray capacitance at a frequency above (e.g., $\sqrt{2}$ times) the highest required frequency. This maintains the load impedance—and hence the gain—up to a higher frequency than for an uncorrected circuit. Response shaping of

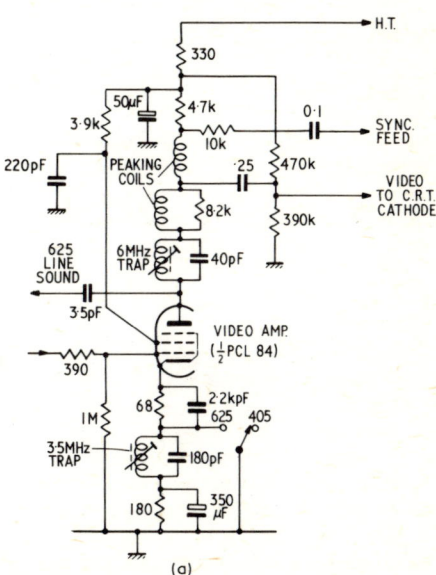

Figure 2.13 (a)—Video amplifier design for a dual-standard receiver where the intercarrier sound is extracted at the anode of the video amplifier. (Note the potentiometer chain across the feed to the c.r.t. cathode to establish the correct d.c. level, the anode peaking coils to obtain a satisfactory video bandwidth, the frequency-selective negative feedback in the cathode modifying the response and the 405/625 switch in the cathode which removes the 3.5 MHz rejector on 625 and alters the bias by shorting the 180 Ω resistor and its 350 µF decoupling capacitor).

66

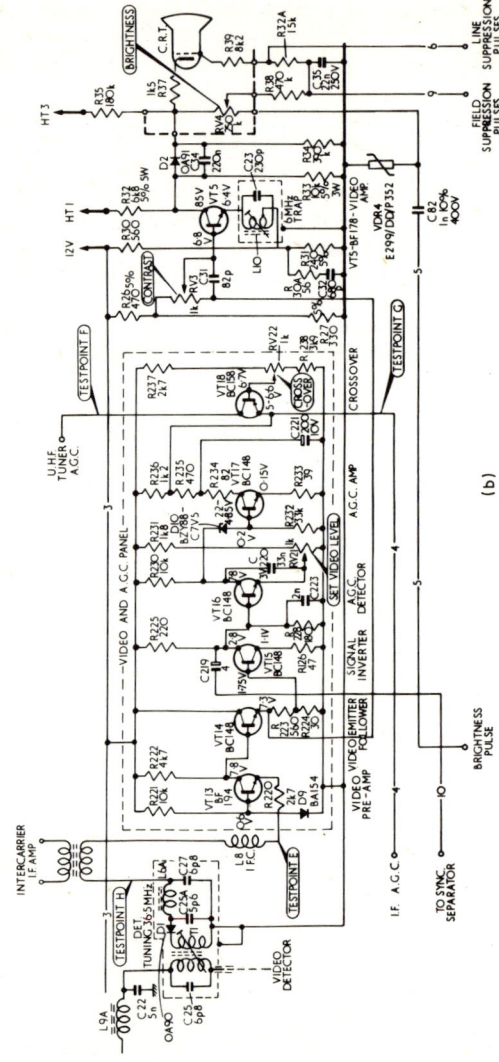

Figure 2.13 (b) Video amplifier and a.g.c. design for single-standard 625-line receiver. (Note the d.c. potential on the cathode of the c.r.t. is settled by the d.c. restorer circuit of D2, C34, R34 and the potential divider R32, R33)

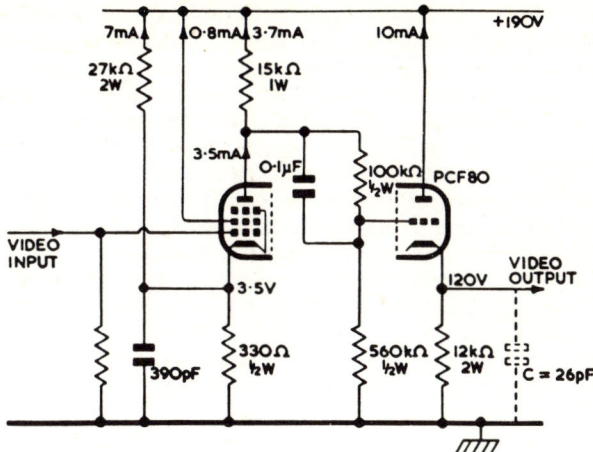

Figure 2.14   Video amplifier stage incorporating a cathode follower
(Mullard Ltd.)

the video amplifier may also be achieved by choice of bypass capacitances in the cathode bias network.

Some valve receivers incorporate a cathode follower in the video output stage (*Figure 2.14*). While a cathode follower stage has a voltage gain of less than 1, the inclusion of such a stage isolates the first video stage from the stray capacitances into which it would otherwise have to work, thus enabling the gain of the first video stage to be increased while maintaining bandwidth.

It was general practice in 405-only receivers to include a vision interference limiter circuit either before or following the video-amplifier stage. This generally took the form of a diode biased so as to conduct and short-out the signal when the input rose above a certain level. In this way interference in the form of sharp pulses, caused for example by car ignition systems, was removed from the signal. In some models a 'black spotter' was used. This consisted of a triode which was cathode-driven from the video-amplifier anode circuit. Its output, consisting of negative-going interference pulses, was fed to the c.r.t. grid so that the c.r.t. was cut off in the presence of interference pulses.

The suppressor grid of the video amplifier can be used to reduce the effect of interference. As the potential on the suppressor grid of a pentode is made more negative, the slope of the valve is reduced, together with the maximum value of anode current. By adjusting the bias supplied to the suppressor grid, the video amplifier can be operated so that interference pulses tend to saturate the valve, which thus acts as an automatic limiter to prevent high peak voltages of interference from being produced in the output circuit. An alternative in some chassis is to switch extra resistors in the anode h.t. feed to the video output stage to limit the maximum anode current.

With negative picture modulation on 625-lines, interference pulses go below black level so their effect is subjectively less annoying on pictures. Interference limiters are not therefore used but precautions have to be taken to prevent sync. mistriggering due to the interference.

On some 625-line receivers, the grid circuit of the video amplifier stage is returned to the a.g.c. line, often via an 'overload diode'. In the presence of strong signals (negative-going) the diode conducts, thus supplementing the a.g.c. voltage and increasing the range of a.g.c. action.

Because of the a.c. coupling between the video demodulator and the video amplifier generally adopted on 625-line operation, d.c. restoration circuits are sometimes found at this point. A simple clamp diode can be used to restore the d.c. conditions.

A black-level stablilisation circuit that has been used is shown in *Figure 2.15*. It is intended to overcome to a considerable degree the effects caused by a.c. signal coupling between the video amplifier and picture tube and the use of mean-level a.g.c., leading to loss of the television waveform d.c. component. The circuit uses a pentode valve

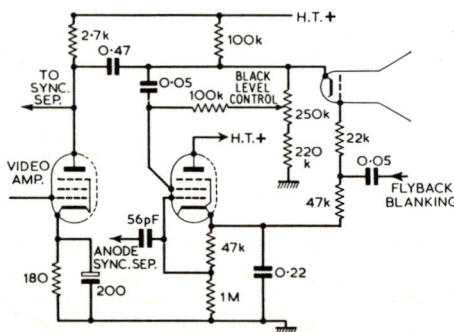

Figure 2.15   Black-level stabilising circuit in the cathode circuit of the picture tube

(the second section of a PFL200) to the screen grid of which the video signal is applied. The d.c. potential on the screen grid is established by the brightness control setting (black level control). Sync. pulses are applied to the pentode grid, the valve conducting on the trailing edge of the pulses. The capacitor in the cathode circuit charges to a d.c. potential corresponding to the black level of the video signal at its screen. This potential, which follows the variations in the video signal black level, is applied to the picture tube grid as the blackness correction signal providing stablilisation of the black level.

Some more modern examples of black level control circuits are shown in *Figure 2.16*. The basic arrangement of any d.c. restorer circuit is that a diode or transistor conducts on the initial signal and charges a capacitor to the peak voltage of the waveform. Should any subsequent period of video change in its reference level (usually the tips of sync. pulses), the diode conducts to short that video to the reference voltage—either chassis or in many cases the brightness voltage—to maintain the same reference. The keyed d.c. restorer or *clamp* performs the same function but is only allowed to react during a specific reference period (normally back porch) when a diode, a number of diodes or a transistor is made to conduct by feeding the device(s) with a line pulse which has been delayed (usually obtained already delayed to coincide with back porch by taking it from the line output transformer). When the switch closes, the video is taken down to the reference voltage and as the porch is the black level the correct reference is always set.

The first example shown in *Figure 2.16(a)* is of d.c. restoration of the video signal with a standard diode/capacitor arrangement, in this case before video amplification. The video signal must then be d.c. coupled to the c.r.t. in order to preserve the d.c. component. In older dual-standard receivers using similar video amplifiers, it was usual to switch out the restoration circuit on 405-line operation.

Figure 2.16(b) shows d.c. restoration used on the chroma output stage of an R.G.B. drive colour receiver. Although only one channel is shown, an exactly similar arrangement is required on each, so that any errors give rise to a general level change rather than one colour being made more predominant than any other. The restorer in this case (the BA145 diode) is keyed every line period by the 'brightness pulse' and the same pulse and d.c. bias (reference potential) is fed to the other chroma output stages.

In a colour difference drive receiver, the luminance must be separately controlled in d.c. level at its final amplifier stage and the circuit of

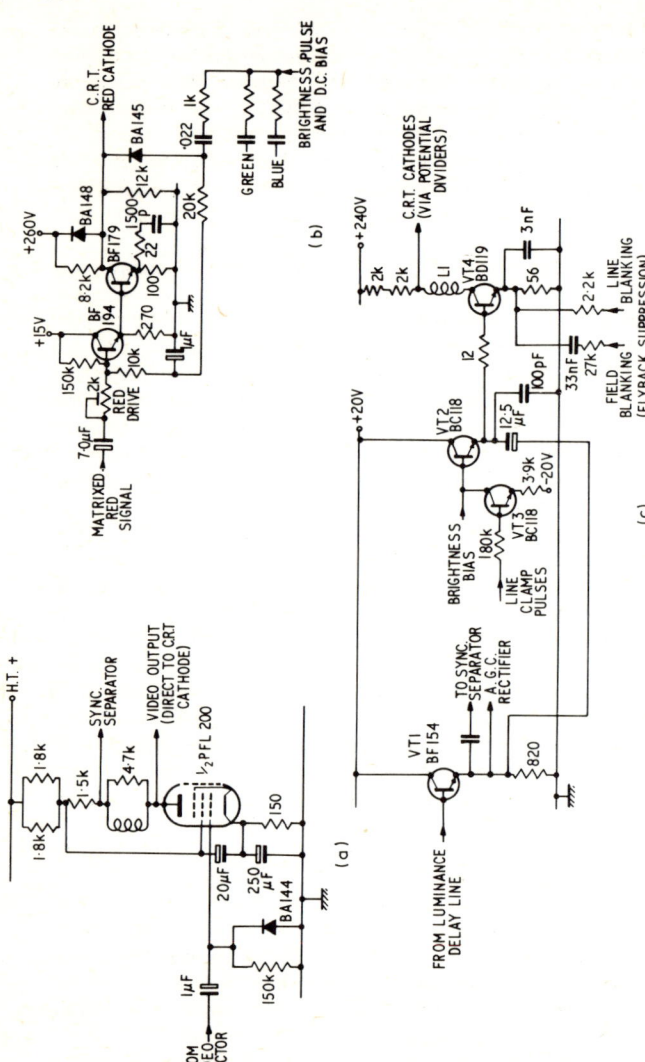

Figure 2.16   Representative d.c. restoration circuits

(a) Monochrome receiver: Philips single-standard G20T300 series u.h.f. receiver (d.c. restoration at output of detector 1 μF, 150 k, BA144 and d.c. coupled from that point to c.r.t. cathode). (b) R.G.B. output colour receiver: Bush single-standard colour receiver series CTV174D (d.c. restoration at output stage, directly on c.r.t. cathode feed—green and blue stages identical). (c) Colour difference output colour receiver: ITT-KB single-standard colour receiver CVC2 (d.c. restoration of input to final luminance stage).

Figure 2.16(c) is reasonably typical. The video passes from the luminance delay line through an emitter follower (VT1) and is taken to the base of the final amplifier (VT4). The d.c. restorer transistor (VT2) conducts every line reference period owing to the clamp pulses derived from the line output transformer and fed to the circuit by conduction of VT3. The brightness bias setting the overall reference level is derived from a potential divider chain across the + 20 V line.

AUTOMATIC PICTURE CONTROL

The incorporation of automatic gain control circuits in the sound and vision channels has been standard practice for many years. In an a.m. sound channel the a.g.c. circuit takes the form of a simple feedback arrangement from the detector to the first sound i.f. amplifier. Delay diodes have sometimes been used but are uncommon today. A.G.C. is not required with the f.m. sound of 625-line operation because absolute gain stability is not required; alternations in gain only appearing as amplitude changes and these are rejected by the a.m. rejection facilities of the f.m. discriminator. In the vision channel the mean-level a.g.c. system described below has often been used, and is almost always used on dual-standard models since the complications arising from the two different sets of parameters make any other system difficult. An alternative approach used on some early Band I/III receivers in order to equalise the gain between the two Bands was to fit independent pre-set sensitivity controls for each Band. Gated a.g.c. systems were used on many Band I/III receivers to overcome the inherent limitations (see below) of the mean-level system and are becoming more common on single-standard u.h.f. receivers; they are described in detail.

Mean-level a.g.c.

The simplest, and one of the most common, forms of a.g.c. is to sample the waveform appearing in the grid circuit of the synchronising separator stage. This signal is d.c. restored by the grid current of the synchronising separator. The control voltage derived from this source is then fed back, as bias, to a controlled i.f. stage and the v.h.f. tuner r.f. stage. The contrast control then takes the form of a potentiometer determining the level at which the a.g.c. begins to operate. With such a system, it is common, as shown in *Figure 2.17*, to include a clamp diode

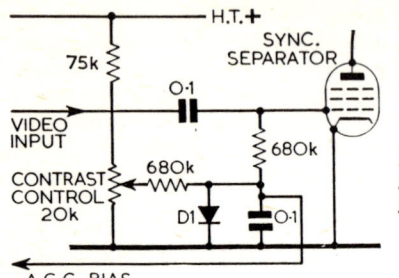

Figure 2.17 Mean level a.g.c. system sampling signal at grid of sync. separator. D1 is a delay diode

to delay the effect of the control voltage so that the gain of the receiver strip is not reduced below the required amplification level and to prevent the a.g.c. line going positive when the contrast control is operated. It is also common to include a diode to provide additional delay to the voltage applied to the r.f. amplifier stage, in order to retain maximum r.f. amplification and thereby obtain the best noise factor. The suppressor grid of a suitable valve, e.g., sync. separator or i.f. amplifier, is occasionally used as a delay diode.

In practice, several diodes may be used to provide different delay levels for various stages, so as to bring the a.g.c. action into operation at different input signal levels. This is done to enable the receiver to operate satisfactorily over a wider range of signal strengths than would be possible with a single fixed delay. For instance, r.f. gain must be reduced only on fairly strong signals in order to maintain good signal-to-noise ratio.

The major disadvantage of mean-level a.g.c. is that the gain of the tuner and i.f. strip vary with picture content. On night scenes, for example, the gain will rise falsely because the mean-level of the picture will be low. This increase in gain is a distortion and will be accompanied with an inevitable rise in picture noise.

Gated a.g.c.

To overcome the disadvantage of changes with picture content, especially in receivers intended for fringe-area reception where considerable picture fading may be experienced, a 'gated' system has been used on many single-standard receivers (both 625-line and earlier 405-line single-standard models). A number of different arrangements have been employed. In gated systems the amplitude of the signal is measured at some time during the picture waveform when it is at a known level, the receiver gain then being controlled accordingly. In the 405-line television system the black level representing 30 per cent modulation is kept constant at the transmitter and is radiated for a few microseconds following the line synchronising pulse and for a few lines following the field pulses. By basing the control bias voltage on these periods, the receiver gain, with a.g.c. applied, can be kept independent of the picture content. Similarly on the 625-line system black level is maintained at the transmitter at 76 per cent, while peak syncs. are maintained at 100 per cent, and either of these may be used as the reference.

It is more common to sample the signal during the 'back porch' period following the line-synchronising pulses than to sample the strength of the carrier at the field frequency so that the system recovers more quickly on, for example, a channel change, and an arrangement for sampling at line frequency is shown in *Figure 2.18*. A negative-going waveform is provided by the cathode-follower (video output). This is coupled to the anode of a 'gating' diode to whose cathode is fed a train of large, narrow pulses, also negative going. These 'gating' pulses are in this system obtained from the line output transformer. They must be narrower than the 'back porch' period and must occur at

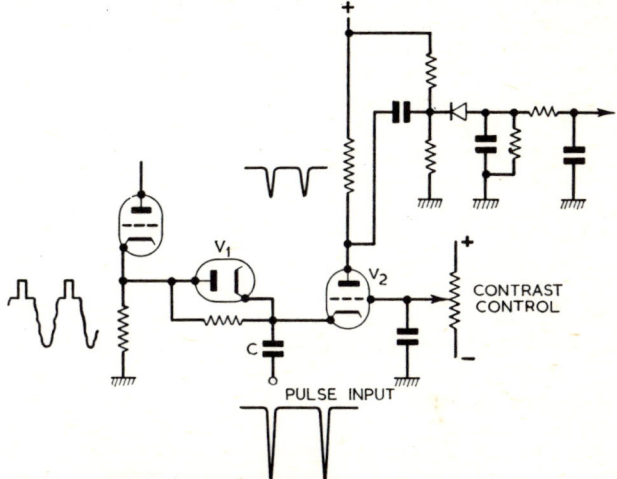

Figure 2.18   Vision a.g.c. circuit with diode 'gate'

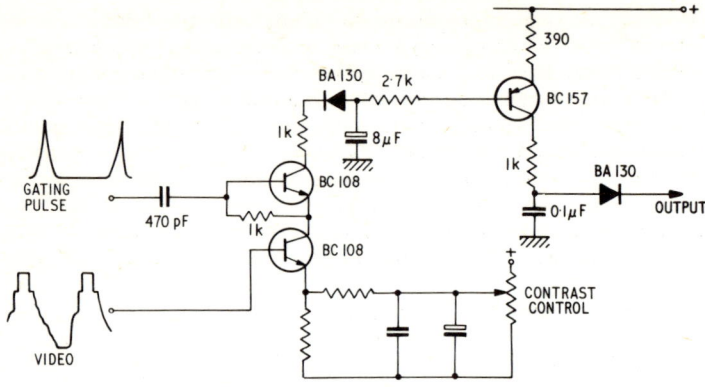

Figure 2.19   Transistor gated a.g.c. circuit

the centre of this period. If the timing is not correct, it is necessary to delay or to advance the pulse train by some suitable means. The pulses must be larger in amplitude than the video waveform. The diode acts as restorer, and the tips of the pulse train are restored to the potential existing at the cathode of the cathode-follower. This potential, if the timing is right, will be the black level. If the signal strength should increase, then the potential of the restored pulse train will fall with respect to earth, and vice versa. The voltage is applied to the cathode of V2. The bias on the grid of V2 is adjusted, by means of the contrast control, so that the tips of the pulses just cause anode current to flow, thus producing a similar train of pulses across the anode load. When the signal strength alters, and the restored pulse train moves its potential with respect to earth, the pulse train in the anode circuit of V2 alters in amplitude accordingly. This information is rectified, by means of the rectifier and smoothing components shown, so producing a d.c. control voltage.

A similar system employing transistors is shown in *Figure 2.19*. The polarity of the output diode in this circuit should be noted for providing forward a.g.c. to the earlier stages. On some single-standard receivers, gated a.g.c. is incorporated in a processing integrated circuit.

Sync.-cancelled a.g.c.

Also used on many 405-only receivers was the simpler 'sync.-cancelled' system, which uses a peak detector to provide a voltage based on the black level of the waveform.

In the usual sync.-cancelled system, the positive-going sync. pulses are cancelled out from a portion of the video signal by mixing with it negative-going sync. pulses of greater relative amplitude from which the picture information has been removed; such pulses can be obtained from the anode circuit of the sync. separator valve. This mixing produces a waveform which is entirely negative, but with the most positive parts of it corresponding to the 'black level' of the incoming signal. This waveform is then fed to a diode (D1, *Figure 2.20*), which acts as a peak detector—i.e. provides a d.c. output which will be related to the most positive peaks of the input but which ignores all the

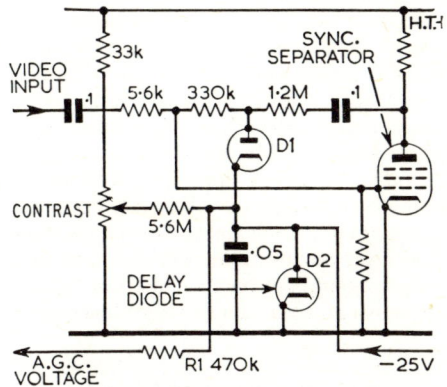

Figure 2.20   Common form of 'sync-cancelled' vision a.g.c.

less-positive parts of the input waveform—and this provides a negative d.c. voltage which follows the 'black level' of the incoming signal, unaffected by the picture content. The rectified output of the peak detector appears at its cathode, and the load resistor has to be returned to a point of negative d.c. voltage of about −25 volts, and this is usually obtained from the grid of the line output valve. It is necessary to introduce a 'delay' voltage so that the a.g.c. system does not operate until the signal reaches a predetermined amplitude. This is done by a delay diode (D2) whose clamping level is varied by the contrast control, which is a potentiometer across part of the h.t. line. Sync.-cancelled techniques were re-introduced on a few dual-standard models.

Automatic picture control on dual-standard models

The differing waveforms for the 405- and 625-line systems would make it difficult to use any of the gated and sync.-cancelled forms of automatic picture control of the type used on a number of single-standard chassis. These systems measure the amplitude of the received waveform at some precise point, such as the peak sync. level or the black-level as indicated by the porches in the transmitted waveform.

As a result, most dual-standard receivers use forms of mean-level a.g.c. which averages out the signal regardless of changes in picture content. This system is far from ideal as already noted.

A complication also arises as a result of the different sense of modulation used on the two systems, positive modulation on 405, negative on 625, which means that changes of the mean level of the waveform derived from the sync. separator stage will tend to affect the gain of the receiver in opposite ways. On 405 a low-key (darker) picture reduces the mean level, resulting in the gain of the receiver being brought up, tending to lighten the picture. But on 625, a low-key picture represents a higher average video signal, and this will tend to darken the picture: this means that on 625 a dark picture will tend to become even darker, a rather more objectionable and noticeable characteristic than the brightening of a low-key picture. These effects can be reduced by using a.c. coupling at some point between the video detector and the picture tube, without any form of d.c. restoration or black-level clamping.

AUTOMATIC CONTRAST CONTROL

Several earlier receivers used small photo-electric cells of the cadmium sulphide type to provide automatic control of contrast or brightness with changes in the ambient lighting of the room. The resistance of these cells in low levels of illumination will be of the order of a megohm, decreasing in a brightly lit room to only a few hundred ohms.

There are various methods of incorporating the photocell into the receiver: one is to include it in a video attenuator network between the video amplifier and the picture tube, so that the video drive and d.c. voltage of the black level are adjusted automatically. *Figure 2.21(a)* shows a typical arrangement for increasing the drive to the picture tube with increasing illumination; the component values of the network are chosen so that the change in drive brought about by the change in the effective load resistance of the video amplifier is roughly

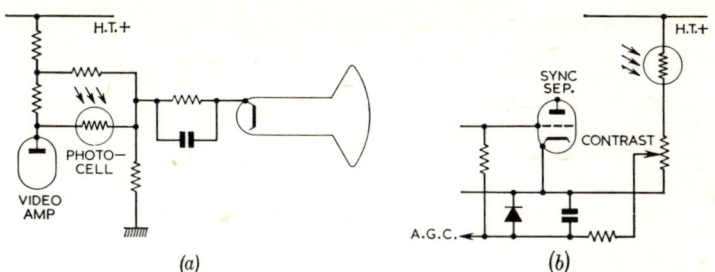

Figure 2.21   Two methods of applying automatic contrast control

equivalent to that required to maintain optimum contrast setting. A disadvantage of this system is that it attenuates video drive in weak signal areas under low illumination conditions; it also adds to the undesired capacitive loading of the video amplifier.

An alternative arrangement is to use the photocell to control the a.g.c. circuit. *Figure 2.21(b)* shows a simple system of this type with the portion of the h.t. applied across the manual contrast control governed by the resistance of the photocell. As the manual contrast control setting is reduced (greater receiver gain) the cell has less effect, and this avoids the difficulty of reducing gain by photocell action in weak signal areas. Automatic contrast control can also be applied with advantage in conjunction with an amplified a.g.c. system.

A.M. SOUND CHANNEL

On 405 with a.m., the general design features of the sound channel up to the detector stage do not differ greatly from the circuits used in the vision channel, except that the inter-stage coupling tends to be less complex. The total gain required is much the same as that for the vision signal—of the order of 100 dB (100 000 : 1). The majority of this gain is supplied by the i.f. stages.

The bandwidth of the sound channel as transmitted will be of the order of 30 kHz, representing the double-sideband transmission of audio frequencies up to 15 kHz, but the receiver bandwidth requires to be much greater than this figure would suggest. This is partly in order to overcome the effect of oscillator drift, which may amount to some 100 kHz, and partly to help in the suppression of impulse interference from car ignition systems and electrical equipment. A broad response of the order of 500 kHz will mean that the interference pulses can be reproduced at the detector without undue lengthening, and this will make for more efficient interference suppression. With a broad-response curve it is sometimes difficult to prevent low-frequency vision-signal components from entering the sound channel (for example, the field sync. signals may be reproduced giving vision-on-sound 'hum').

With the high sound i.f. of modern receivers—usually 38·15 MHz—i.f. stage gain with conventional transistors or high-slope pentodes tends to be limited to about 30–40, in order to ensure stability. Considerably higher stage gains can be obtained with frame-grid valves, although it may be necessary to use some form of neutralisation to maintain stability.

It is common practice to control the gain of a sound i.f. stage by an a.g.c. voltage obtained from the detector stage. However, in some designs no separate sound a.g.c. system is incorporated.

The detector

As in the case of radio receivers, the diode detector has long been popular in television-receiver designs. The value of the load resistance will, however, be much lower than in radio practice, and the i.f. filter may consist of chokes and very small capacitor values (10–20 pF) in order to prevent the undesirable integration of interference pulses. The

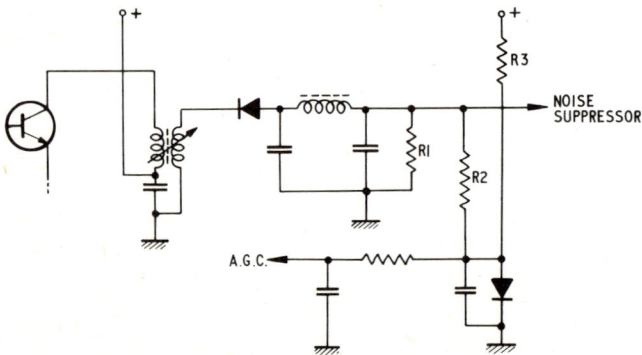

Figure 2.22   Sound detector and a.g.c. circuit

use of germanium crystal diodes has ousted the thermionic diode on the counts of low cost and low self-capacitance.

An a.g.c. potential may be taken from the detector load, a suitable arrangement being shown in *Figure 2.22*, in which a delay diode is incorporated. The detector circuit has, of course, to produce a signal sufficient to cause a current flow in R2 equal to the current normally flowing from the h.t. line through R3 and the delay diode before the delay is overcome. Delay circuits are seldom encountered today, and a compromise method is to apply only a part of the detector d.c. component to the a.g.c. line. Whichever method is used, it is necessary to reduce the impedance of the a.g.c. line to the minimum so that blocking does not occur in i.f. circuits connected to it.

Interference suppression

The level at which the detector operates is determined largely by the performance of the audio-output stages, which follow it, and also by the circuit used to provide suppression of impulsive interference. While audio amplifiers to provide the required output (normally 1–3 W) can easily be designed to accept very low inputs, the noise-suppression circuit operates best at fairly high levels. It is placed immediately after the detector, so that the need for large bandwidths can be dispensed with after the interference has been reduced sufficiently, and it is not difficult to operate the detector to provide a demodulated output of 5–10 V peak-to-peak (80 per cent modulation).

One of the most successful noise-suppression circuits* is shown in

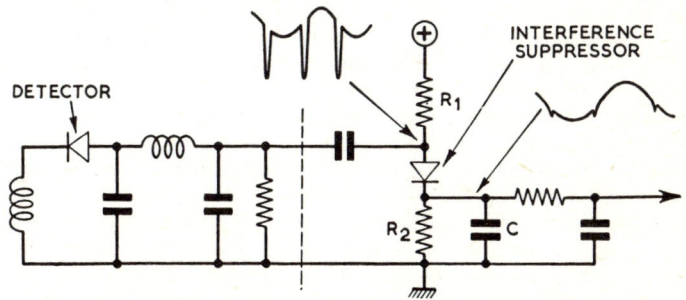

Figure 2.23   Sound interference suppression circuit

*British Patent no. 605,206.

Figure 2.23. Resistances R_1 and R_2 are the order of megohms, so that a steady current of, say, 50 μA flows through the diode. The capacitor C is a critically large value chosen so that the time constant formed with R_2 is just capable of being charged and discharged by the current flowing through the diode at the highest audio frequencies it is desired to reproduce. From the waveforms it is apparent that at audio frequencies the cathode of the diode follows the anode. If, however, a negative-going interference pulse of very short duration appears, the diode will immediately cut off and the cathode circuit will commence to discharge. Before any great change has taken place, the pulse ends and the audio signal reappears at the output bearing only a small triangular 'pip' instead of the original large pulse. The further smoothing is provided to smooth out this remaining pip.

It is now possible to see why so much attention has been given to preserving the shape of the interference pulses and to the prevention of blocking.

V.H.F./F.M. sound reception

A number of earlier television receivers incorporated facilities for the reception of Band II v.h.f./f.m. sound broadcasting stations.

One of the simplest methods used was to fit three extra sets of coils in the turret tuner, tuned to the local Radio 1/2, 3 and 4 channels, and to arrange to switch into circuit when required an f.m. discriminator, usually a ratio detector, in place of the television a.m. sound detector. H.T. and heater supplies to the picture tube are switched off during f.m. reception to extend its life. Similarly, the timebases and sometimes the video strip are taken out of service.

Figure 2.24 shows a representative sound channel using dual-channel i.f. stages. The anode circuit of the mixer contains circuits tuned to 10·7 MHz and 38·15 MHz, and both channels are amplified by V9 and V10. The 10·7 MHz circuits offer negligible impedance at 38·15 MHz, and vice versa. D3 is a conventional sound detector, while V11 is a ratio detector for the f.m. signals. The pre-set resistor P9 is adjusted to provide minimum output of a.m. signals. The two-stage a.f. amplifier (V17A, B) is common to both channels. On f.m. reception the heater supplies to the picture tube and the timebase valves are switched off.

81

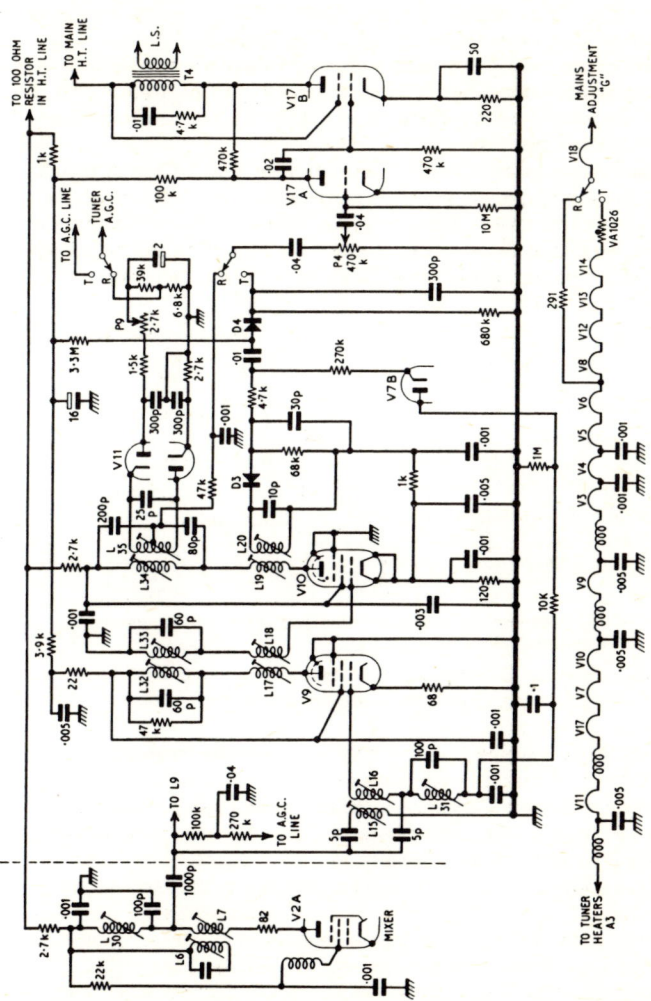

Figure 2.24 Dual channel sound receiver for a.m./f.m. reception L_7, L_{15}, L_{16}, L_{17}, L_{18}, L_{19}, L_{20} are tuned to 38·15 MHz; L_{30}, L_{31}, L_{32}, L_{33}, L_{34}, L_{35} to 10·7 MHz; D3 is the television sound detector, V11 the f.m. ratio detector

F.M. SOUND CHANNEL

On the British 625, the C.C.I.R. 625, East European 625 and American 525 systems, the sound transmissions are frequency modulated. A u.h.f. receiver must therefore be able to cope with f.m. transmissions which have a maximum deviation of 50 kHz and a pre-emphasis time-constant of 50 microseconds.

The general use of intercarrier techniques for f.m. reception means that the f.m. sound signals are presented to the f.m. discriminator at 6 MHz, preferably with some pre-limiting to remove amplitude variations. This limiting may sometimes be achieved by the reduction of the screen voltage on the final i.f. stage in a valve receiver, or by

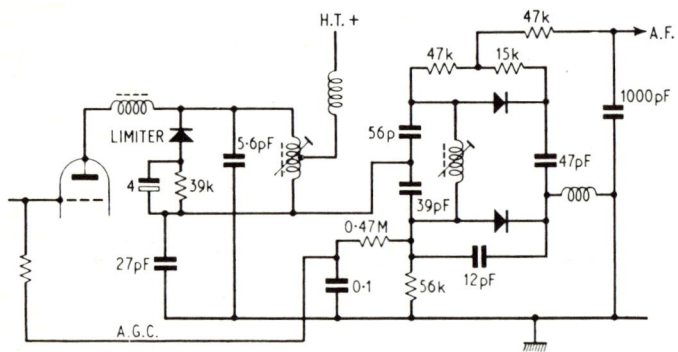

Figure 2.25   Form of unbalanced phase discriminator with separate limiter diode used on some dual-standard models

setting the bias conditions on the final i.f. transistor amplifier, or by 'soft' limiting of the signal over two or more stages, for instance by the use of envelope feedback forming virtually an a.g.c. loop. The detector should also provide inherent self-limiting action, unless a high degree of limiting is provided independently.

The most common form of f.m. discriminator is the ratio detector, similar to that used widely on f.m. sound radio receivers. This has self-limiting action which can be further enhanced by the incorporation of a pre-set a.m.-rejection balancing control (see also *Figure 2.27*).

A rather different form of two-diode phase discriminator with a diode limiter has been used on some models: this is shown in *Figure 2.25*. This circuit incorporates an a.g.c.-type limiting loop. Diode limiters such as this may also be used in transistorised i.f. strips on one or more stages.

Quadrature f.m. detectors

A form of f.m. detector not found in sound radio practice has been widely used in the United States and Europe, and is sometimes found in British dual-standard models. This is the quadrature f.m. detector which, provided that it is accurately aligned, has very good a.m. rejection and provides a high-level a.f. output which can be fed directly to the sound output stage. The same valve may be used during 405 reception as a first a.f. amplifier, thus enabling a single pentode to be used as sound output rather than the triode-pentode arrangement commonly used with the ratio detector.

Two resonant circuits tuned to the same frequency but in which the alternating currents have a phase difference of 90 deg are said to be 'in quadrature'. In the quadrature detector two such circuits are used to control the anode current of a valve, thereby converting an f.m. input signal into a.f. output. There are two main forms of quadrature detector. One uses a special 'gated-beam' valve (*Figure 2.26(a)*) and is common in the United States. The other, termed the quadrature-grid, locked-oscillator discriminator (*Figure 2.26(b)*), uses a pentode or heptode of more conventional design.

In the gated-beam detector the required quadrature relationship between the two tuned circuits is developed by space-charge coupling in the electron stream of the valve, the f.m. input signal being applied across a tuned circuit in the input circuit. When the anode current of a

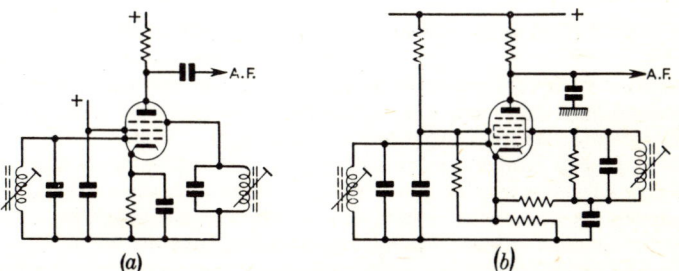

Figure 2.26   Quadrature f.m. detectors (a) uses gate-beam valve (b) uses heptode or pentode

single valve is controlled by two sine waves having a nominal 90 deg phase difference, the magnitude of the signal appearing across the anode load will depend upon the phase angle between the two signals; this is the basic requirement for an f.m. discriminator, as the phase angle of an f.m. signal varies in accordance with the a.f. modulation. Since it is possible to make such a detector relatively insensitive to amplitude variations of the two control signals, the detector will be self-limiting, and can provide fairly high frequency/amplitude conversion gain.

The action of the quadrature-grid, locked-oscillator discriminator is basically similar to that of the gated-beam detector, but requires a less expensive valve, though rather more complex coupling arrangements since the two tuned circuits operate in a self-oscillating mode. The tuned circuits are connected to the control grid and third grid of a heptode, and the space charge coupling again produces a voltage which lags 90deg behind the control-grid signal. The tuned circuit connected to the third or suppressor grid draws energy from the electron stream, and will oscillate provided that a feedback path exists. This oscillator locks to the frequency of the incoming signal (but with a phase difference of 90deg) so that an f.m. input signal produces an a.f. signal across the load as in the case of the gated-beam valve. Self-limiting is provided at low input signal levels by the gating action of the oscillator, and at high input signal levels by grid current action between grid and cathode. A requirement for valves in this application is that they must provide appreciable space-charge coupling to the third grid, and in Europe the EH90 heptode valve was developed for this purpose.

DUAL-STANDARD SOUND I.F. AMPLIFIERS

Whether valves or transistors are used, the sound i.f. stages are made considerably more complex by dual-standard working. Not only are the intermediate frequencies different for the two systems, but the use of amplitude modulation on one system and frequency modulation on the other means that two demodulation circuits are needed. Fortunately the use of the intercarrier sound technique simplifies things to the extent of eliminating much of the switching that would otherwise be necessary in the coupling and detector stages. This is because the f.m. intercarrier frequency at 6 MHz and the standardised a.m. intermediate frequency at 38·15 MHz are sufficiently well separated

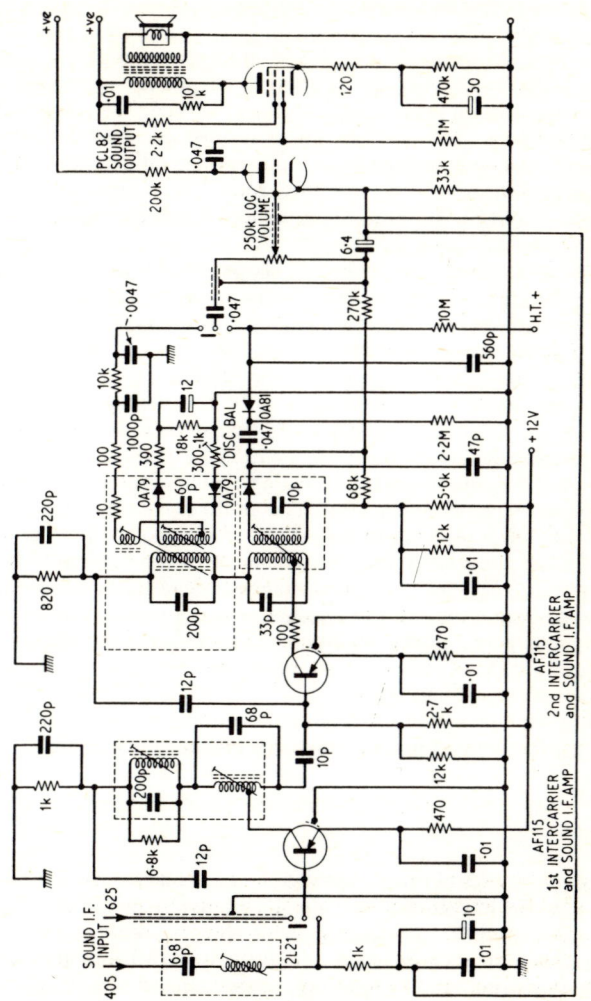

Figure 2.27   Two-stage sound i.f. amplifier and audio stages in the Bush TV135

to allow the two sets of tuned couplings to be connected permanently in series without serious interaction between them.

The resulting simplification of switching does place some restriction on neutralising arrangements. If two sound-only transistor i.f. stages are used, and this is really the minimum acceptable, then it may be necessary to neutralise to get adequate gain at 38 MHz. The unswitched series connection of the coupling circuits, however, does not readily lend itself to providing anti-phase voltages so that a different neutralising arrangement, involving the formation of a capacitor bridge circuit, will probably be used. This is shown applied to sound i.f. amplifier stages in *Figure 2.27*.

As with any method of neutralising, component replacement should be made with great care to preserve balanced conditions, but there is the added danger here of a service engineer assuming that one of the high-capacitance components of the bridge is merely serving as a decoupling element, in which case he may unwittingly replace with a wide-tolerance capacitor or perhaps even with a deliberately chosen higher value one in the mistaken belief that he is improving things.

It is common practice to use one or more of the vision i.f. stages also as sound i.f. amplifiers by arranging that their coupling circuits have sufficient bandwidth to embrace both the vision and sound information. The a.m. sound take-off has to be made before the final vision i.f. stage because of the danger of cross-modulation. This is not so serious on 625-line f.m. sound channels because a certain amount of amplitude modulation can be rejected by the ratio detector, and it is common practice to retain the sound carrier right through to the vision detector where it beats with the vision carrier to produce the 6 MHz intercarrier sound signal.

On single-standard (u.h.f.) receivers it is now common practice to use an integrated-circuit amplifier and detector for the f.m. chain.

Sound a.g.c.

By commoning stages in this way, vision a.g.c. will also have an effect on sound level, which is not unreasonable because many of the factors causing carrier amplitude variations will apply to both carriers. But there is a need for separate a.g.c. derived directly from the sound carrier and operating on a sound-only stage to correct any changes in sound-to-vision carrier differential and also to buffer the sound gain from the effects of changes of contrast control setting by the user.

The obvious stage to control is the first sound i.f. stage, and if forward a.g.c. is used, as in the vision stages, then a fairly powerful d.c. amplifier stage will be essential to amplify the small amount of control power available from the sound detector to the level required for forward control. However, by paying special attention to the selectivity of the sound take-off circuits it is possible to produce sufficient rejection of vision i.f. to enable reverse a.g.c. to be safely used without risk of cross-modulation. This allows direct control from the d.c. component of the signal at the detector, using a very simple circuit. Many receivers use a d.c. connection between the detector and the audio amplifier so that the first audio stage, be it valve or transistor, can be used to improve the a.g.c. loop gain. With a valve stage the a.g.c. take-off will probably be at the cathode, as in *Figure 2.27*, while with a transistor it will usually be at the emitter.

AUDIO OUTPUT STAGES

While it can be argued that any reasonable quality audio stage(s) will suffice the needs of television sound, it is as well to remember that considerable concern goes into the production of high-quality sound from the television studio. Some of the quality of this sound is lost between studios and transmitting site because of poorish circuits (apart from the London area) but in future years this will become less true as vision circuits are also used for the high-quality carriage of sound using the B.B.C. sound-in-syncs system.

Unfortunately the very size and space allocated for the loudspeaker in a television receiver is very much smaller than that recognised for 'hi-fi'. With one compromise, other factors also pertaining to sound reproduction are also compromised including the output power being used and the design of the amplifier. The output stages shown in *Figures 2.24* and *2.27* are typical of those used for many years in valved receivers delivering between 1 and 3 watts into an elliptical speaker.

Generally, all-transistor receivers are to be found using an integrated circuit as sound i.f. amplifier and discriminator, and the output level may be very low (even when use is made of an i.c. with a built-in pre-amplifier), perhaps less than 100 mV. A typical audio amplifier for such use is shown in *Figure 2.28*. This is a fairly straightforward Class A amplifier with high impedance loudspeaker directly coupled—i.e. without output transformer. The two output transistors are operated as a push-pull pair in which the collector current of one drives the other.

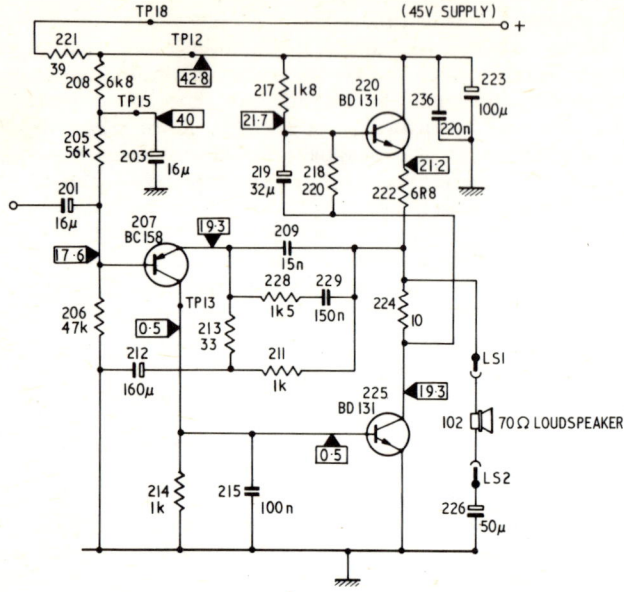

*Figure 2.28   Representative transistor audio amplifier: Philips 520 series
single-standard colour receiver*

The driver transistor (BC158) drives the output pair and also stabilises
their d.c. potentials by using the output centre point for its emitter
supply. The amplifier gives 2 watts output.

Although more efficient output stages using Class B could be employed in the all-transistor receiver, the regulation requirements of the
supply rail are high because of the 'pulsy' current drain; unless a separate supply point is used with its own smoothing and regulation—
relatively costly—Class B is unlikely to be extensively used.

3

BASIC TIMEBASE AND POWER
SUPPLY CIRCUITS

As we have seen, timebases are required in a television receiver to generate the 50 Hz (field) and 10 125 Hz or 15 625 Hz (line) scan waveforms needed to produce the cathode-ray tube raster. The waveform required is basically of sawtooth shape, so as to deflect the beam across and down the tube face linearly and then, in the much briefer flyback period, allow the beam to return for the commencement of the next line and field.

It should be emphasised that because beam deflection using magnetic fields is due to the *current* flowing through the line or field coils, the sawtooth waveform referred to is a *current sawtooth*. The voltage necessary to produce a sawtooth of current through a perfect inductance must be a constant level.

Timebases consist basically of two stages, first an oscillator producing a regularly varying control waveform (at 50, 10 125 or 15 625 Hz), and secondly a power stage switched by the control waveform to supply the deflection coils with the necessary sawtooth current waveform. The oscillator may generate a sawtooth voltage waveform to control the power stage, and this is in fact the general practice with field timebases. Because of the differing mode of operation of the line output stage, however, a somewhat modified control waveform is required, and this is generally shaped to the necessary form by the interstage network.

The essential difference between line and field requirements is caused by the different operating frequencies. At field frequency (50 Hz) the scan coils used are more resistive than inductive so that the drive voltage waveform is quite closely allied to a sawtooth; the output stage can therefore be considered almost as an amplifier with some necessary distortion added to correct for the scan coil inductance and transformer effects.

At line frequency (10 125 or 15 625 Hz) the scan coils are more inductive than resistive and the line output stage must provide an almost constant voltage during the scan period and another value during the flyback time. This requires a stage that switches between two states and the drive waveform must be of a form that will perform the switching action cleanly.

All timebases operate on the basis that they will provide an output even under no-input conditions. This ensures that in cases where the signal fails (e.g. a receiver left on after the end of a day's programmes) there is no damage to the c.r.t. by an undeflected electron beam, or that a power circuit (e.g. the line output stage) is not left with abnormal bias conditions, which might cause valve, transistor or transformer damage. All timebase oscillators are therefore free-running oscillators triggered in the correct position by the incoming sync. waveform.

Generating a sawtooth waveform

The simplest method of producing a sawtooth waveform is through the slow charge and rapid discharge of a capacitor. *Figure 3.1* shows

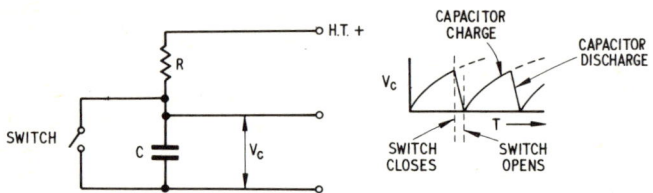

Figure 3.1   Basic principle of the sawtooth generator

a simple arrangement with a capacitor (C) charged through a resistor (R), with discharge initiated when the switch is closed short-circuiting the capacitor. The resultant waveform is also shown.

The rise time of the sawtooth (which will generally be used for the scan period) will depend on the time constant of $R \times C$. The fall-time (which will generally be used for the flyback period) will depend on the time constant of $C \times$ the resistance of the closed switch (which is theoretically zero but a practical, electronic switch must have some resistance).

Although the charging of a capacitor is not a linear function, the rate of charge reducing with time, providing that only a fraction of the charge time is used before flyback is initiated, the waveform can be considered reasonably linear. If this is not the case in a practical circuit because a required voltage level is demanded by the output stage, some correction to the waveform shape may be necessary.

The blocking oscillator

In a practical timebase an electronic means must be used in place of the switch to initiate the discharge of the capacitor. A circuit commonly used for this purpose is the blocking oscillator, generally employing

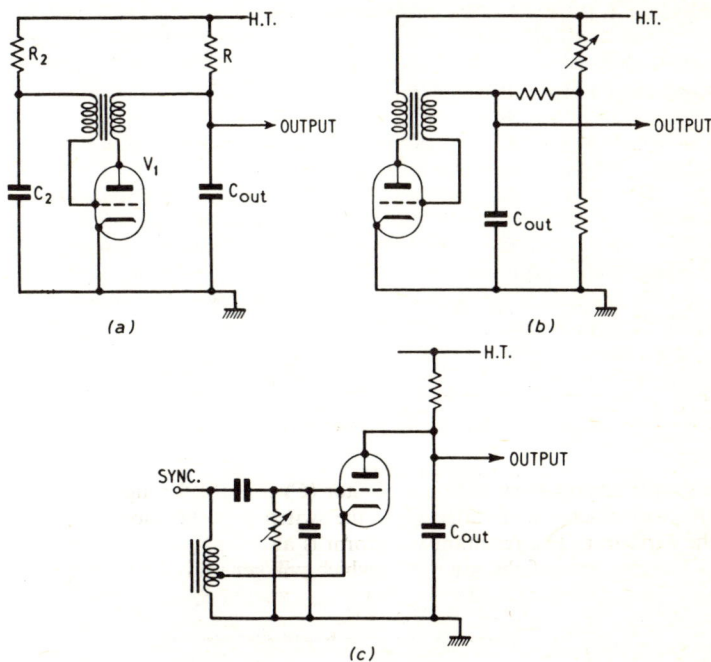

Figure 3.2   Basic blocking oscillator circuits
(*a*) Common form in which the time constant C_2R_2 controls the conduction of the valve; (*b*) variant using a single time-constant network; (*c*) circuit sometimes used for field timebases, in which an autotransformer is connected in the grid-cathode circuit.

the triode section of a multiple valve (pentodes were frequently used in early receivers). The basic arrangement is shown in *Figure 3.2(a)*, C_{out} being the capacitor across which the waveform is produced. When V1 is cut off by the negative charge on the upper plate of C_2, C_{out} charges through R. The charge on C_2, however, gradually leaks away via R_2, and when it has leaked away sufficiently V1 grid is driven sufficiently positive for it to conduct. When it conducts, positive feedback via the transformer secondary rapidly drives V1 into saturation, its anode voltage falls sharply and C_{out} discharges. When V1 has reached saturation, the field around the primary of the transformer collapses, and a negative-going pulse is reflected back via the secondary, cutting V1 off. When V1 saturates, C_2 is charged negatively by grid current and this keeps V1 cut off or 'blocked'—hence the name blocking oscillator—until the charge leaks away through R_2 and the cycle of operations is repeated. The values of C_{out} and R are chosen so that their time constant is suitable for the frequency being generated.

It will be noticed that the blocking oscillator is a free-running circuit, i.e. it is not necessary to apply triggering pulses to it in order to bring it into operation. To synchronise it, however, control pulses can be applied in order to regulate the precise moment when the valve cuts off. A negative-going series of sync. pulses for this purpose may be applied to the anode or grid.

If R_2 is made variable it will act as hold control to control the frequency of the stage by varying the time constant of the grid network.

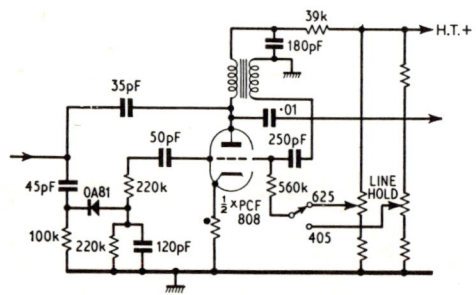

Figure 3.3   Line blocking oscillator with sync. inputs to grid and anode
The 180 pF charging capacitor is charged through the 39 k resistor. Output to the line output valve is via the 0·01 coupling capacitor. The 250 pF grid capacitor and associated resistors determine the basic frequency of oscillation (*Thorn/B.R.C.*).

Making R variable provides a means of controlling the amplitude of the sawtooth waveform.

A number of variations of the basic circuit has been used. *Figure 3.2 (b)* shows a rearrangement in which the output is taken from the grid circuit and a single time constant is involved. *Figure 3.2(c)* shows an arrangement sometimes found in field timebases, employing an auto-transformer connected in the grid-cathode circuit. *Figure 3.3* shows the line timebase blocking oscillator used in several recent Thorn receivers: an interesting feature of this is the application of the sync. pulses to the anode and, after being fed through a delay network, the grid to increase the range of sync. control over the operation of the stage.

The charging curve of the capacitor C_{out} is of course exponential, being more linear at its commencement than towards full charge. To obtain maximum linearity it is discharged at an early point on its charging curve and it is advantageous to use as high as possible a voltage for charging. For this reason the boost h.t. voltage is generally used in field timebases for the oscillator stage.

The multivibrator

Perhaps the most commonly used type of timebase oscillator is the multivibrator, which may also be used to discharge a capacitor so as to produce a sawtooth waveform. Two basic forms of multivibrator, the anode-to-anode cross coupled and the cathode coupled varieties, are shown in *Figure 3.4*, used in conjunction with a charging capacitor C_{out}.

The operation of the anode-to-grid coupled version is as follows. Assume that V1 is starting to conduct. Its anode voltage falls and, as

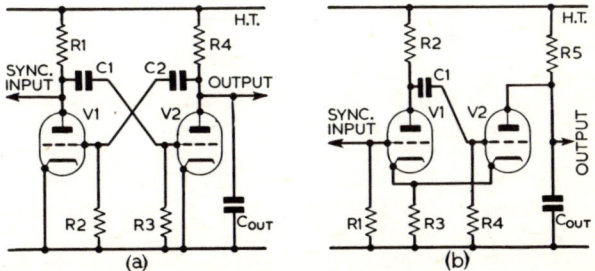

Figure 3.4   Basic multivibrator circuits, (a) anode-to-grid coupled, (b) cathode coupled

a result of the capacitor C_1 linking V1 anode to V2 grid, a negative voltage spike occurs at V2 grid, cutting it off. Thus V2 anode voltage rises towards h.t. potential and, with C_2 coupling V2 anode to V1 grid, a positive potential appears at V1 grid driving it into saturation. Thus we have V1 fully conducting, V2 hard off, a low voltage at V1 anode and high voltage at V2 anode. Under these conditions C_1 will discharge exponentially via R_1 and R_3, and C_2 will charge via R_4 and R_2. Then at a time determined by the RC time constants of the cross-coupling networks, V2 grid will be sufficiently positive for it to begin to conduct, while V1 grid will be sufficiently negative for it to cut off. The initial situation is thus reversed, V2 will rapidly be driven to saturation and V1 driven hard off, while C_1 will recharge and C_2 discharge. The process will continue indefinitely. C_{out} charges through R_4 when V2 is cut off, and discharges when V2 is conducting, to produce a sawtooth output waveform. The waveform is made the right shape by making V2 conduct for about 10 per cent of the time and V1 conduct for about 90 per cent of the time. The conductive periods of V2 thus act as an electronic switch discharging C_{out}. To initiate the changeover of conduction between the two valves positive-going sync. pulses can be fed to V1 anode, or negative-going sync. pulses to V1 grid or V2 anode. Making R_3 variable provides control over the frequency of oscillation. R_2 and R_3 are sometimes connected to h.t. positive.

As in the case of the blocking oscillator, when used in the field timebase R_4 may be fed from the boost h.t. line to improve the linearity of the waveform, and making R_4 variable provides control over the amplitude of the waveform (height control).

The operation of the cathode-coupled multivibrator (*Figure 3.4(b)*) is as follows: Starting with V2 cut off, V1 conducting and C_1 discharging exponentially so that V2 grid is becoming less negatively biased, eventually V2 grid potential will rise above the cut-off value and V2 will conduct. When V2 conducts the current through R_3 increases so that both cathodes become more positive with respect to their grids. The current through V1 immediately decreases as a result, and its anode voltage rises accordingly—V2 continuing to conduct because C_1 is still discharging. The sudden rise in V1 anode voltage, greater than the change at its cathode because of the amplification V1 provides, causes a momentary positive-going grid drive at V2, so that heavy saturation current, with grid current, flows through V2. C_{out} discharges rapidly, and V2 grid current begins to charge C_1 so that the positive grid drive is reduced to nearly zero. Thus the current through V2 and R_3 falls, decreasing the bias applied to V1. V1 anode current rises, its anode voltage falling, and quickly driving V2 grid negative so that it cuts off. C_{out} then begins to charge gradually. C_1 begins to discharge and the sequence of events repeats. Synchronisation is simply effected by feeding negative-going sync. pulses to V1 grid.

Double-triode valves (or alternatively the triode sections of two multiple valves) are generally used for multivibrator circuits, but triode-pentodes have also been widely used and at least one chassis employed a triode-heptode. In the case of a triode-pentode a technique once favoured in line timebases was to take the feedback to the triode section from the screen grid of the pentode section, with the charging capacitor C_{out} connected to the 'free' pentode anode. The purpose of this was to reduce the loading effect of the output stage on the running of the oscillator.

A refinement often encountered in multivibrator circuits is the inclusion of a variable reactance in either the cathode circuit or one of the anode circuits. This acts to stabilise the frequency of operation of the stage (see *Figure 6.9*).

Although transistors can be used in multivibrators using similar circuit techniques (renamed of course as: collector/base coupled and common emitter) they have not been favoured in recent designs because of the basic temperature instability of transistor oscillators in this format.

Self-oscillating timebases

In the self-oscillating timebase the output stage also forms part of the oscillator circuit. The circuit is a development of, and functions in the same way as, the multivibrator, comprising a triode RC crosscoupled to the output stage. Originally developed for the line timebase, and still used for this application, the self-oscillating timebase has also been widely used for the field timebase. An example of its use in the line timebase is shown in *Figure 3.5*, from which it will be seen that the feedback to the triode stage is via a capacitor connected to a suitable tapping point on the line output transformer. This is an early example—the system has been in fairly wide use for many years—but is basically the same as that in current use.

An alternative arrangement that has been used is to take the feedback to the triode from the screen grid of the line output valve. In several models—including current ones—a combination of these two systems is used, the screen grid-cathode section of the output valve forming one section of the multivibrator during the warming up period before the efficiency diode is brought into action, the mode of operation then changing and the main feedback then being via a capacitor connected to the line output transformer.

An arrangement used in several old models was the use of the screen and control grid of the line output valve in a blocking oscillator arrangement, using a separate transformer or a section of or over-winding on the line output transformer.

Two examples of self-oscillating field timebases, using a PCL85 triode/output pentode valve, are shown in *Figure 6.11*, and in *Figure 3.6*. Feedback to the triode section of the valve is from the anode of the pentode section. The frequency of oscillation is determined by the time constant of the components in the triode grid circuit, and the charging capacitor is connected in the anode circuit of the triode ($0.05\,\mu F$ capacitor to chassis in *Figure 6.11*. In *Figure 3.6* it is C_{81}, which is taken to the cathode of the output pentode, a common practice in recent sets). A problem with the use of the self-oscillating arrangement for the field timebase is that line pulses picked up by the field deflection coils are fed back to the coupling components and may affect the operation of the oscillator, resulting in poor interlace. A bypass

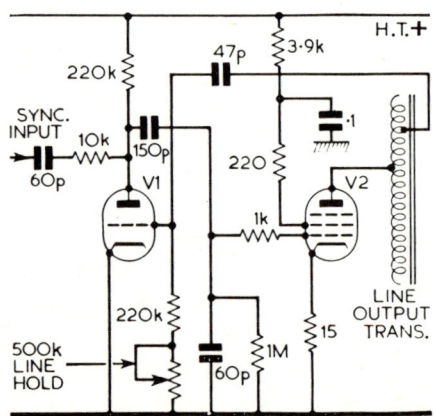

Figure 3.5   A common variant of the multivibrator in a line timebase
The feedback to V1 is via the 47 pF capacitor connected to a tapping on the line output transformer.

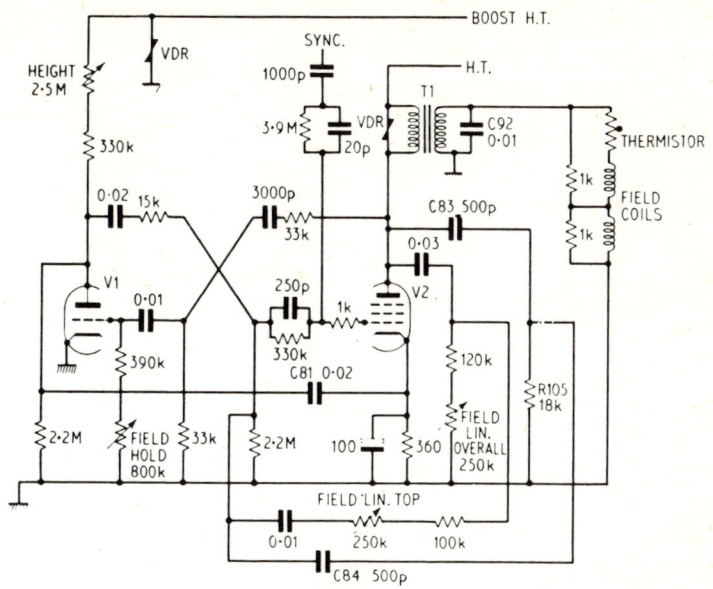

Figure 3.6   Self-oscillating field timebase

V2 is the output valve and also forms part of the multivibrator circuit. C_{81} is the charging capacitor (*Thorn/B.R.C.*).

capacitor is generally fitted in the pentode anode circuit—often across the primary or secondary of the output transformer—to overcome this. In addition, in the circuit shown in *Figure 3.6*, a second feedback circuit in the output stage (C_{83}, R_{105}, C_{84}) is used to provide protection against this.

Sinewave oscillators

Also used for many years but much more common today is the *LC* sinewave oscillator as line timebase generator. This very stable oscillator produces a sinewave output which is shaped to be nearly a square wave and used to control the operation of the line output stage. *Figure 3.7* shows a representative valve example. The pentode section is the oscillator, with the tuned feedback circuit connected between the

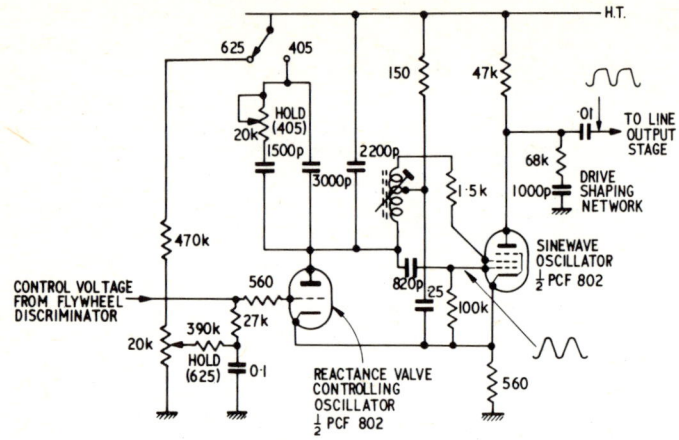

Figure 3.7   Sinewave oscillator used as a timebase generator. The frequency is controlled, via the reactance valve, by a control potential from a flywheel sync. circuit

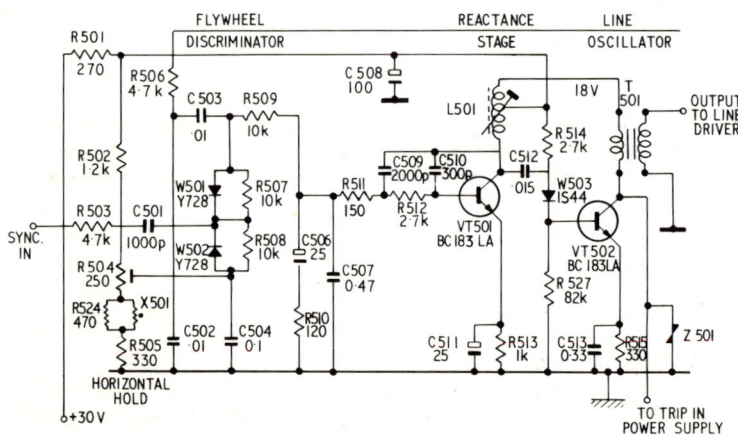

Figure 3.8   Line oscillator (sinewave): B.R.C. 3000 series colour receiver (single-standard)

screen and control grids. A PCF802 triode-pentode, which is intended for this application, is generally used and provides a ready indication of whether a set incorporates this particular type of circuit (though other triode-pentodes have been used in this position). This type of oscillator is found in conjunction with flywheel sync. circuits (which are generally used on dual-standard receivers), the control voltage from the flywheel sync. discriminator being applied to the grid of the triode reactance valve which acts as a variable capacitive reactance in the tuned circuit, controlling the frequency of oscillation. In the example shown, on 405 lines a variable resistor in the anode circuit of the triode provides hold control, while on 625 lines a variable d.c. control potential is applied to the triode grid via the 20k potentiometer to provide hold control. The output waveform is made the required shape by the interstage network coupling the output to the line output stage.

The reactance of the control valve in this circuit is determined by the input from the flywheel discriminator and the waveforms applied to its cathode as a result of the shared, unbypassed 560 ohm cathode resistor and the coupling via the $0.25\,\mu\text{F}$ capacitor to the oscillator tuned circuit.

Single-standard, u.h.f. receivers generally employ this form of circuit. If the line output stage uses a valve then the stage is generally similar to *Figure 3.7* but if solid-state output is used then a transistorised version of the circuit may be used as shown in *Figure 3.8*. Here, the basic arrangement is as before: the reactance stage coupled across the oscillator section so that changes in reactance cause a change in the tuned circuit frequency and a change in oscillator frequency. The pre-set line hold in this case is the oscillator coil itself, L501. The output is transformer coupled by VT501 so as to match the input impedance of the following driver stage.

Silicon controlled switch (s.c.s.) oscillator

The sinewave oscillator is impractical for use in a field timebase because of the large inductor that would be required for it to operate at 50 Hz. The blocking oscillator is practical (the transformer in the blocking oscillator is used as such and not used as an inductor), and the multivibrator or self-oscillating multivibrator is practical. Because receiver manufacturers prefer, if possible, to avoid the use of wound components due to their relatively high cost and the expense of having them set to

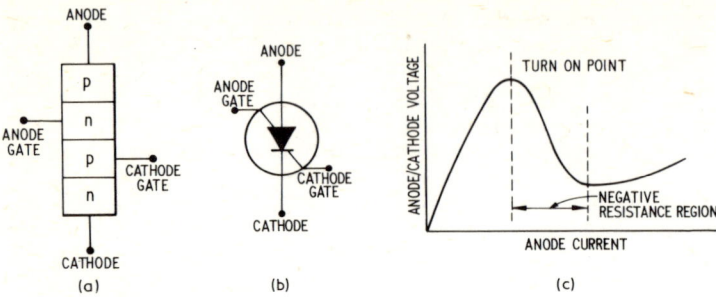

Figure 3.9 (a) Layer structure, (b) nomenclature, (c) plot of anode/cathode voltage versus anode current in silicon controlled switch

their correct tuning position on the production line, the majority of earlier receivers used the multivibrator or the self-oscillating multivibrator. As already pointed out, however, the transistor is not very temperature stable in the multivibrator connection and should therefore be avoided.

If the receiver is fully solid-state, one way of avoiding these problems is by the use of an oscillator employing a silicon controlled switch. The s.c.s. is a four-layer semiconductor device similar to the thyristor but with all four layers brought out to terminal connections. *Figure 3.9* shows the layer arrangement and the symbol of nomenclature used with the device.

With the two gates left unconnected (i.e. zero voltage) a very small leakage current will flow with a positive anode/cathode voltage. As this voltage is increased a breakdown point is reached where the device enters a negative-resistance region (similar to the tetrode valve)—*Figure 3.9(c)*. This is a low impedance condition. The device will continue this conduction for as long as the anode current is allowed to remain above a 'holding current' level. Below that level it returns to being a high impedance device between anode and cathode.

With the two gates connected to suitable supplies, the s.c.s. can be made to turn on below its normal threshold with forward biasing of the intermediate junctions in the device.

A representative circuit for an s.c.s. field oscillator is given in *Figure 3.10*. To understand its operation we must assume that initially C_1 is discharged and that the s.c.s. is in the 'off' condition. The diode D1 will then be forward-biased through the resistor R_1. The junction of C_1 and R_1 will therefore be virtually at chassis potential and C_1 will charge from the $+40$ V line through R_2 and R_3. The anode gate is biased to a positive potential set by the potential divider chain R_4, R_6 and the frequency control R_5. When the anode voltage is just greater than the anode gate (which will occur while the capacitor C_1 is charging) current will flow in the s.c.s. turning it 'on'.

This low-impedance state passes current derived from the anode gate potential divider and the discharge current of C_1 and the increase in cathode voltage turns on the transistor VT1. Across VT1 is the charging capacitor for the timebase, C_2, which is charged from the 21 V rail through the height control R_7 and R_8. The charging and discharging of C_2 is therefore controlled by the switching on and off of VT1.

As the capacitor C_1 is discharging through the s.c.s. there comes a point where the voltage on the R_1 side of the capacitor is greater than that on the R_3 side (R_3 is larger than R_1). At that point the diode D1 suddenly conducts and the cycle is completed as the s.c.s. falls below the holding current and turns off. The scanning period time is dependent on the values of R_3 and C_1 whilst the flyback time is dependent on the values of R_1 and C_1. Both times are also dependent on the setting of the frequency control, R_5, under the free-running conditions. When synchronised by negative pulses appearing on the anode gate, R_5 has no control.

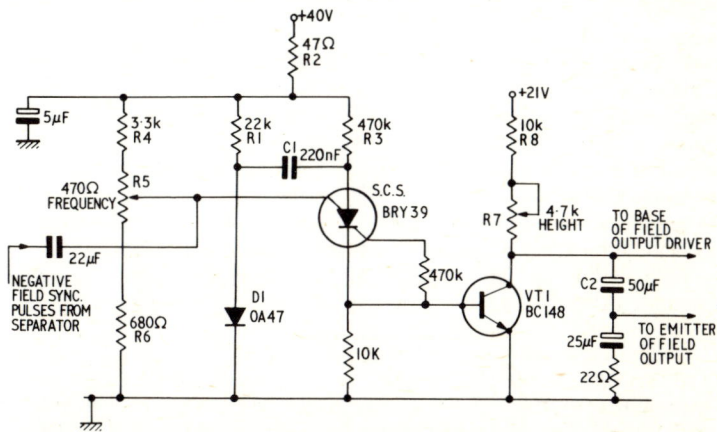

Figure 3.10   Silicon controlled switch oscillator for field timebase (Philips and Bush)

Waveform shaping

As we have seen, the 'basic' scanning waveform is of sawtooth shape. Because, however, of the mode of operation of the line output stage, and in both the line and field timebases the electromagnetic characteristics of the output transformers and scanning coils, and also because of the correction necessary to take into account the nonlinearity required for correct scanning of modern flat-faced picture tubes (the distance the electron beam travels varies at different points on the screen of a flat-faced tube), the timebase output current waveforms are not of exactly sawtooth shape. In the case of the field timebase the output waveform required is a mixture of parabolic and sawtooth components, and the energy stored in the transformer at the end of a scan period is used to contribute to the deflecting current following the flyback period. The line timebase output current waveform is more nearly of sawtooth shape.

A number of factors can be made use of in obtaining the required waveform. Frequency conscious networks in the coupling circuits play a part; the transfer characteristic of the output valve or transistor can be employed; and feedback circuits incorporating frequency conscious networks can be used. The basic frequency conscious network relied upon is the simple *RC* combination. The shaping effect is provided by the time it takes for the capacitor to charge and discharge.

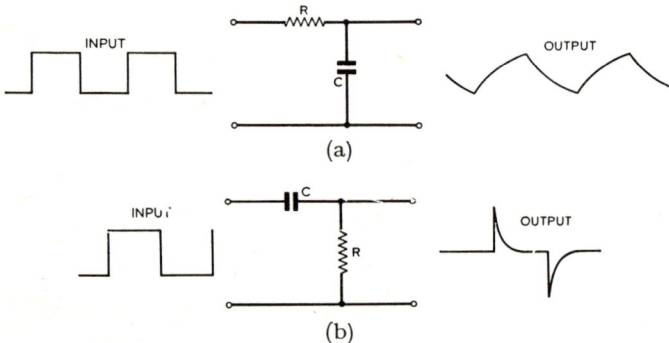

Figure 3.11   (*a*) *Basic RC integrating circuit, showing its effect on a square-wave input when its time constant is approximately a half-cycle of the input.* (*b*) *Basic RC differentiating circuit, have a time constant very short compared with one input pulse*

The two basic alternatives, called integrating and differentiating circuits from the effect they have on the input waveform fed to them, are shown in *Figure 3.11*, together with waveforms. The time constant of the networks is given by the component values chosen, so that by suitable choice the required wave or pulse shape is obtained. *RC* shaping networks are used to select and shape the sync. pulses, to shape the feedback pulses or waveforms used in multivibrator-type circuits, and in the feedback circuits used to adjust the linearity of the field scanning waveform.

Field output stage

The valve field output stage consists of a pentode transformer coupled to the field scanning coils. An example has already been illustrated in *Figure 3.6*. A further example is shown in *Figure 3.12*, in which the output stage is controlled by a separate cathode-coupled multivibrator circuit. The basic charging capacitor is the 0·05 μF capacitor connected between the anode of the second triode and, as in the previous example, the cathode of the output stage, the waveform being fed to the grid of the pentode by the 0·1 μF coupling capacitor. As shown in both examples, feedback networks are included in the output stage from anode to grid, to obtain waveform linearity adjustment. The networks

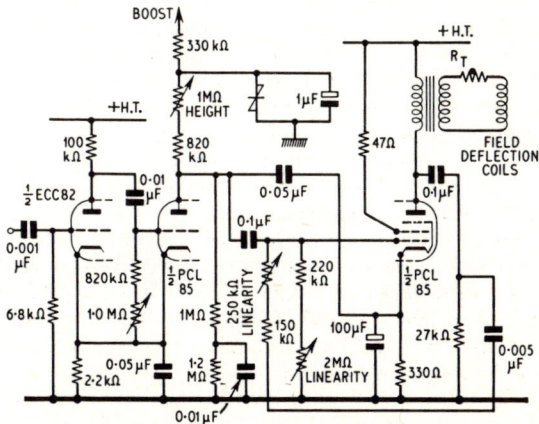

Figure 3.12   Field output stage driven by a cathode-coupled multivibrator

are generally designed to be the *RC* equivalent of the *LR* circuit comprising the output transformer and scanning coils, the feedback voltage then being proportional to the deflection current. Numerous variations in detail exist in practical designs to take into account such factors as the characteristics of the output valve, output transformer, etc. The feedback circuit commonly takes the form of a *CR* differentiating network followed by an *RC* integrating network. The former provides low-frequency attenuation of the feedback voltage, to take into account losses in the transformer; the latter averages and shapes the feedback voltage, and also provides attenuation of the voltage peaks that occur at the anode and the effects of line pulses that may be fed back via the field deflection coils.

To overcome the problem of picture height shrinkage with rising temperature of the deflection coils, it is common practice to include one or more thermistors to provide temperature compensation. The thermistor is mounted close to the deflection yoke and connected either in the field output h.t. line or, more often, in series with the secondary winding of the field output transformer: sometimes both methods are used. As the temperature of the deflection coils increases so does their resistance, whereas that of the thermistor falls.

As a protection against mains-voltage variations a voltage-dependent resistor can be used to stabilise the boosted h.t. applied to the field oscillator, and a v.d.r. may also be found in some circuits across the output transformer primary (as in *Figure 3.6*) partly to stabilise the output level and partly to limit the amplitude of the flyback pulse.

The approach to transistor field output stage design follows the same considerations as the design of an audio output stage. Class B amplifiers cannot be easily used because of the 'pulsy' loading of the power supply, the accuracy required in components for Class B, and the problem of crossover distortion. Class A is therefore preferred and directly-coupled variations give additional advantages in simplification of the drive requirements.

The circuit of *Figure 3.13* uses the s.c.s. oscillator of *Figure 3.10* to drive it. The stage consists of an emitter follower (VT1) feeding a push-pull pair of transistors (VT2 and VT3). At the start of the scan stroke the base of VT1, and therefore of VT3, is at chassis potential and VT3 is cut-off. No current will therefore flow in R_1 and about half the rail voltage will appear across the capacitor C_1. The value of R_2 is chosen so that VT2 is conducting when VT3 is cut-off. Apart from specific components in the circuit, the base/emitter potential of VT2 is determined by the effective voltage across C_1.

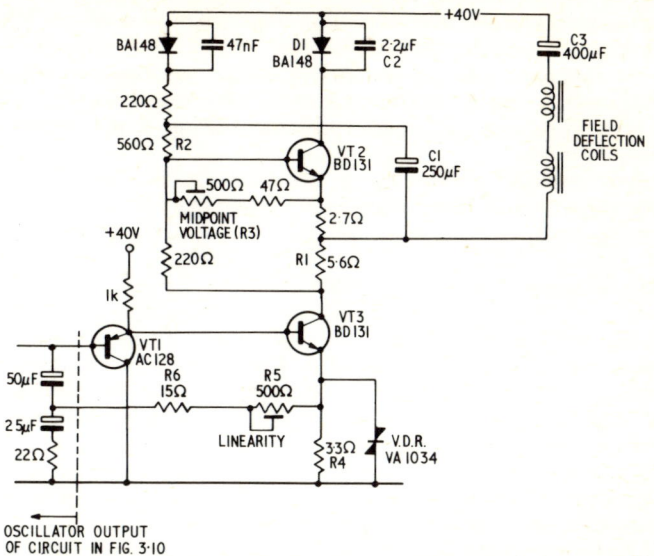

Figure 3.13   Representative transistor field output stage which would be connected to the oscillator of Figure 3.10

As the scan progresses, VT3 gradually conducts harder and the collector voltage gradually falls, slowly turning VT2 off. The voltage across the deflection coils therefore swings about the rail voltage and R_3 is set up to make sure that this is so.

During flyback, VT3 is cut-off and stored energy in the deflection coils causes the voltage across them to rise positively taking the base and emitter of VT2 positive.

However, as these potentials approach the rail voltage D1 ceases conduction and the voltage across the coils goes into a half-cycle of oscillation (*ringing*) because of resonance between the deflection coils and C_2. As soon as the half-cycle is complete—bringing the potential back to the rail voltage—D1 again conducts clamping the potential until the coil current has completely reversed. At the start of the next scan VT3 begins to conduct slowly once more to complete a cycle of operation.

Stability of the output level is maintained by the v.d.r. across the emitter feedback resistor R_4, and linearity correction is made by feedback through R_5 and R_6 to the 50 μF charging capacitor of the oscillator.

The dissipation in the two BD131 output transistors is about 3 watts each. There is little dissipation in the driver transistor (VT1) because of its emitter follower operation and its bottoming during flyback periods.

C_3 prevents unwanted d.c. passing through the field deflection coils. Its value is large because of the 50 Hz signal it must pass.

Valve line output stages

The mode of operation of the line output stage is very different. The line output valve acts more like a switch than like the waveform amplifier used in the field output stage. The valve provides only part of the scan, the first part being derived from recovered flyback energy. The line output stage is also used to generate the e.h.t. required for the final anode of the picture tube, and provides a boost voltage of 400–1 000 V which is used for the picture tube first anode and focusing electrodes and for the charging circuit in the field timebase.

The basic arrangement used in valve line output stages is shown in *Figure 3.14*. V1 is the line output valve, V2 the efficiency diode—also called boost, recovery or reclaim diode—and V3 the e.h.t. rectifier. There are three main sections of the line output transformer to be

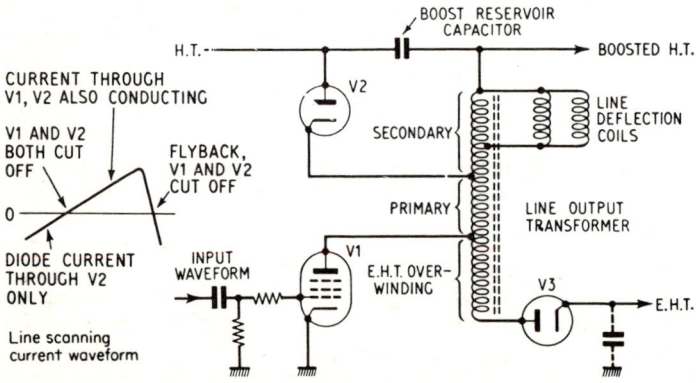

Figure 3.14   Basic line output and e.h.t. circuit

considered, the primary and secondary, which are connected as an autotransformer, and an overwinding across which the e.h.t. is developed. The primary and secondary form a voltage step-down transformer, with the deflection coils tapped across the secondary.

When V1 conducts, it drives a current through the transformer and deflection coils and this deflects the picture tube electron beam. The operation of the line output stage depends mainly, however, on what happens when the flyback period arrives. The electron beam in the picture tube being an almost weightless mass, little power is required to deflect it, so that at the end of the scan considerable power is stored in the field established around the deflection coils. When the line drive ceases at the end of the scanning stroke, this field collapses rapidly and a high positive voltage pulse is reflected back through the transformer. This is stepped up by the transformer since, for power reflected back from the scan coils, the autotransformer is now a step-up transformer. The voltage is further stepped up across the e.h.t. overwind and, since it is in the correct polarity, V3 e.h.t. rectifier conducts and charges the e.h.t. reservoir capacitor (shown dotted) to the e.h.t. voltage required by the picture tube final anode (16–20 kV). The e.h.t. reservoir capacitor is shown dotted since a separate component is not used in modern receivers, which make use instead of the capacitance formed by the internal and external cathode-ray tube conductive coatings with the glass bulb as dielectric (for this reason it is important that the external coating has a good earth connection and that the capacitance is discharged before any service work is done).

The high-voltage pulse reflected back cuts V2 off and, as a result of the negative grid drive, V1 is also cut off. Following the positive pulse the circuit tries to swing negative. V2 then starts to conduct, damping the tendency towards oscillation and charging the boost reservoir capacitor. The current flowing through V2 as a result of the collapse of the scanning field gradually decreases, and it is this flow of diode current that forms the first part (see *Figure 3.14*) of the scan waveform. Hence the terms efficiency, recovery or reclaim diode: the power that needs to be dissipated during the collapse of the field around the deflection coils following the flyback period is recovered by the efficiency diode circuit and made use of to provide the initial drive for the following scanning stroke while the potential developed across the boost reservoir capacitor, being in series with the h.t. line, increases the effective h.t. supply to the stage. V1 is held non-conducting during the first part of the scanning stroke by the waveform applied to its grid. When the waveform is sufficiently positive, V1 conducts and takes over from the efficiency diode circuit the supply of current to the deflection coils. Thus V1 is held cut off for $\frac{1}{4}$ to $\frac{1}{2}$ of the scanning period, then switching on until the flyback part of the drive waveform arrives and cuts it off again.

This part of the drive waveform has to go to a substantial negative potential to avoid the valve being switched on again by the positive flyback pulse which then appears at its anode.

The charge delivered by V2 to the boost reservoir capacitor is in addition to that already acquired by the capacitor from the h.t. line, so that on the other side of the capacitor a boosted h.t. of 400–1 000 V is present which is used for the purposes mentioned at the beginning of this section. The efficiency diode circuit also provides the damping required to prevent the stage going into oscillation.

It is common practice to operate the line output pentode valve 'below the knee' of the I_a/V_a characteristic, in which condition the grid potential has little effect upon anode current. The main advantages of this mode of operation are that the e.h.t. regulation is improved, and the fall off in scan power during the life of the output valve is reduced. For example, a drop of 25 per cent in the peak anode current of the valve need result in a loss of scan power of only about 0·25 per cent. In this case, the valve is being operated as a switch.

Representative valve circuit

A representative B.R.C. dual-standard line output stage circuit is shown in *Figure 3.15*, and it can be seen that it closely follows the basic arrangement shown in *Figure 3.14*. Three additional features require explanation: the need for switching to compensate for the slightly different characteristics required for deflection at the two different line frequencies; the use of a feedback circuit to stabilise the operation of the stage; and the need for a means of providing scan linearity adjustment.

The dual-standard switching in this particular example is relatively simple: S1A adjusts the third harmonic tuning (see later) of the transformer, and S1B alters the linearity compensation (a capacitor in series with the scan coils is used to provide the linearity compensation needed, as previously mentioned, in scanning modern flat-faced tubes: the capacitor is resonant with the scanning coils at half line frequency).

Feedback stabilisation circuits are used on all current models, and generally take the form shown in *Figure 3.15*, a pulse being fed back

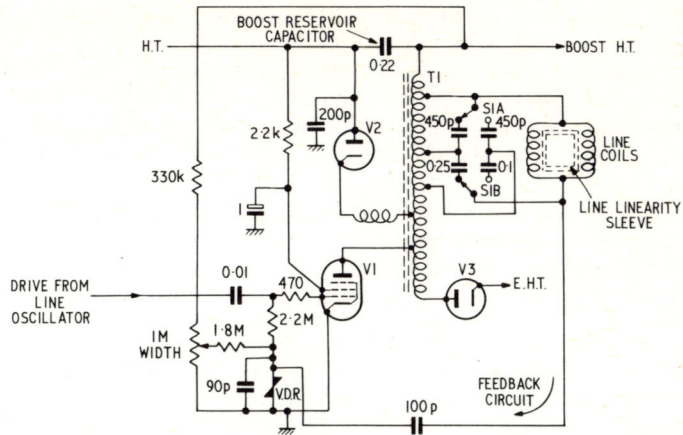

Figure 3.15   Representative dual-standard valve line-output stage, incorporating an e.h.t. stabilising circuit

via a capacitor from the deflection coils or a suitable point on the line output transformer and applied across a nonlinear v.d.r. (voltage-dependent resistor) to the line output valve grid circuit. The v.d.r. presents a lower resistance to voltages more positive than a certain critical value. Since more current flows through it on positive than on negative peaks, the averaged value across the resistor is slightly negative, and this potential, in conjunction with the pre-set control, determines the working point of the valve. Should the output tend to rise, the negative voltage across the nonlinear resistor rises rapidly and reduces the current flowing through the valve. A fall in output has the opposite effect. Thus the system is really a form of a.g.c. which maintains the line output fairly constant. The pre-set control is used as a width control.

De-saturated transformers

Since about 1955, some line output transformers have had de-saturated cores, which is to say that the magnetising flux arising from the direct current flowing in the windings is cancelled out.

A representative example of a dual-standard line output stage using a de-saturated output transformer is shown in *Figure 3.16*. It will be seen that the main change is the splitting of the transformer winding and the insertion of the boost reservoir capacitor at this point, and also that the h.t. supply to the stage is fed into this point. With this arrangement the line output valve anode current flows through the two

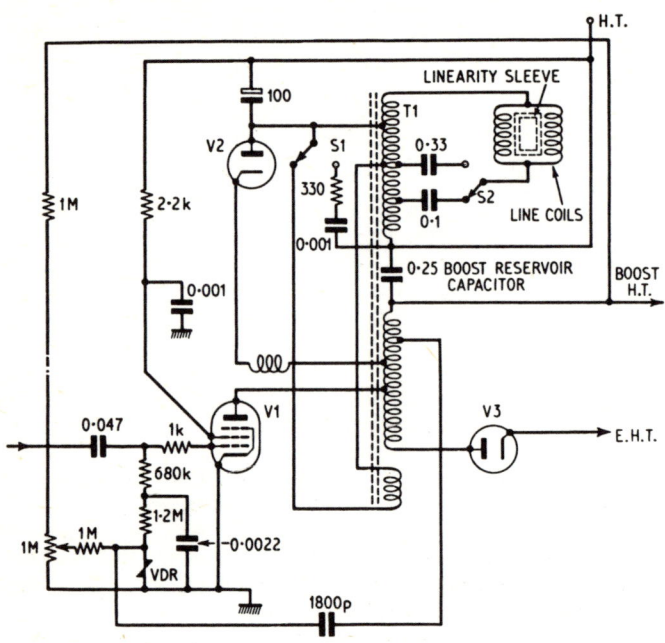

Figure 3.16   Dual-standard line output stage using a de-saturated line output transformer

separate windings in the opposite sense, so that the magnetic flux due to the d.c. component is largely neutralised, resulting in greater efficiency of operation for a given core section (10 and 15 kHz whistle is also reduced).

S1 alters the third harmonic tuning, in this case by introducing a coupling winding on the transformer into circuit on 625 lines to reduce the leakage inductance between the primary and the e.h.t. overwinding.

This technique is used on a number of chassis. S2 adjusts the linearity and amplitude of the scanning for the two standards.

Transistor line output stages

The line output is still the most sensitive stage of a receiver to transistorise. In a full size monochrome receiver using an e.h.t. of 20 kV, the power requirements for e.h.t. and line deflection are about 25 watts. In a colour receiver the figure is even higher because of the larger e.h.t. and the greater current drain on the e.h.t. supply. The transistor requirements are therefore severe in terms of switching speed, peak voltage, peak current and thermal dissipation.

The first transistorised line output stages were found in portable receivers where the smaller e.h.t. and scanning requirements were within the capability of the transistors available.

Because the line oscillator stage will be transistorised (if it were not there would be no point in making the line output solid state) the output voltage will usually be quite low and the stage must be reasonably isolated from the switching impedance of the output transistor—to prevent damping. These two factors involve the introduction of a line driver stage usually with transformer coupling from the oscillator and transformer coupling from the driver to the output stage.

Timing of the drive is very important with transistors because of their inability to switch very quickly due to 'minority carrier storage'. In the case of a *p-n-p* transistor the minority carriers are electrons. To prevent high collector dissipations during flyback the transistors must always be turned off relatively slowly. In earlier designs this was always done by including in the base circuits a series inductance (typically 20 μH) which reduces the speed of the turning-off pulse edge (effectively a low-pass filter). While this technique is still used in some circuits, it has been shown that with some other transistors the charge storage effects are very much reduced and the inductor can be neglected. Output devices are also gaining in the power dissipation of which they are capable, at low-cost.

Examples of each form of timing are given. In *Figure 3.17* the inductor is present using the transistor BU105/02 whilst in *Figure 3.18* there is no inductor when using the transistor R2008. In both circuits it will be noted that the transistor is protected against flashover by the capacitor connected between collector and base. Additional protection is included

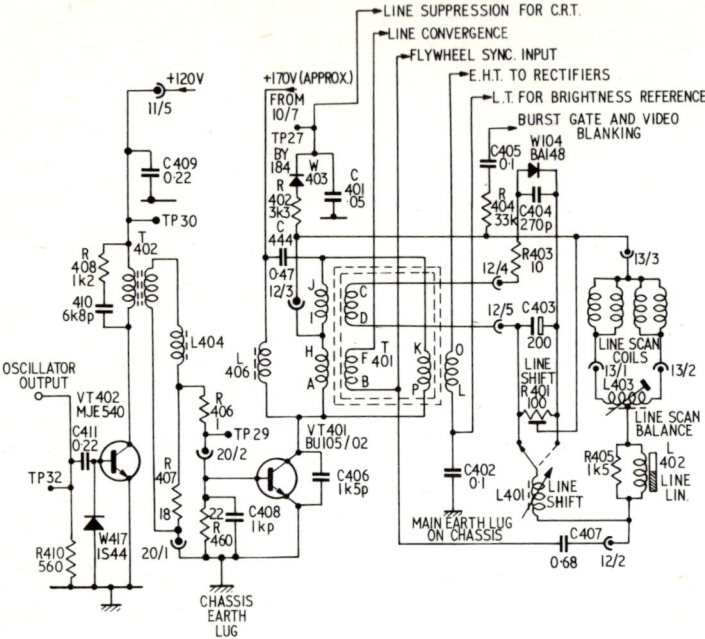

Figure 3.17   Single-ended line output stage (BRC 8000, single-standard, colour) with inductive timing correction (Note: BU 105/02 is a higher rated version of BU 105)

in one circuit (*Figure 3.17*) at the base/emitter junction by the parallel R_{460}/C_{408} combination presenting a low impedance path when the transistor is cut off. On both circuits, too, the primary of the driver transformer is heavily damped by series resistance and capacitance. Phase reversal is set up in the driver transformer so that the output transistor is off when the driver is on and vice-versa: this gives better isolation and it is sometimes known as the *non-simultaneous mode*.

Apart from these basic differences essential for the correct protection and operation of the output transistor and maintenance of the oscillator, the transformer effects and waveforms are similar to those encountered in the valve output stage. One notable exception is that there is no need for the inclusion of a boost diode because the reverse current can be handled by the transistor. If a diode is included it suggests that the transistor is operating at its limits of dissipation and the diode should always be checked in any servicing on the line output.

Present economical transistors for use in the line output stage are quite able to handle the power requirements but there tends to be little margin for error. In some designs the value of the base stopper resistor (e.g. R_{406} in *Figure 3.17*) is critical and may have to be set up to give specified current drain whenever the line output stage is serviced.

An alternative version of the receiver in *Figure 3.18* is available with a *pair* of output transistors (*Figure 3.19*). These work together to produce the scan and e.h.t. as they act effectively as a series switch whilst reducing the individual power dissipation requirements.

In the future, designs may well make use of other devices, such as the thyristor, in this stage of the receiver. This would be required to allow operation of the stage from rectified mains voltage. With transistors this is not possible in the forseeable future because the device would have to be able to withstand peak voltages across it of the order of 4–5 kV.

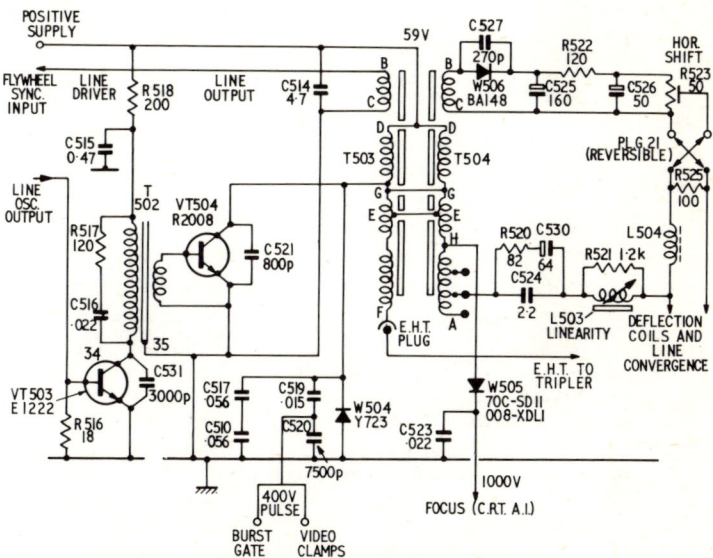

Figure 3.18   *Single-ended line output stage (BRC 3000, single-standard, colour) with direct timing and efficiency diode (W504)*

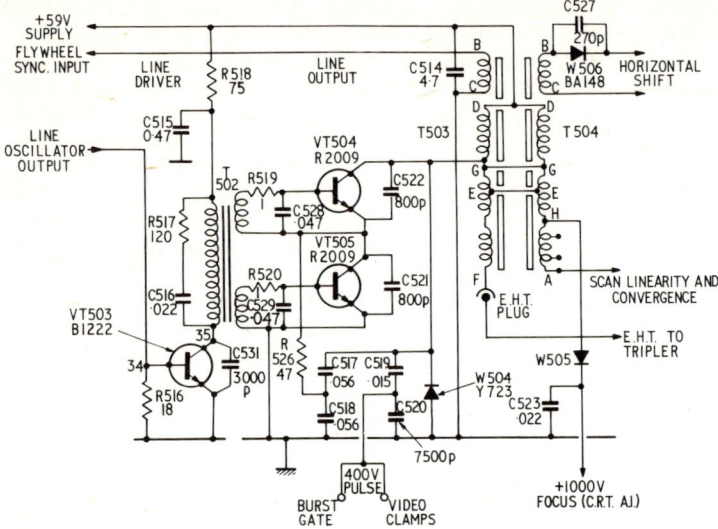

Figure 3.19 Alternative version of line output stage for BRC 3000 series receiver using output transistor pair (note, compare with single-ended version of Figure 3.18)

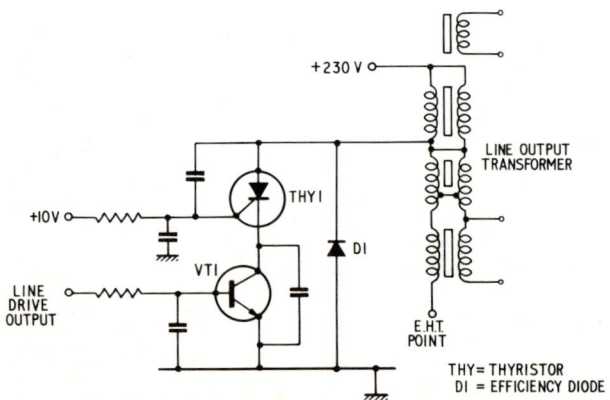

Figure 3.20 Probable form of thyristor line output stage

A thyristor output stage would be capable of handling the power requirements and it would be switched by a driver transistor with its collector connected in series with the cathode of the thyristor. With the cathode gate of the thyristor held at the h.t. voltage (*Figure 3.20*), the output of the line drive stage would pulse the control transistor to turn the thyristor on and off. The current would flow through the cathode gate of the device. Because the thyristor is such a good switch, there would be no return path back through it for the reverse current and a separate efficiency diode D1 is needed. The first experimental thyristor output stage was produced by R.C.A. some years ago and it is expected that they will begin to find their way into new designs.

The line output stage is often called on to supply other voltage sources. In hybrid receivers for example there may be a voltage tapping off the line output transformer which is rectified and smoothed to provide the l.t. for the transistor stages. In particular pulse outputs are used for a number of purposes in receivers: flywheel sync. (e.g. *Figure 3.8*), burst gate (in colour receivers), keyed d.c. restorers or clamps and other functions. All these feeds increase the loading on the output stage. It is expected that eventually a device such as the thyristor will be used in the line output with mains rectified h.t. being used to feed it and then, provided the power capability was there, the rest of the receiver could be fed from a rectified tapping on the line output transformer. With the introduction of 110 deg colour tubes, the thyristor becomes an almost essential device if the line output is to be solid state.

Line output harmonic tuning

There is always some leakage inductance between the main winding

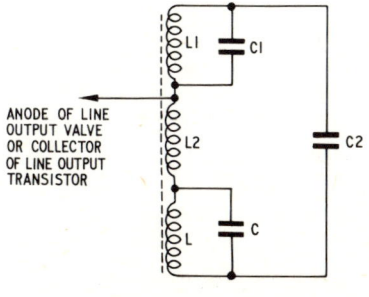

ANODE OF LINE
OUTPUT VALVE
OR COLLECTOR
OF LINE OUTPUT
TRANSISTOR

Figure 3.21   Equivalent circuit of line output transformer—leakage inductance L is represented in series with overwind, L_2, because overwind is made narrow in comparison to main winding, L_1

and the e.h.t. overwinding on a line output transformer because of their different physical sizes, all the magnetic flux that is produced being unable to link them both completely. To get better control of this inductance modern transformers have the two windings mounted on separate limbs. The leakage inductance although not physically existing as a winding must be represented in any equivalent circuit of the transformer. Indeed the inductance must also have some stray capacitance. These are L and C in the equivalent circuit of *Figure 3.21.* Also in this circuit the main winding and its stray capacitance are L_1 and C_1. These include the effect of the line deflection coils, any linearity coil and the stray capacitances across them; it may also include some physical capacitance. The e.h.t. winding is indicated by L_2.

The capacitance C_2 represents the load capacitance of the circuit which will include the e.h.t. rectifier, any e.h.t. smoothing or regulation and the c.r.t. capacitance at the final anode.

During the active line period (the scan time), the line output valve will be conducting all the time or just for the second period (with boost diode operation) and a current will build up in the main winding L_1. As soon as the conduction ceases during the flyback period, a large half-sinusoid of voltage is induced into L_1 and L_2—the flyback pulse. The pulse is taken to the e.h.t. rectifier for supply to the final anode of the c.r.t.

None of this flyback pulse comes from the leakage inductance because it does not physically exist and cannot therefore have a voltage induced into it. However, the pulse reaches L as a shock wave and sends L and C into a damped oscillation. If the damping is low, the oscillation will continue into the next scan causing vertical striations

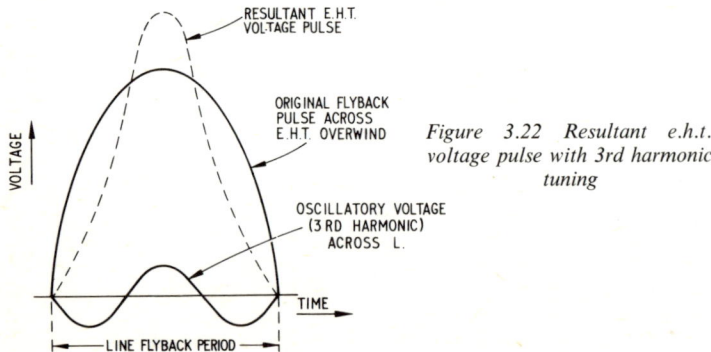

Figure 3.22 Resultant e.h.t. voltage pulse with 3rd harmonic tuning

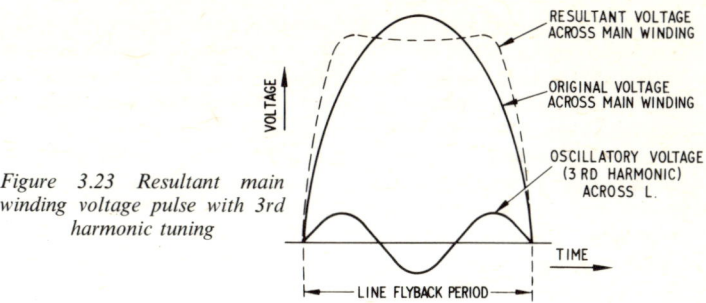

Figure 3.23 Resultant main winding voltage pulse with 3rd harmonic tuning

on the left-hand edge of the picture. This is inevitable because some of the oscillatory voltage will appear across the main winding.

The best method of damping is to allow the conduction of the boost diode to achieve it. This is possible if the oscillatory voltage is at zero potential at the instant the new scan begins. This will be so if the ringing frequency is a harmonic of the flyback pulse frequency. This ringing frequency is determined by the values of L and C. Additional benefits are gained if the harmonic chosen is an odd harmonic and also if it is the third harmonic. *Figure 3.22* shows how the voltages across the circuit add to produce the e.h.t. voltage pulse in the third-harmonic case. Because of the winding directions, the oscillation voltage of L and C begins in antiphase to the flyback pulse across the overwinding so the resultant of the two added together produces an e.h.t. pulse which has a higher maximum value.

Across the main winding the voltage is modified in a different way. *Figure 3.23* shows the relevant voltages: the oscillatory voltage is now in phase with the flyback pulse and gives a resultant which has a lower maximum with a 'flatter' top to the pulse. As this is the voltage present at the anode of the line output valve (or collector of the line output transistor) it means that the valve (or transistor) ratings can be maintained more easily or that slightly more scanning power can be obtained from a particular valve (or transistor) without exceeding the voltage ratings. This is very important with the higher power requirements of modern 110 degree scanning angle picture tubes.

The third-harmonic tuning adopted in this case therefore gives a larger e.h.t. voltage with a greater scanning power being available whilst the control of the leakage inductance (which must exist) prevents ringing in the line scan time.

To control the value of L, the position of the overwind on the second limb of the transformer can be varied. Fine tuning is generally provided by external capacitances on the other windings. In dual-standard receivers it is usual to switch an additional coupling coil in and out of circuit to change the value of L to give the different frequencies of oscillation required for the two line standards (*Figure 3.16*).

As the use of e.h.t. multiplier circuits increases (as we will see in the next section), the demands on the line output stage have also changed. The size of the e.h.t. pulse required from the line output transformer overwinding is considerably smaller (approx. 8·5 kV for the 25 kV supply for a shadowmask tube) and it would be quite possible, simply from insulation requirements, to lay the e.h.t. winding *on top* of the main winding, so reducing the leakage inductance to minimal proportions and ending the need for third harmonic tuning. Unfortunately, this system gives a flyback pulse which, as we have seen, is poor in shape and which gives poor voltage regulation. Whilst this is not so important with monochrome receivers it is very important with

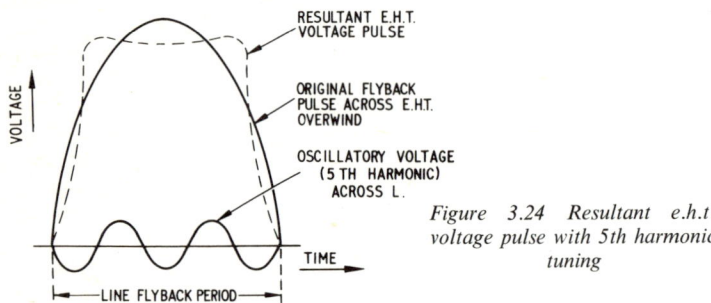

Figure 3.24 Resultant e.h.t. voltage pulse with 5th harmonic tuning

colour receivers where the changes of beam current during operation are larger. To improve the regulation the primary capacitance of the line output transformer must be reduced, and this can only be achieved by once more removing the overwind to a separate limb on the transformer.

Once the overwind is on a separate limb, the need for tuned leakage is again introduced to reduce or remove the ringing on line scan. However, we have seen that third-harmonic tuning produces an e.h.t. pulse which is peakier than the flyback pulse; this would rather negate the point of moving the overwinding to the separate limb because the peakier the pulse the poorer the regulation will be. This difficulty can, however, be overcome by employing *fifth* harmonic tuning instead of third. The result on the e.h.t. pulse shape is indicated in *Figure 3.24*. It can be seen that the top of the e.h.t. pulse is much flatter than before indicating a more constant voltage and hence better regulation. The oscillation voltage is still zero at the end of the flyback period so conditions to prevent faster ringing are still present. A secondary effect of the use of fifth-harmonic tuning, however, is that the voltage on the anode of the line output valve (or collector of the line output transistor) is made more peaky with a higher maximum. This must be taken into account in the design using a particular device—particularly where an output transistor is being used.

Shorted-turn linearity and width controls

Owing to the varying impedance of components in the line timebase during the scanning cycle, an uncorrected timebase would cause the picture to be stretched on the left and cramped on the right. One common remedy is to connect a saturated reactor in series with the

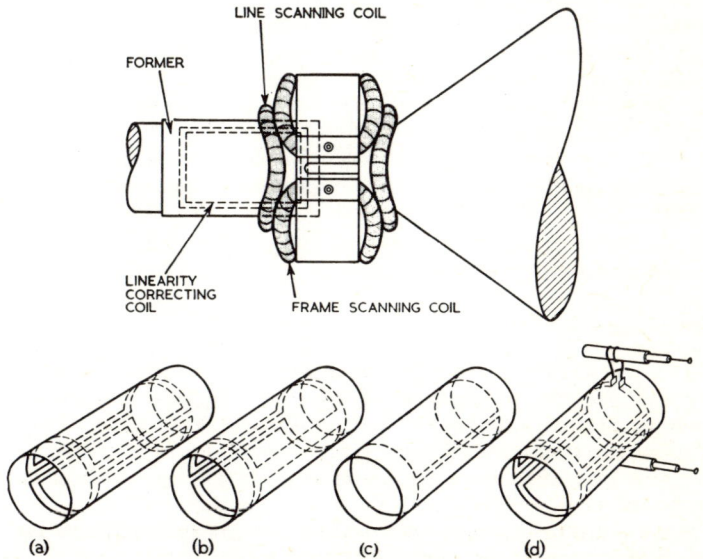

Figure 3.25   Linearity sleeve in position around neck of picture tube, and various forms taken by sleeve showing foil loops inside non-conductive former

line-scanning coils; this can be arranged to compensate for the impedance variations so that a relatively undistorted picture is obtained. Linearity controls of this type are found on many modern receivers. Any tendency to ringing can be reduced by damping the saturated reactor by means of a parallel resistor.

Also used has been the shorted-turn line linearity sleeve which consists basically of two rectangular loops of metal foil fixed to a cylindrical former and placed around the neck of the picture tube beneath the line-scan coils. There is no direct electrical connection betwen the loops and the scanning coils, but a current is induced in the loops by the magnetic field, and this in turn affects the current in the scanning coils. By adjustment of the physical position of the loops the linearity of the line scan may be controlled. There are several possible versions of this form of linearity control (see *Figure 3.25*).

The adjustment of the loops is critical, and may be made more difficult by the effect of other assemblies on the neck of the tube. For this reason the sleeve carrying the linearity control loops may be cemented to the scanning-coil assembly, and in such cases it is advisable not to disturb the adjustment when changing tubes.

In a number of models this sleeve technique was extended to act as both linearity and width control. This is possible because modern line output stages are designed to provide a nearly constant output throughout the life of the output device.

Where the shorted-turn sleeve is used for width and linearity adjustments, the usual method of setting, using a transmitted test card, is as follows:

Picture shape.   Adjust the sleeve to provide symmetry of the horizontal and vertical lines and corners by *rotating* the sleeve about the neck. This adjustment is for minimum coupling to the field coils which would cause poor interlace and geometric distortion.

Width and linearity.   Slide the sleeve *along the neck* without rotating it until the correct width is obtained. With 110 degree tube masks this may require some horizontal overscan. Correct width setting should also provide good linearity. It should be noted that a sleeve inserted too far into the deflection-coil assembly may cause overheating of the deflection coils.

On older tubes having an ion trap, the ion-trap magnet should be checked after any adjustment of the sleeve.

Width control by means of a tapped line output transformer was also once common.

Sync. pulse separation

To synchronise the receiver timebases with the timebases used to control the scanning of the television camera, the sync. pulses inserted in the television waveform must be separated from the picture information and used to control the operation of the receiver's timebase oscillators. The portion of the transmitter carrier wave used for the sync. pulses has been shown in Chapter 1. It follows that an amplitude limiter stage can be used to separate the sync. pulses from the rest of the received waveform. Time conscious circuits can then be used to separate and shape the short (10 μs on 405 lines, 4·7 μs on 625 lines) line sync. pulses and the relatively longer field sync. pulses. The output from the video amplifier on both standards consists of negative-going (towards peak white) picture information, the sync. pulses being positive-going. Due to the phase inversion occurring in the normal grounded-cathode (or common-emitter) amplifier stage, the sync. pulses at the output of the sync. separator will thus be negative going. Where it is desired to use a positive-going sync. pulse to control a timebase oscillator, the sync. pulses are passed through a second amplifier stage which acts as a pulse inverter.

A triode or pentode valve was generally used as the sync. separator, the operation of the stage being as follows. The positive-going sync. pulses applied to the grid drive the valve into grid current, charging the coupling capacitor negative with respect to the grid. The time constant of the grid circuit components is made such that the coupling capacitor charges to the mean-level of the input signal (hence the use of the potential across the grid leak resistor as the a.g.c. voltage in mean-level a.g.c. systems). If in these conditions the grid base is kept short as shown in *Figure 3.26* (this is done by using a low anode voltage and, in the case of a pentode, low screen voltage) the stage will be kept cut-off when picture signal information is present but will be driven into conduction by the sync. pulses so that only the sync. pulses appear in the anode circuit. Optimum results are obtained when, as shown in *Figure 3.26*, the sync. pulses drive the valve into saturation, thus clipping both the negative and positive sides of the input pulses to give a train of constant amplitude output pulses. With a triode there is the possibility of h.f. components of the picture appearing at the output via the grid-anode capacitance of the valve, so that a pentode is generally preferred for use as a sync. separator. Transistor sync. separator stages operate on the same principles. In the input circuit a resistor of about 5–20 kΩ is usually connected in series with the coupling capacitor to reduce

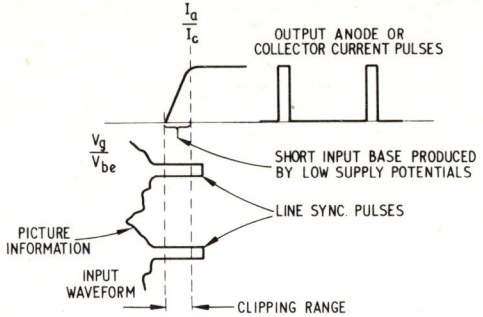

*Figure 3.26 Operating conditions of a triode, transistor or pentode sync.
separator stage acting as an amplitude limiter*

Negative-going output voltage pulses are produced as anode (or collector)
voltage falls when increased anode (or collector) current increases voltage
across anode (or collector) load resistor.

loading on the video amplifier and to reduce the effect of noise and
interference on the sync. separator; there is a practical limit to the
value of this resistor, as if too great it will distort the shape of the sync.
pulses. In some 405-line circuits the conditions of the sync. separator
stage may be changed slightly in areas of weak signal strength to
minimise 'line tearing' caused by poor synchronisation.

In the usual form of pentode sync. separator both the field and the
line sync. pulses appear at the anode of the valve, and it is then necessary
to separate the two sets of pulses before using them to control the
appropriate generators. (In a few circuits, however, the line pulses are
taken from the anode and the field pulses from the screen circuit.) It is
important that as much as possible of the line sync. pulses should be
removed from the field synchronising signals, as otherwise false
triggering of the field generator may occur and impair interlacing.

Since the line and field sync. pulses are of the same amplitude but
differ in duration and repetition frequency, they can be separated by
being passed through fairly simple 'time conscious' filters. These
filters consist of resistor-capacitor integrating or differentiating networks. Line sync. pulses normally pass through a differentiating network: field sync. pulses may pass through either an integrating or a
differentiating network or a combination of the two.

The calculations in the following paragraphs relate to the parameters of the 625-line system.

In *Figure 3.27(a)* the time constant C_1R_1 is short compared with the length of the line-synchronising pulses (e.g. about 1·5 μs $C_1 = 50$ pF and $R_1 = 30$ k). The output voltage (across R_1) rises sharply to approximately the full value, but, as C_1 charges rapidly, the voltage across R_1 falls to give the waveform shown. For the integrating circuit (*Figure 3.27(b)*) the time constant is long compared with a line-synchronising period (about 25 μs and, consequently, the voltage appearing across C_2 is small for line pulses, but when the five broad 27 μs field pulses arrive, the voltage across C_2 builds up. The state at the end of even and odd fields is shown. This appears to produce a flat-fronted pulse, but when the scale is considered, the five broad pulses total about 160 μs and the rest of the field period 20 ms, a ratio of about 1:125, so that the pulse may be steep enough to ensure good interlace. Further shaping, however, is often applied.

The field pulses can also be separated by means of a differentiating network of selected time constant (about 10–15 μs) and some designers consider this provides sharper pulses.

To reduce the effect of noise and residual line pulses on field sync., a series-connected clipper diode is frequently incorporated in the field-sync. circuit. This gives a short charging and slow discharging time constant to the field-sync. separation circuit, operating on similar lines to the series sound interference limiter circuit previously described (see *Figure 2.23*). Alternatively, a shunt clipper diode may be used in conjunction with the time-constant network arranged so that noise and residual line pulses are removed from the output.

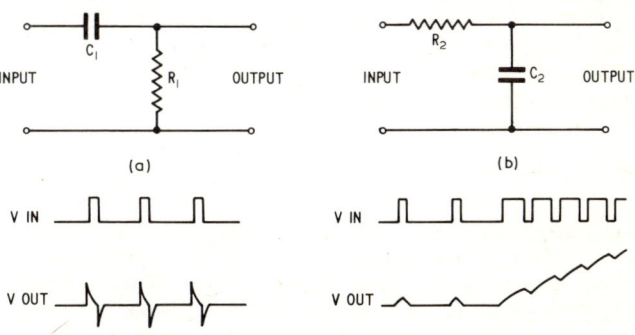

Figure 3.27   Waveforms produced by: (a) differentiating circuit; (b) integrating circuit

On 405-line standards, the separation of the field pulses from the waveform does not result in precisely the same waveform. This is because the odd and even fields are not identical—a problem not encountered in the 625-line system because of the use of equalising pulses.

Because of this problem on 405-lines, it is quite possible for the field timebase to trigger at slightly different times on alternate fields. This gives rise to the alternate lines being closer and further spaced than they should be; a problem which in its worst case gives rise to *line pairing*.

A wide variety of circuits have been used in the attempt to obtain improved interlace, i.e. avoid line pairing. One technique was to use an additional valve as a discharge device. The valve is cut off by the sync. pulses so that the charging capacitor it controls only charges during these times. When a narrow line pulse cuts the valve off the capacitor will only charge a little; when the valve is cut off by the broad field pulses, however, it will charge to a much higher value. Clippers and pulse shapers are also occasionally used between the sync. separator and the line generator.

Line pulses can leak into the field timebase by various indirect routes—for example, by inductive coupling between the line and field deflector coils—and precautions are taken in the design to reduce the effects of this coupling.

Interference to the picture (faint white lines, slightly tilted, across the picture) is sometimes caused by the field flyback, and suppression, in the form of a negative flyback pulse applied to one of the picture-tube electrodes, is generally incorporated to cut off the scanning beam during the field flyback period. Line flyback suppression is also generally used, with, occasionally, a clipper diode included to suppress positive excursions of the waveform.

FLYWHEEL SYNCHRONISATION

Fringe reception is often rendered more difficult by the line oscillator being triggered by noise pulses reaching the oscillator just before the arrival of the line-sync. pulse, producing ragged vertical edges to the picture. This is particularly so on 625-line reception where the negative modulation used causes any interference present on the signal to be *below blacks*, i.e. in the sync. region. Several systems have been developed in which the frequency of the line oscillator is not controlled

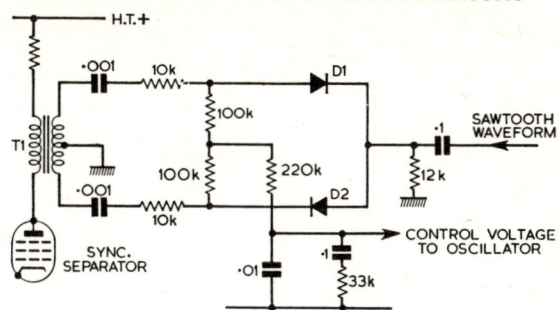

Figure 3.28   Commonly used flywheel line-sync. circuit

D1 and D2 form a phase discriminator circuit. The components in the output side smooth the output to provide a d.c. control potential. The long time constant of the smoothing components gives the 'flywheel' effect.

directly by the sync. pulses but indirectly by means of a discriminator circuit which compares the phase relationship between the line oscillator frequency and the incoming sync. pulses. Should the frequency of the oscillator begin to 'wander' from that of the pulses, the discriminator develops a control potential which modifies the biasing of the oscillator circuit in such a way as to bring the frequency of the oscillator back into step with the sync. pulses.

A common form of line flywheel synchronisation circuit is shown in *Figure 3.28*. From this it can be seen that the sync. pulses are fed, through a phase-splitter transformer, T1, to the phase discriminator circuit, D1 and D2. A sawtooth waveform, from the line output transformer, is also fed to the discriminator circuit. The integrated output is applied as a bias voltage to control the frequency of the

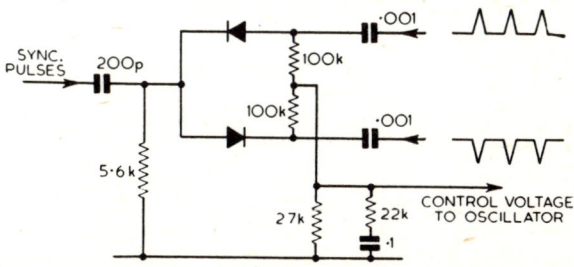

Figure 3.29   Alternative flywheel synchronisation circuit

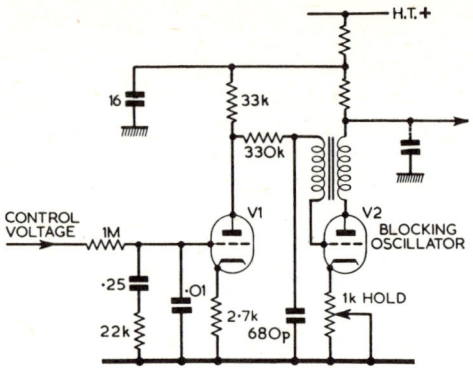

Figure 3.30　Blocking oscillator (V2) controlled from a flywheel circuit via a control valve (V1)

oscillator. In some receivers a valve phase splitter replaces T1. A double diode, such as the EB91, may also be used in place of D1 and D2.

An alternative system is shown in *Figure 3.29*. The synchronising input consists of positive-going pulses, which are compared with a series of short pulses of positive and negative polarity derived from the line output transformer during the flyback period. A separate winding on the output transformer provides these pulses.

The distinction between the circuits of *Figure 3.28* and *3.29* is that the sync. feed is phase split in one (*Figure 3.28*) and the feedback signal has to be phase split in the other (*Figure 3.29*). The circuits are, in essence, phase discriminators with a long time constant.

The two circuits so far mentioned are suitable for controlling a line oscillator of the cathode-coupled multivibrator type. Where the line generator is a blocking oscillator, however, the control voltage has to be applied via a separate control valve. A typical circuit is shown in *Figure 3.30*, the control voltage being derived from a discriminator circuit as shown in *Figure 3.28*. The state of conduction of V1 controls the time constant of the oscillator grid circuit components.

POWER SUPPLIES

Because of the original provision of either a.c. or d.c. mains supplies in the U.K. the television receiver has had a history of operation on either supply. This prevented the general use of transformer supplies (some of the early receivers were a.c.-only and used a mains transformer) and led to the 'live chassis' technique where one side of the receiver chassis is connected directly to the mains supply; although an installer should normally arrange that this is the neutral side of the supply it must always be assumed that this is not so and that the chassis is at full mains potential. This may be true even if the receiver is switched off if a single-pole mains switch is fitted or if the switch is faulty. To satisfy British Standards, the receiver is designed so that no live metal can be touched by the user. The standards are not however a sign of safety for the service technician and appropriate safety precautions should be taken when working on any 'live chassis' technique receiver. In particular a receiver should never be installed in a home in a condition where the user might be left in jeopardy.

In recent years, d.c. supplies have been completely phased out, and receiver manufacturers have been able to employ some forms of mains transformers. In general, however, the live chassis technique is still used—mostly on economic grounds. There is also a general progression in the U.K. towards a standardised mains supply of 240 V and a number of present receivers will only operate satisfactorily on this standard. Older receivers usually have some form of mains tapping changeover system to allow operation on any supply from 200 to 250 V.

H.T. supplies

For some time h.t. supplies have been derived directly from the mains supply. In order to standardise the h.t. voltage with different possible mains inputs to older receivers, suitable dropper resistor(s) are included in series with the input. There may be up to four sections of wire-wound dropper mounted on the same former; a tapping or soldered lug system is often employed to change over the number of sections in circuit for the particular mains supply.

On newer receivers designed for operation on 240 V mains only, no switching is of course necessary. This means that the h.t. can also be higher.

Half-wave rectification is universally used for h.t. supplies. In older receivers using valve diodes such as the PY32 the output voltage may be as low as 190 V. With more efficient silicon diode rectifiers and 240 V only input receivers, the rectifier output may be as high as 270 V.

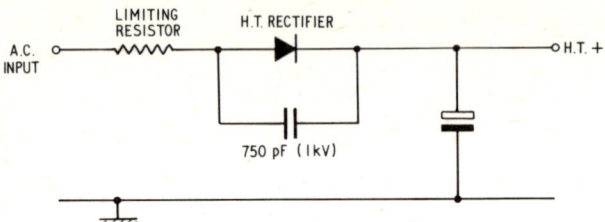

Figure 3.31   Protective resistor and capacitor used to reduce current and voltage transient risks in a silicon rectifier

After smoothing and filtering this will probably be reduced to 200/220 V for distribution within the receiver. In some receivers valve rectifiers may be paralleled in order to be able to supply sufficient current and operate with reasonable efficiency; on receivers fitted with early silicon rectifiers it was sometimes necessary to have two units connected in series because of their low peak inverse voltage (p.i.v.) ratings.

Unlike valve or metal (selenium) rectifiers, silicon rectifiers must be protected against voltage and current surges (*Figure 3.31*).

The current surge is prevented from doing any damage by the inclusion of a limiting resistor in series with the diode; this will be at least 5 Ω. The prevention of damage by voltage surges is by a shunt capacitor of a value small enough to pass fast-edge transients but ineffective at mains frequency. A typical average value used is about 750 pF. The capacitor may bridge either the rectifier, or the rectifier and limiting resistor, although in the latter case the protection is for a larger resistance so its value can be higher (perhaps 5 nF).

Conventional smoothing techniques are generally used for the h.t. with resistive filtering—some earlier and a few recent receivers used the more costly inductive filtering.

In some chassis, an autotransformer technique has been employed for tapped mains inputs to maintain, by step-up action, an h.t. voltage higher than that obtainable with dropper resistors. The higher h.t. voltages have become more and more necessary as the demands of the receiver—particularly in the line output stage—have increased.

Heater supplies

Receivers using valves either completely or in part require a separate supply system. Television receiver valves are all designed for operation from a particular *current* through the heaters rather than a particular voltage *across* them; this is to simplify the design by the use of a series chain of valve heaters. The standard current is 300 mA (0·3 A) normally indicated by the prefix 'P' in the valve number. In this system the power consumption of the heater chain will always be the same, the product of the mains voltage and the heater current; for example $240 \times 0·3 = 72$ watts.

As with the h.t. supply, different values of dropper must be included in series with the chain in order to operate the receiver from any mains input voltage. A tapping changeover system will also operate on this part of the dropper chain.

Because of the heater current requirements, additional resistance must be added in series with the chain if fewer valves are employed. When there are only a few circuit valves the power dissipated in the droppers is very considerable. For these hybrid receivers it is usual to employ a form of wattless reduction of power by including a diode rectifier in the chain (*Figure 3.32*). With only half cycles being passed there is an automatic reduction in the heater current with no dissipation problems.

In such a system the waveforms in the chain are not sinusoidal and they are not d.c. The results of voltage measurements must therefore be interpreted with some care and note should be taken of the sort of test instrument used by any particular manufacturer when a heater chain voltage is quoted in a service manual.

In a heater chain it is conventional to place the c.r.t. last in the chain (i.e. closest to chassis) so that the chances of breakdown of it or other heaters due to flashover or other fault conditions are smallest. Protection must also be given in a fully valved receiver against the

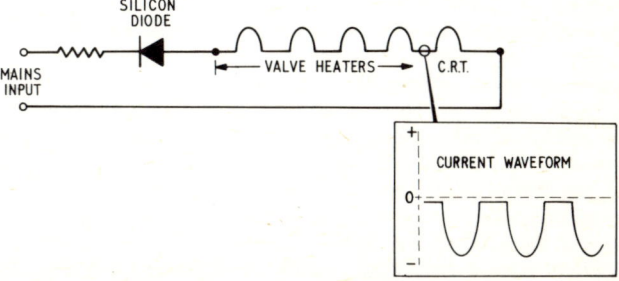

Figure 3.32   Diode fed heater chain giving 'wattless' reduction of heater current

overdriving of the heaters on initial switch-on. This is due to the very low resistance of a cold valve heater which increases quite rapidly after powering. In order to protect these heaters, a thermistor is included in series with the chain; the thermistor has an opposite temperature coefficient that exhibits a high resistance when cold which falls as current flows; the Mullard VA1015 thermistor, for example, changes resistance from 930 to about 100 Ω when heating from 24°C to 100°C. With only a few valves in the receiver the dangers of damage due to this switch-on current surge are very much smaller and no thermistor is usually included in circuit.

In diode-fed heater claims there is a danger that the silicon diode employed might go short-circuit with no apparent effects until one or more valve heaters burn out. On some chassis this is provided for by taking a feed from the heater chain to either the sync. separator or field oscillator stage as a smoothed d.c. bias. If the heater diode should fail, a.c. will be fed to this point and cause an interference with the field waveform (which for some years has not been tied to the mains frequency). The result is a field fault which is unrelated to the field hold control; the effect is something like the results of excessive sound-on-vision where the picture collapses and expands regularly.

Low-voltage supplies

In the same way that a feed to the field stages can supply a warning of rectifier short-circuit in the heater chain, such a chain can also supply the low-voltage requirements of the semiconductor parts of a hybrid receiver by taking a feed of the unidirectional voltage on the chain and smoothing it to provide a d.c. voltage. With such a system the point of feed must be after the heaters in the chain, otherwise a danger exists that should a valve be removed with the receiver powered, the low-voltage rail would jump up to the full mains potential.

Another method of providing low-voltage supplies for transistor stages is to take a feed from an additional winding on the line output transformer, smooth it and distribute it. This should be remembered in a receiver using this technique when servicing is taking place, for some line output faults can give very misleading effects.

In all-transistor receivers there is generally insufficient margin of safety for the low-voltage rails to be provided from the line output stage. In this case a separate, usually regulated, supply is provided. An example is given in *Figure 3.33*.

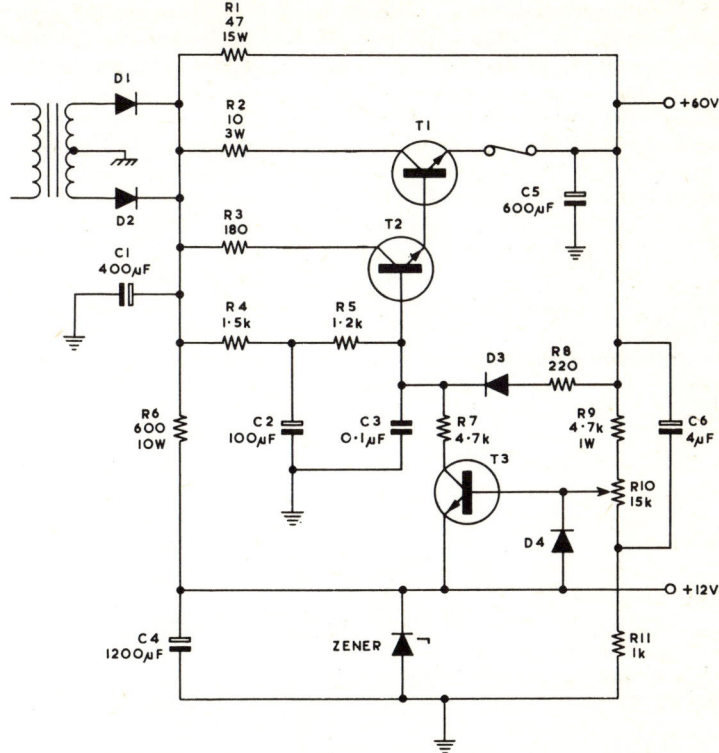

Figure 3.33   Stabilised power supply unit in an American Westinghouse 19-in transistor receiver

This circuit is in an American Westinghouse 19-in 27-transistor receiver, and is particularly interesting because it uses a voltage regulator. Obviously it is an expensive circuit but, in addition to rendering the receiver immune from mains voltage variations, the regulating action also assists the filtering process. Full-wave rectification by diodes D1 and D2 produces about 75 V across the reservoir capacitor C_1. A proportion of the load current passes through R_1 but the rest passes through the series regulator transistor T1, a power device mounted on a heat sink. The base current, and hence the regulating action, of T1 is controlled by transistor T2 which is itself controlled by the sensing transistor T3. A zener diode clamps the emitter of T3 and provides the voltage reference, but the base potential is directly proportional to the d.c. output voltage and is, therefore, sensitive to any change. Such a change thus precipitates a correcting action at T1. The pre-set resistor R_{10} is used to establish initially the exact desired output voltage, in this case 60 V. A subsidiary supply at 12 V is obtained directly from across the zener diode but is only suitable for feeding small-signal stages.

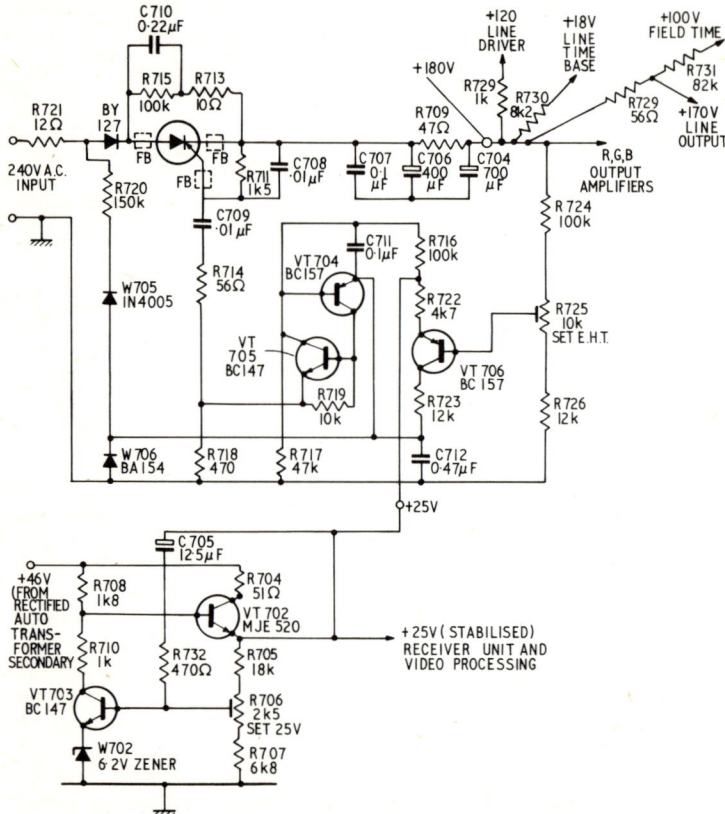

Figure 3.34   Part of regulated power supply of B.R.C. 8000 series colour receiver using secondary stabilised supply of 25 V

For less critical stages where regulation of the voltage is not as important, the low-voltage supply may be taken from an autotransformer or transformer secondary winding and full-wave rectified with conventional smoothing. The reverse may be true as well; colour receivers using RGB output stages and a transistorised line output require a very accurately controlled supply voltage. One method of providing this is shown in *Figure 3.34* which is part of the supply circuitry of the B.R.C. 8000 series colour receiver.

This power unit supplies all three colour output amplifiers and the line oscillator and line output stage and the current requirements would make it a little difficult to achieve economically with a series transistor regulator. Instead a thyristor control is used with feedback from the output of the unit. A proportion of the output voltage is sensed by the potentiometer R_{725} and passed to the emitter follower VT706. The output is amplified by VT704 and the d.c. signal is then coupled via a further emitter follower VT705 to drive the gate of the thyristor. The required potential for the operation of the three-transistor amplifier is derived from the secondary, 25 V power unit. As R_{725} controls the output voltage of the power supply and this in turn dictates the operating conditions of the line output stage and the magnitude of the e.h.t. voltage, it is termed 'SET EHT'.

The secondary power unit uses transformer rectified 46 V and with the zener W702 stabilises the output to $+25$ V in a series regulator format, VT702 being the regulator with VT703 the d.c. amplifier.

E.H.T. supplies

The e.h.t. may be derived by at least three methods: from the line flyback, by rectifying the output from a separate oscillator operating at about 40 kHz, or by the 'ringing choke' method. The very early receivers also used a step-up mains transformer but this is costly and with present levels of e.h.t. voltage it is extremely unsafe to have a circuit where the current that can be drawn is not strictly limited by the circuit design.

For some time all receivers have used the line flyback to derive the e.h.t., either using a transformer pulse directly or by means of a smaller amplitude pulse followed by a voltage tripler.

Figure 3.35 shows the form of circuit that has been conventional for deriving e.h.t. voltages up to about 20 kV for modern picture tubes. The e.h.t. rectifier in this thermionic form requires heater voltage,

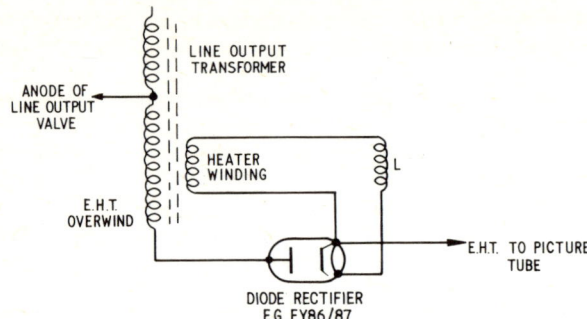

Figure 3.35   Conventional form of e.h.t. rectifier circuit L is included in dual-standard receivers to reduce heater voltage on 625-line operation and so reduce e.h.t. output (greater repetition frequency of flyback pulse on 625-lines would otherwise give higher average amplitude

but this cannot be derived from the main heater supply because of the dangers of flashover from the cathode to the heater chain. Instead, a directly heated cathode system is employed, and the heater voltage is derived from a further winding on the line output transformer. In this way the insulation problems are very much reduced. On dual-standard receivers it is also conventional to include a small inductance (which may be formed by the leads to the valve base) which exhibits a higher reactance at the higher 625-line frequency and so reduces the e.h.t. output to the same value as that for 405-line operation.

Early receivers used voltage regulation devices for smoothing the e.h.t. voltage but tubes have for years now employed a conductive coating of *Aquadag* around the final anode cavity connector which, with the internal final anode electrode, forms a capacitor to smooth the supply reasonably well. This coating is often earthed via a static discharge resistor of perhaps 2·2 M. It is obviously important that for correct operation of the receiver this resistor is replaced in any servicing or tube change and that any earthing springs across the tube flaring are similarly cared for. The coating should be kept clean in order to prevent arcing and to preserve the capacitive effect. Any deficiency in these respects will usually show itself as 'blooming' of the picture with any increase in brightness or contrast.

The development of suitable silicon rectifiers has meant the almost universal use of these devices for rectification of the flyback pulse of the line output transformer. In 25 kV supplies for colour receivers, two rectifiers may be connected in series in order to obtain a sufficiently high p.i.v. rating.

It is however more conventional in colour receivers, and in a good number of monochrome receivers as well, to employ a form of voltage tripler to obtain the high e.h.t. now needed. This greatly reduces the demands of insulation on the line output transformer because for a 25 kV supply the input from the transformer need only to be about 8·5 kV. Also it was conventional to employ a shunt regulator on the rectifier output and these were not only expensive circuits, because of the valve requirements, but they also posed problems in screening because of the production of X-radiation. With the amount of screening required, there were then ventilation problems because over 100 W was dissipated inside the screened area. With voltage triplers and fifth-harmonic tuning of the line output transformer (see earlier section) the only e.h.t. exposed is the small lead from the tripler output to the final cavity, and regulation is not required because of the flat shape of the top of the flyback pulse presented to the input of the tripler.

Although the process of voltage tripling is fairly well known, it is given here because the process is slightly different for television purposes. The system is based on the *addition* of voltages rather than multiplication. The basic arrangement is composed of two voltage doublers connected together, each of the form shown in *Figure 3.36*.

The load capacitance of this circuit and of each of the circuits given here is assumed to be present in the aquadag capacitance of the picture tube. With a positive input pulse (the flyback pulse from the transformer) D1 conducts and C_2 charges to the peak value of the pulse.

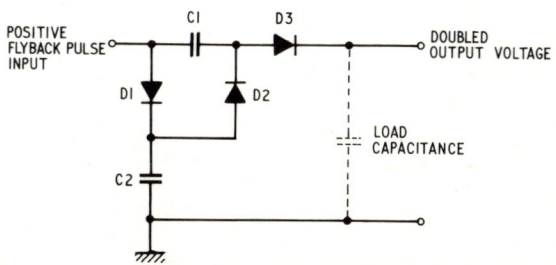

Figure 3.36   Basic voltage doubler

During the following scan period the input is very nearly at chassis potential and as the junction of D1 and D2 is positive, D2 conducts and charges C_1 to the peak value of the original pulse.

During the next flyback period, another positive pulse appears at the input and the low potential side of C_1 is raised to the peak value of this pulse and so is the opposite side of the capacitor. However, there is already a voltage on the D3 side of the capacitor so that it now raises to *twice* the peak value of the input. D3 rectifies this combined result and the load capacitance is now charged to a d.c. potential of twice the pulse input voltage.

In a voltage tripler, the process is repeated again in a further section of circuit to give an output equivalent to three times the original pulse amplitude. The basic circuit of the tripler is shown in *Figure 3.37*.

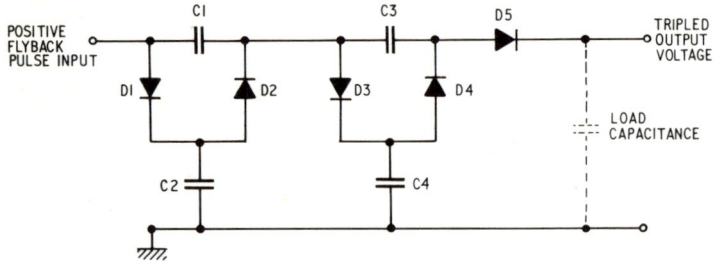

Figure 3.37   Basic voltage tripler

In this basic circuit the peak voltage across C_4 is twice the input potential whilst that across the other capacitors is only equal to the pulse input potential. This is undesirable because of the increased cost and lack of standardisation of that one component. In a practical circuit this is overcome by taking C_4 to a different point without affecting the operation of the circuit as a whole. Also in this practical circuit (*Figure 3.38*) C_2 is taken to the line output transformer where it can be connected to three alternative positions: either chassis as before or a positive potential, so increasing the output level, or a negative potential, so reducing the output level. In this way a reasonably fine control can be obtained for the e.h.t. Also, the focus potential for a colour receiver can be conveniently taken from the tripler unit— either from point C or point D. The latter may often be employed if a large voltage rated v.d.r. focus control is used but it is generally more convenient for point C to be used. *Figures 3.39* and *3.40* show two

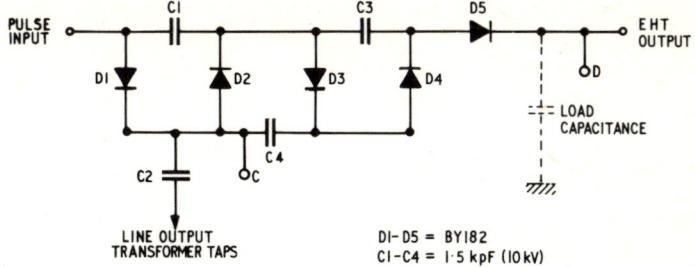

Figure 3.38   Practical e.h.t. tripler (Mullard LP1174/1 module)

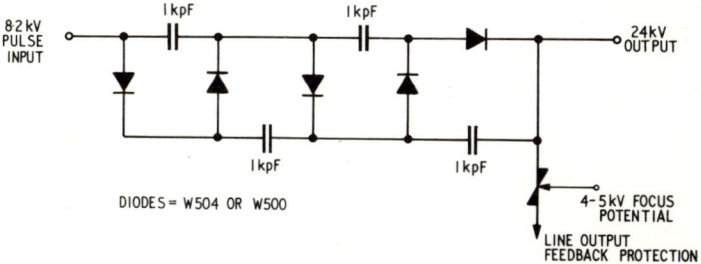

Figure 3.39   E.H.T. tripler used in B.R.C. 2000 series colour receiver

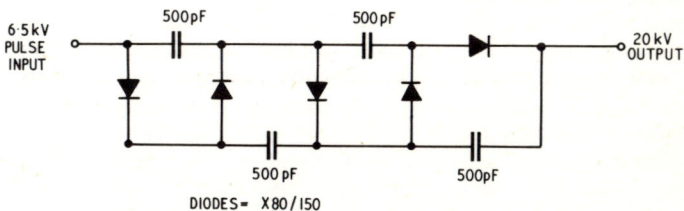

Figure 3.40   E.H.T. tripler used in B.R.C. 970 and most 1400 series monochrome chassis

additional circuits for triplers slightly rearranged but operating on the same principles. Note should be taken of the different capacitor values used with the different current demands of the monochrome and colour picture tubes.

Supply protection

The number and location of fused circuits in a receiver differs according to manufacturer, but it is conventional to always have at least a mains input fuse and a separate h.t. fuse. There is a movement towards protecting other circuits individually—such as the line output stage—and this can be done with either a fuse or a thermal protection device.

Fuses are basically either *quick-blow* or *anti-surge* in their action. A circuit normally taking 100 mA might be protected by an anti-surge fuse of 150 mA or a quick-blow fuse of 250 mA, for example, either sort providing sufficient protection against fusing with switch-on surges. It is obviously important that the correct size and type of fuse is used in any replacement.

Thermal protection devices, usually known as either *drop-off resistors* or *thermal cut-outs*, consist of a heating resistance soldered in a spring fashion such that if the circuit it is protecting exceeds its current rating the solder melts and the resistor springs off to break the circuit. These elements should always be resoldered *without* any mechanical joint—the wire end should not be wrapped around a tag, for example—and any advice given in the manufacturer's manual regarding the type or amount of solder to be used should always be followed so that the element will operate at its correct current rating.

Bright-spot suppression

Unless precautions are taken, the switching off of a receiver results in a white spot appearing in the centre of the screen for several seconds. In most older sets the spot rapidly defocused and did not cause damage. However, with electrostatic focusing, the focus may be relatively little affected by the disappearance of the focusing voltage (if any), and the spot may remain bright and sharp; this could give rise to screen damage. To overcome this, various methods are used to ensure that the e.h.t. falls rapidly, despite the presence of the e.h.t. reservoir capacitor, when the set is switched off.

The usual method is to ensure that the bias applied to the grid of the picture tube is automatically raised immediately the set is switched off, producing a heavy beam current in the moment before the raster collapses. This may be done by connecting the 'earthy' end of the brilliance control network to the neutral mains supply lead—at a point beyond the main on/off switch—instead of to chassis. The control network is thus returned to chassis only when the mains switch is 'on', and the grid potential rises immediately to h.t. positive when the network floats. *Figure 3.41* shows a typical arrangement. An alternative technique is to return the brightness control to chassis via a v.d.r.; since the action of a v.d.r. is to increase the time constant of the circuit, the grid will be held positive after switching off so that the beam current discharges the e.h.t. capacitor.

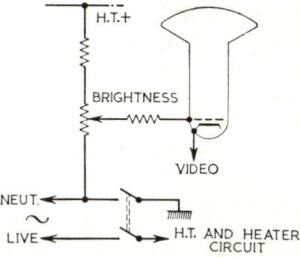

Figure 3.41   One method of bright spot suppression

A long time-constant network is needed in the h.t. supply to the first anode of the picture tube to accelerate the beam after switching off. The timebases should also have long time constants so that the raster is maintained while these processes take place.

The alternative approach is to suppress beam current on switching off, allowing the e.h.t. capacitor to discharge slowly. As the cathode cools slowly this technique can lead to a spot appearing shortly after switching off.

4

TELEVISION INTEGRATED CIRCUITS
by J. I. Sim

The integrated circuit (i.c.) is a development in electronics that has come about because of military requirements. Equipment for military purposes must be reliable, low weight and low volume. With an enormous number of manufacturers moving into the field in the 1960s the costs of integrated circuits plummeted by enormous amounts—a particular circuit dropping, for example, from £10 in 1965 to 60p in 1971. In this respect, advancement has been too rapid and a proportion of manufacturers have been priced out of this aspect of the market. The situation has now become reasonably stable and will probably remain so.

The first i.c.s were digital devices for the performance of straightforward logic operations such as inversion and gating of various kinds. These were in standard formats and allowed great flexibility of use in many sorts of equipment, so permitting large production volumes to be built up; this is an important aspect with i.c.s—unless the market is a large one the selling price must be high to cover the development costs.

Indeed it was not really until the digital i.c. market began to saturate that the manufacturers turned towards the analogue processes of a system like television in order to produce specific circuits; the market is not as large, the devices are inevitably specialised for the purposes of television and so the costs of production are higher.

Any receiver manufacturer must be satisfied that the costs are not too high; he knows he can save production money using i.c.s because they are cheaper to wire up than the equivalent circuit using discrete components, but the capital cost of a new i.c. may be higher than that of discrete components it is replacing, there might be delivery problems with a new device, and so on. If this is so there must obviously be some encouragement from the i.c. manufacturer that costs will drop as production increases and that perhaps other receiver manufacturers will begin to use the same i.c. This is basically what has happened in practice.

We call an integrated circuit developed for a particular circuit purpose, handling signals rather than pulses, a *linear i.c.*

Nature of an integrated circuit

Well, what is an i.c.? As the name implies it is the condensation of a complete circuit into one *integrated* package. This could be understood to mean any micro-electronic circuit but this would be a wide interpretation of the term; in particular we always mean the production of a complete circuit on a *single chip of semiconductor material*. This leads to the colloquial use of the word *chip* when talking about i.c.s.

The material used for the base (or *substrate*) of our i.c. is silicon so that we are in general always using a *silicon integrated circuit (s.i.c.)*. We will have a look at the manufacture of an s.i.c. a little later but first we have to be able to understand how we can obtain our circuit components on a single chip of silicon.

Obviously with such a base it is easy to construct transistors and semiconductor diodes by the correct infusion of doping elements to form p and n regions giving suitable p-n junctions. These can be extremely small and are relatively simple to produce.

Capacitance is not easy to produce in a small area and in general capacitors cannot be economically constructed in an s.i.c. If we want small amounts of capacitance these *may* be obtained using the capacitive effects of a reverse-biased diode—but the absolute limit is about 200 pF. If the capacitor must be larger or indeed must have special properties—very low leakage or electrolytic properties, for example—then there must be a component connected external to the chip.

Inductance is virtually impossible in an s.i.c. using the monolithic construction that is current. Inductors may be constructed using thin-film techniques but their size is not really comparable to the chip size and they cannot be incorporated. Again, for television purposes, an inductor must be an externally connected component.

Resistance is another of our electrical properties that in its normal form requires well-defined parameters—of length, cross-section and resistivity—always needing more area than we care for! In practice resistance values up to 20 k are *possible* but should be avoided and present thinking suggests an upper limit for design work of about 5 k.

At the lower end, it is difficult to control the fabrication of resistances of less than 100 Ω so again these are avoided. It should be pointed out though that some manufacturers seem to be ahead of others in the development of resistances in practical i.c.s. In some cases resistances can be formed by transistors but there are then limitations on the circuit positions where they can be used—any semiconductor junction resistance being nonlinear, i.e. not obeying Ohm's law. The tolerance on resistance values is generally poor, $\pm 25\%$ being typical.

Fabrication of silicon integrated circuits

The number of processes in the formation of s.i.c.s is quite large and any description of it makes the process appear extraordinarily cumbersome. But it should be remembered that production-line techniques are employed and that, after the initial design and prototype stages have been passed, the production can be virtually automatic.

The best description of the fabrication is to take a circuit example and follow the basic processes through: for example, let us imagine that the simple circuit in *Figure 4.1* is to be integrated. This could be a simple amplifier with resistive collector load (R_1) and a variable load in the emitter formed by D1 and D2 which may be controlled by an external voltage at pin 4. The transistor in this case is *n–p–n*.

The process begins by growing an *n* type epitaxial layer on to a *p* type silicon substrate. This is done in a very similar manner to the manufacture of epitaxial or planar transistors, by heating the chip to about 1 200°C and passing it through a mixture of impurity gases. Silicon dioxide is then formed on the surface of the *n* growth by passing the heated chip through oxygen. This silicon dioxide is a good insulant and also has the property of preventing the penetration of impurities through to the semiconductor layers (*Figure 4.2*).

A photoresist is applied to the silicon dioxide surface and exposed to ultra-violet light passing through a photographic master. After photographic development, where the oxide is to remain it hardens—elsewhere it is unaffected by the light and a solution of hydrofluoric acid etches the silicon dioxide away (*Figure 4.3*).

Impurities of *p* type are now diffused into the epitaxial layer through the gaps that have been made in the silicon dioxide layer (*Figure 4.4*). Because the *n* regions are now separated by *p* regions, if we put a reverse bias on the junctions—i.e. a negative bias on the substrate with respect to the *n* regions—we get quite a high degree of electrical isolation

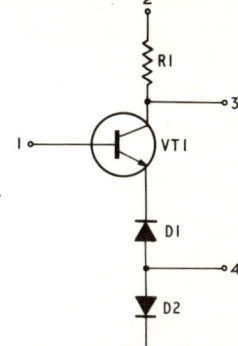

Figure 4.1   Simple circuit to be fabricated as an i.c.

between the *n* regions, and for that reason they are often known as *isolation islands*.

The anodes of the diodes in the circuit and the base of the transistor must all be *p* type and these areas must now be introduced into the chip. This is done by inserting them as 'wells' in the isolation islands and the photo-etching process is repeated to enable this to be achieved—reforming a complete silicon dioxide layer on the surface, coating with photoresist, exposure to ultra-violet light, development and etching to leave the required spaces. The *p* type impurities are then diffused into those regions (*Figure 4.5*). This process, because it is providing the *n–p–n* transistor base, is known as *base diffusion*.

The diffused areas in *Figure 4.5* have been marked 1, 2 and 3. Area 1 will be used for the resistance R_1, area 2 for the transistor, and area 3 will be used for *both* diodes. For the resistance we only need to provide contacts on to the well of *p* material. For the transistor and the diodes we need to introduce further *n* regions suitable to give us our

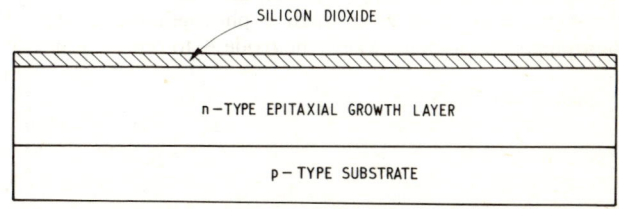

Figure 4.2   Epitaxial growth and surface oxidation

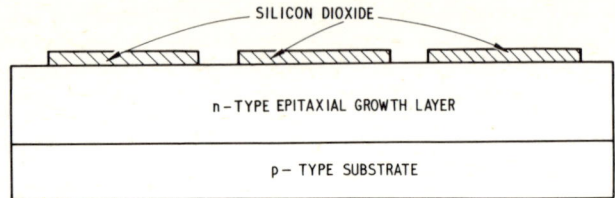

Figure 4.3   Silicon dioxide removed by photo etching in selected areas

p–n junctions. This is achieved by again repeating the silicon dioxide and photo-etching process and then diffusion of the *n* materials required can take place (*Figure 4.6*).

As can be seen this *n* diffusion material is marked n^+. This indicates a very heavy doping, which must be used because the final contact

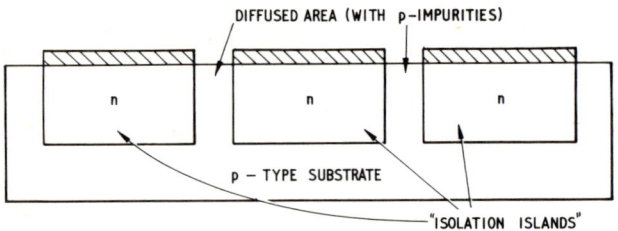

Figure 4.4   Diffusion of p-type impurities to form 'isolation islands'

material, which is aluminium, is in fact a *p* type impurity for silicon. If a lightly doped diffusion of *n* were used the aluminium would react with the *p* doped silicon to form an unwanted *p–n* junction.

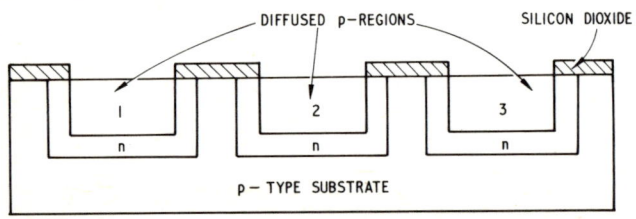

Figure 4.5   Diffusion of p type impurities into isolation islands—base diffusion

Finally, the silicon dioxide and photo-etching process is repeated to form 'windows' in the silicon dioxide where contacts to the chip are needed. These windows and the whole surface of the chip are metallised with aluminium and then this surface is photo-etched to give the required shapes and positions of contacts (*Figure 4.7*).

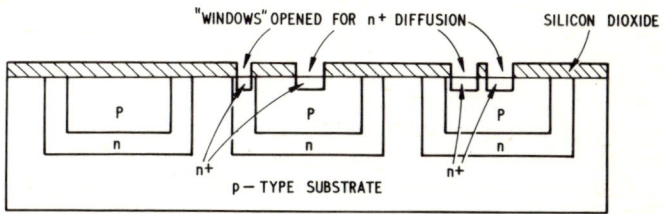

Figure 4.6   Re-etching and diffusion of n⁺ material to form transistor collector and emitter and diode cathodes

In this final form we have our resistance formed by the passage through the *p* material in the left well (the size and doping of the well give the required variations in circuit resistance that might be required); the transistor formed by the centre well, with the collector on the left connected externally through aluminium strip to the resistor,

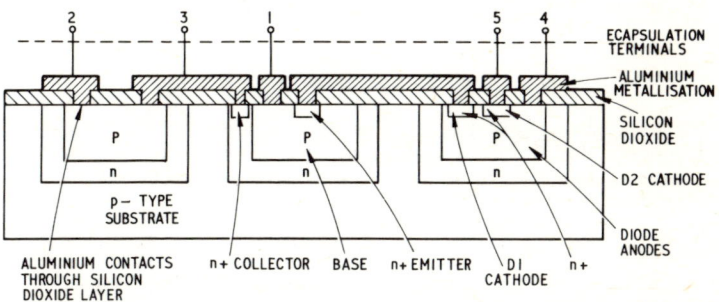

Figure 4.7   Final form for 'chip' of circuit in Figure 4.1

the base being the well of *p* material and the emitter in the centre of the well; the diodes are formed by the right-hand area with the common anode connections of the circuit being formed by the central *p* region and the two separate cathodes being formed by the two n^+ diffusions.

To enable the chip to be used in a practical circuit, this final form has leads attached to the appropriate aluminium points and these are taken to the outside of the package. The rest is then completely encapsulated in plastic to form an environment-proof unit.

Encapsulations

In our simple example we have only considered the fabrication of a four-component circuit. Manufacture of such a circuit would be very uneconomic and it is unlikely that any circuit requiring less than fifty components could be produced economically.

The encapsulation of the s.i.c. comes in a variety of forms but common are the transistor-type package using a TO5 or similar style can or 'header', and the *dual-in-line* (D.I.L.) package. The transistor-type

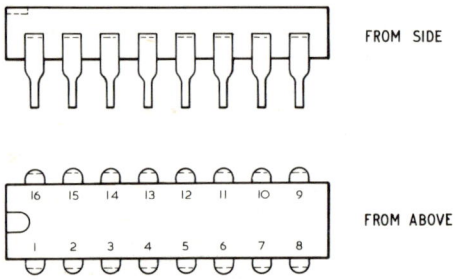

Figure 4.8   Typical appearance of dual-in-line integrated circuit with identification notch

package, with up to 12 lead connections, was initially the most common in i.c. work but the D.I.L. package offers the advantage of being able to extend the number of connections simply by lengthening the package. It is rectangular in shape and the connections (up to 20) are brought out from both sides at right angles, with the mounting in printed-circuit boards eased by 0·1 in (2·54 mm) pin spacing. There is usually some coding, such as a notch, made on the top surface of the package to identify the direction in which it should be mounted (*Figure 4.8*) or the pins may be staggered along the body.

Power and instability

Although the packaging of a chip suggests that it is quite large in size, the majority of the space within the i.c. package is taken up with the encapsulation and the lead-outs. The chip itself, sitting in the middle of all the encapsulating plastic may be only a few cubic millimetres in volume.

Obviously the power dissipation abilities of such a circuit are going to be severely restricted, unless we encapsulate only a small circuit and use some form of *heat tab* within the encapsulation to carry off heat to an external heat sink. For the majority of television applications this is not done, although a small heat tab may be provided. But certainly for audio i.c.s reliable performance has been obtained up to 5 W r.m.s. and there have been claims for up to 50 W! In our circuits, wherever i.c.s may be used to reasonable advantage, these powers cannot be economically sought because there are also voltage limitations.

Currently, few i.c.s are rated above 30 V and the majority are limited to 20 V. In any situation where, for example, flyback pulses appear, the s.i.c. is of no use.

There is also, unfortunately, a restriction in the amount of overall gain that can be obtained from an s.i.c. Audio enthusiasts who have experimented with audio i.c.s often know the difficulties of using a high gain i.c. without the whole circuit going into oscillation due to positive feedback. Circuit layout is often critical.

These difficulties fall into two basic parts. Firstly, the size of the chip itself is so small that a high level output signal (compared to the input) inevitably causes some input due to simple a.c. coupling. This small input is amplified, part of it is fed back to the input again and so on until oscillation takes place. This problem increases with frequency. The i.c. designer takes this into account and the package itself will be stable; but as soon as there is some external encouragement to assist the feedback—due to stray capacitances on a printed circuit board for example—oscillation occurs. It is particularly important that in an i.c. where there are a number of earth pins that the ones suggested for use with inputs or outputs are used for just that. The manufacturer assists further with the external circuitry design by putting input, output and other critical terminals as far displaced from one another as possible. A good circuit layout should prevent all problems.

The second difficulty is the manufacturer's and is due to the design of the chip itself. We have already said, in discussing the fabrication, that the p substrate of the chip must have negative bias on it to give reverse-biased junctions between it and all the 'wells' in which the components are formed. Unfortunately this reverse bias does not isolate the junctions completely and there is always some reverse capacitance. If the values of this leakage capacitance are great enough at the particular gains being used oscillation may occur. As these oscillations are parasitic the cause is often known as the *parasitic capacitance*. Again this must limit the gain that can be obtained from the i.c.

Field output stages will be practical in i.c. form in the near future and it is possible that sufficient advance will have been made during the next five years to allow integration of video output stages—monochrome or RGB. It is not foreseeable that the line output stage will ever be in i.c. form. The rest of a receiver, apart from the u.h.f. tuner and necessary selectivity units, is wide open to integration and the present breed of receivers is quickly approaching this point. Before we look at some practical s.i.c.s from receivers it is as well to examine the philosophy that goes into circuit design within an integrated package.

Circuit arrangements

In the television receiver it is not terribly important to know the exact circuits used in any particular s.i.c. In fact we will give an example of such a circuit and the difficulties involved. It is, however, of great interest to know the basics of the circuit techniques, to understand how some circuit functions are realised and of course what outputs we expect or get from an s.i.c.

The circuit arrangements are based completely on the advantages and disadvantages inherent in the format of the i.c. Remembering that we explained that inductors and capacitors cannot be formed—for all intents and purposes—and that semiconductor diodes and transistors are both easy to form and cheap to provide, the essentials are therefore to reduce to an absolute minimum any a.c. coupling (because this requires external capacitors), to use circuits that do not depend on resonance techniques (inductors and capacitors having to be external) and to reduce to an absolute minimum the use of resistors (because of the difficulty in accurately forming them and the limits in value). These features are not universally possible to achieve but the circuit designers go a long way towards them by using direct coupling completely, quadrature detectors and their like, and the use of further transistors rather than introduce a resistance.

D.C. amplifier circuits are renowned for their instability and difficulty in setting up. These problems do not occur in the i.c.—instead of a straightforward d.c. amplifier the differential amplifier (long-tailed pair) is used—*Figure 4.9*. This gives stability balance and the cost of the additional components for each stage is minimal when the

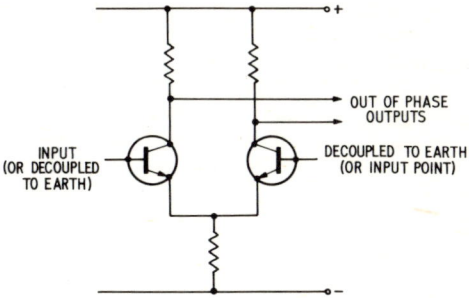

Figure 4.9   Basic differential amplifier circuit: operating on principle that signal voltage is common to both transistors across emitter resistor. Any drift in collector current of one transistor causes change of equal magnitude in the other

i.c. is being produced anyway. The differential amplifier also lends itself to use as a quadrature detector, limiter, frequency multiplier, modulator, etc., with little complication. Problems in setting up cannot occur, of course, because this is a function of the chip design and not a problem of the user. The properties of the differential amplifier are only gained if the devices match one another; this is no problem in an s.i.c. when an error in the doping of one transistor also occurs in every transistor on the chip.

PRACTICAL S.I.C.S

It was thought at one time that when the all-integrated-circuit receiver began to be common that they would all look the same inside—that the receiver would be divided up in precisely the same ways such that this and that function were always combined. It has now become apparent that this will not be the case and that i.c. designers have their own individual ideas about the division of the receiver functional blocks.

These arguments apart there are some basic problems in combining some particular functions in an i.c. and we are unlikely to ever see the line oscillator and the field timebases combined, for example, because of the interference problems that would arise.

We will look at one example from each of the sections of a receiver that currently uses an i.c. to show how they are employed.

Intercarrier sound channel

The intercarrier sound channel was the first to be found in an i.c. package in a receiver and it is the one that all manufacturers agree on as being the ideal integration. The channel presents few problems for the i.c. designer; the frequency of operation of the channel is relatively low (6 MHz), the gain required for this frequency is moderate (about 65 dB), f.m. detection can be handled reasonably easily and audio preamplification at the output is simple.

In the Mullard example of the TBA750 given (*Figure 4.10*), the 6 MHz amplifier and limiter chain is formed from five stages of differential amplifier, TR1–TR10, and all five stages receive their collector voltage through TR11, which provides a self-limiting action; this is very similar to putting a fairly large resistance in series with each collector load so that the voltage swing on the collector is limited.

The detector in an s.i.c. could, of course, still use conventional techniques—such as the ratio detector—but this would mean that every connection to an external transformer would have to appear on an i.c. pin. Instead of this, a *quadrature detector* is used. This operates on a switching basis in a very similar way to the synchronous detectors used in colour. Instead of diodes, however, transistors are used and the switching is between alternate pairs of transistor amplifiers of which one transistor in each pair has a common load resistor. The input signals used are the limited f.m. signals appearing out of phase on alternate sides of each differential amplifier and a reference signal, to perform the switching operation. The reference is derived from the output of the limiter chain and provided it is fed in exact *quadrature* (i.e. 90 deg out of phase) to the differential amplifiers, efficient demodulation will be effected; as the input frequency changes, the periodicity of the output signal changes with constancy in magnitude and the average level of the output signal varies.

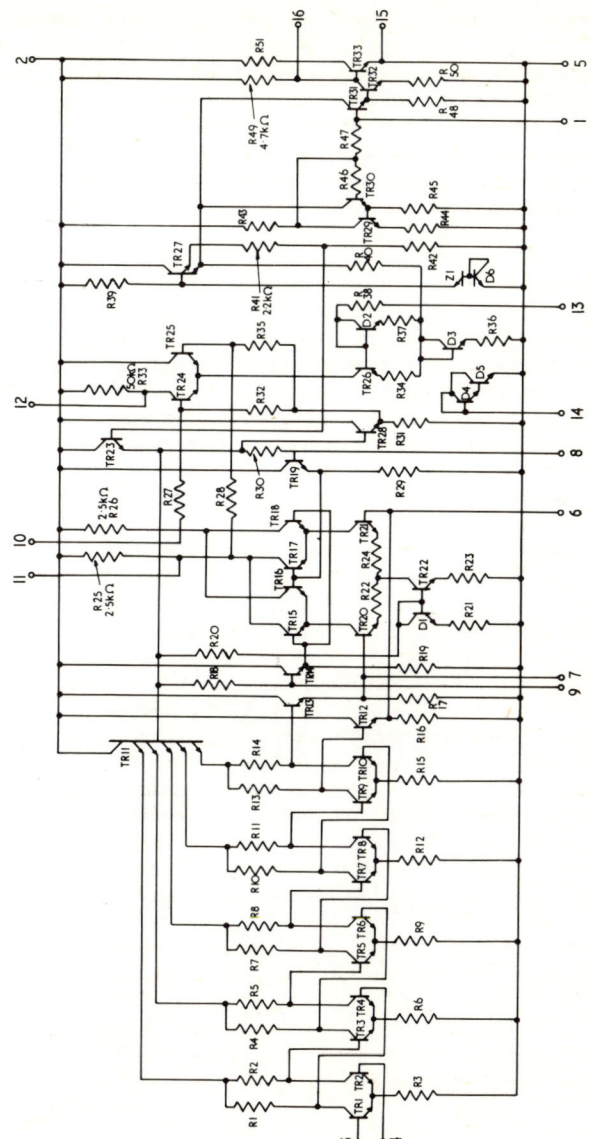

151

Figure 4.10   Circuit used in Mullard TBA750 limiter amplifier i.c. (for intercarrier sound channel)

A preamplifier chain is added to the detector output (TR24–TR32) to give a maximum output signal of about 1 V—sufficient to drive a triode-pentode or transistor Class A output stage directly.

As the circuit of *Figure 4.10* indicates, the i.c. is not a straightforward arrangement and in fact there are also features, such as parasitic and distributed capacitance, which are used to obtain the correct circuit operation but which cannot be readily indicated on a conventional circuit diagram. Having shown this point we will not now go into circuit detail in the next parts of this section.

The external circuit requirements for the TBA750 are:

(1) Input bandpass filter (tuned to pass 6 MHz and to give high rejection in the region of colour subcarrier).
(2) A 90 degree phase-shift circuit to provide the reference signal for the quadrature detector.
(3) Supply rail and other decoupling.
(4) Volume control (which varies the supply voltage to TR26 through D2).
(5) Normal de-emphasis capacitor.

These external circuit features are indicated in *Figure 4.11*. The TBA750 is a 16-pin D.I.L. package. Again it is pointed out that this

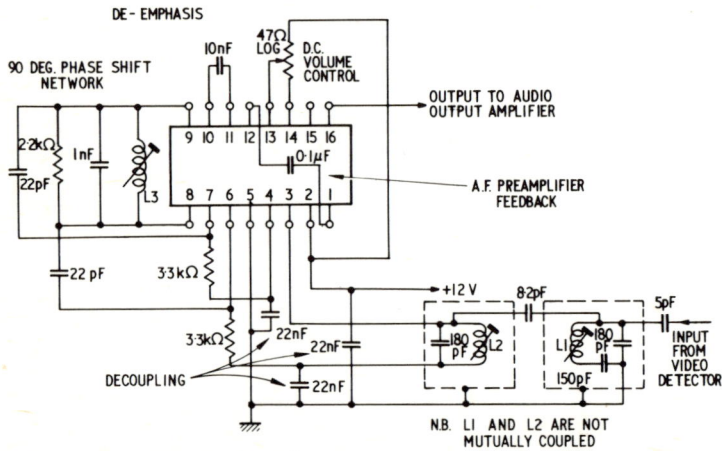

Figure 4.11   External requirements for Mullard TBA750 i.c.

is only one version of an intercarrier sound s.i.c. and that other manu-facturers naturally have their own ideas about circuit implementation.

Video processing

The name of this i.c. section could really mean anything and indeed in the future there is likely to be considerable variety in what the manufacturer considers should be in a video processing i.c. The one we have chosen to glimpse at has been used in U.K. receivers although it has now been superseded by an improved version. The considered one is the Mullard TAA700.

This s.i.c. contains functions for video preamplification, line gated a.g.c. with feeds for i.f. and r.f. stages, sync. separation, line flywheel phase detection, and video amplifier blanking. It is encapsulated in a 16-pin D.I.L. package and a simplified operational block of the device is shown in *Figure 4.12* together with typical values of the external components required and the general arrangement of the i.c. in a monochrome receiver.

The video input, straight from the detector at about $2 V_{p-p}$, is amplified to give about 4·2 V at the output (pin 12) from peak white to black level. With flyback blanking pulses of line and field on pin 11 no video comes from the i.c. during the flyback periods so that we have flyback suppression. The amplitude of these gating pulses at pin 11 will be between 1 and 5 V. The video output level is sufficient to drive a transistor or valve video amplifier/output stage or, in a colour receiver, it may be used to drive the luminance buffer stage through the luminance delay line. When the TAA700 is used in a colour receiver flyback sup-pression is not applied and pin 11 is earthed.

Sync. separation is shown as one block but in fact includes divided field and line separation both preceded by a low-pass filter to remove noise and chrominance information. The integrator capacitor for field separation is connected to pin 14 and after shaping into a square wave drive it is led off (pin 15) to the field oscillator.

The line separation makes use of peak detection using external components on pin 13. The output is connected directly to the line phase detector used for flywheel sync. but it is gated by line pulses on pin 3 to make it operational only during line flyback. This gives greater protection against noise, a larger pull-in range, and less disturbance during field flyback.

The line phase detector or discriminator operates with a positive voltage of about 7 V on pin 1, and a sawtooth reference of about $7 V_{p-p}$

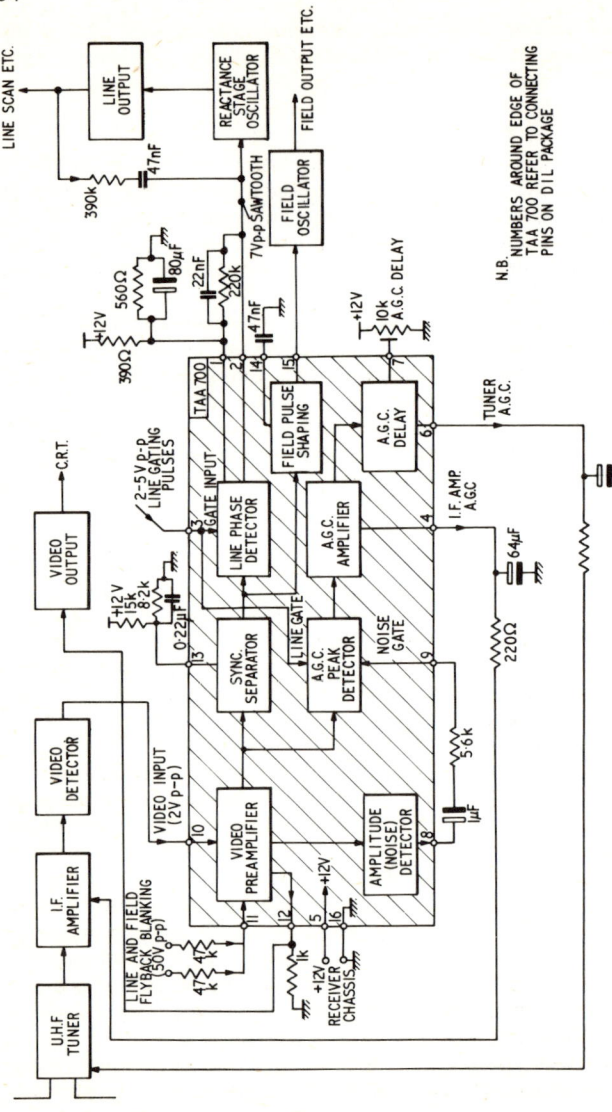

Figure 4.12   Simplified block of Mullard TAA700 video processing i.c. and its arrangement in a monochrome receiver

taken from the line output stage and connected to pin 2. During the reference period (set by the gating as being during line flyback) conduction of the 7 V d.c. takes place from pin 1 to pin 2. Added or subtracted from this voltage is another voltage dependent on the centring of the sawtooth reference. If the phase and frequency are exactly correct only 7 V d.c. is passed. If there is an error the voltage will be greater or less than the 7 V. This d.c. potential controls the frequency of the reactance oscillator to bring it into the correct running conditions.

The a.g.c. units of the i.c. provide forward a.g.c. potential for the i.f. amplifier and delayed a.g.c. (i.e. a.g.c. operative only at the higher levels) for the r.f. stage in the u.h.f. tuner. The a.g.c. detector looks at the amplitude of the line sync. pulses and passes the signal to the a.g.c. amplifier. The detector is gated during line flyback by the pulses from pin 3 to prevent operation during active line time and field flyback. Gating is also applied from a noise detector through a.c. coupling (between pins 8 and 9) to prevent a.g.c. operation on impulse interference. The noise detector consists of a comparator sensing signals occuring at levels greater than the video signal.

Colour processing

The production of R′, G′ and B′ signals from modulated chroma inputs using discrete components involves unfortunately tight tolerances in matrix components and d.c. levels. Rank-Bush-Murphy pioneered the use of an s.i.c. to perform these functions and since 1968 have used the SL901, and later the SL901B, in their colour receivers. The s.i.c.s are manufactured for them by Plessey.

A block diagram and most of the external circuitry required for the SL901 is shown in *Figure 4.13*. The internal arrangements of the s.i.c. are quite straightforward with quadrature type detectors used for the synchronous demodulators (the required 90 degree phase shift for quadrature detection is already arranged by the reference subcarrier phasing, of course). The outputs of these detectors, which have been fed from pins 3 and 18 with the resultants from the delay-line circuitry (see Chapter 5 on PAL-D decoding), go direct to the red and blue matrices, with a secondary feed from each through an inverter to the G′–Y′ matrix. This is formed by two resistances and while they cannot be accurate in value in an s.i.c., they are accurate in *ratio*.

The red, green and blue matrices are formed in a similar way but because the resistances are separated by a transistor they cannot be

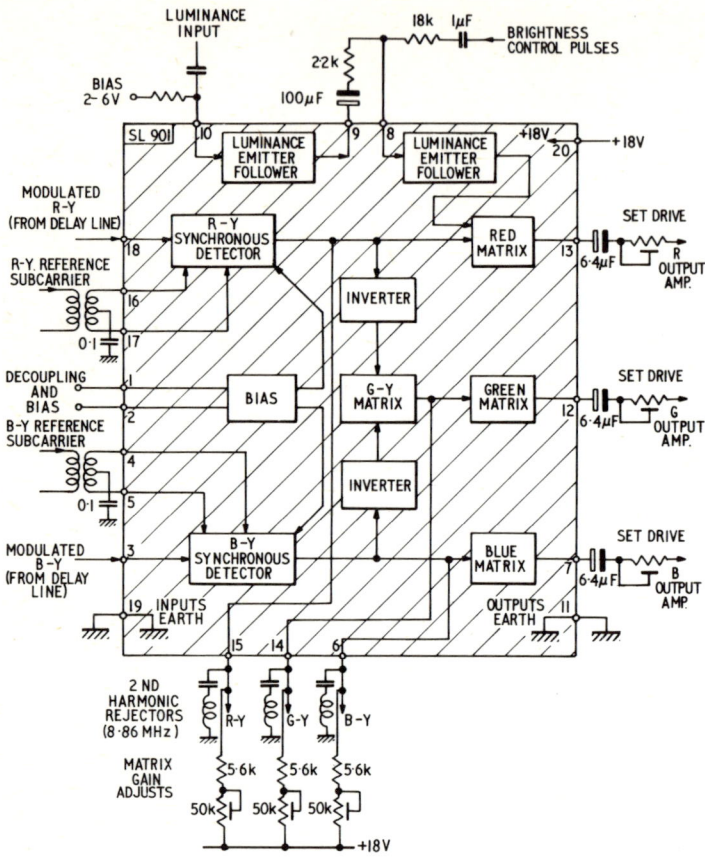

Figure 4.13　Bush/Plessey chroma demodulator and RGB matrix i.c. SL901 with external arrangement

guaranteed to be accurate in ratio; a small amount of balance control is given by a preset bias voltage on pins 6, 14 and 15. The colour-difference signals are also available at these pins should they be required in a different receiver arrangement and second harmonic rejectors at these points remove the 8·86 MHz components on the signals. Those components are caused by the switching form of the demodulator.

Rank-Bush-Murphy/Plessey have also produced an s.i.c., the SL917A, which integrates a large proportion of the rest of the colour decoder. This integrated circuit, in a 24-pin flat pack, uses a *passive subcarrier regenerator* instead of the normal subcarrier oscillator frequency controlled by an error voltage. This is done by using the high Q property of a crystal and the fact that such a high Q will maintain an oscillation for a relatively long period of time if it is driven by a signal at its resonant frequency. The resonant frequency is set to be subcarrier, the drive signal is the PAL burst amplified. When driven in this way subcarrier is available from the crystal filter arrangement for the line period after a burst has triggered it and until the next PAL burst has triggered it again.

In practice the circuit is not quite so simple because sidebands at both 7·8 kHz and 15·625 kHz are present on the burst—the 7·8 kHz is produced because of the swinging burst rate and the 15·625 kHz is present because of the burst occurrence at line rate.

A crystal can quite readily reject the line sidebands but a crystal which can reject 7·8 kHz sidebands would be uneconomic. Because of this the 7·8 kHz components are removed prior to the regenerator circuit by a phase switch driven by the ident signal circuitry. The sideband components at line rate can also reappear in the output because of the leakage across the crystal due to the holder capacitance. This is neutralised in the practical circuit.

The circuit is external to the i.c. which itself provides chroma amplification to drive the delay line, colour killer, subcarrier amplification and ident switching and bistable.

Vision i.f. amplification

The Motorola MC1352P is the first s.i.c. to be used in any U.K. receiver in a position *before* the video detector. In format it contains both i.f. amplification and a gated a.g.c. system. We therefore see an immediate conflict between manufacturers about the positioning of various circuits in their i.c. form. The video processing i.c. we looked at earlier also has an a.g.c. system in it. One would therefore never expect a receiver to have both i.c.s (!) and it leads to receiver manufacturers using one 'brand' of i.c.

Two stages of i.f. gain are given in the MC1352P with a total of 53 dB of power gain. This is an extremely high gain for the frequencies of operation. The differential amplifiers used are balanced from the input right through to the external transformer where the signal is coupled off to the video detector.

The a.g.c. circuit is a line-gated system with direct a.g.c. to the input amplifier in the i.c. and delay a.g.c. available for the r.f. stage of the u.h.f. tuner. The a.g.c. maintains the output to within 0·3 dB for a 60 dB change at the input of the amplifier. The value of R in the circuit

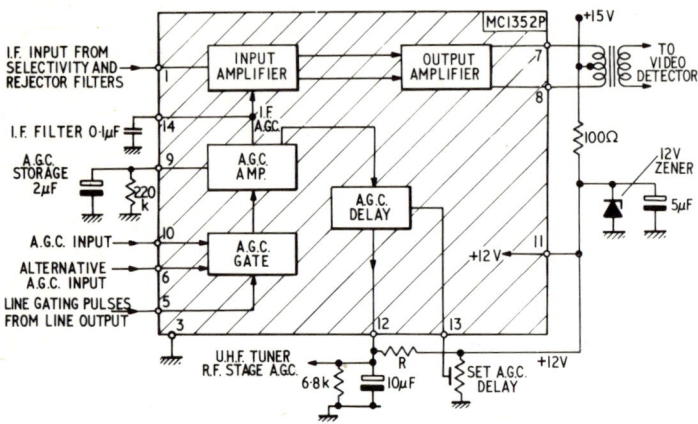

Figure 4.14   Simplified block of Motorola MC1352P and a.g.c. i.c. with basic external corrections

of *Figure 4.14* is chosen to give the correct maximum gain in the r.f. stage of the tuner when the a.g.c. output transistor is just cutting off.

To prevent instability in the external a.g.c. circuits the power supply to the i.c. must be of very low source impedance. This is achieved with the 5 μF decoupling shown; at the same time the rail voltage is held steady by a zener diode.

Other stages

Integrated circuits are available for other functions although they are not yet used in current receivers. Examples which exist are for field output and audio output where powers up to 5 W can be successfully handled. One example is the General Electric (G.E.) PA246 for audio work with a sensitivity of 12 mV for 5 W into 16 Ω. This· kind of circuit should begin to find its way into the receiver quite shortly. All that will then remain for integration will be the video and chroma outputs, which will not be for some years, the line output, which it is difficult to imagine can ever be integrated, and the power supply, some of which, in all transistor receivers, it might be possible to integrate, but because of the range of voltages required this would probably be integration in a number of separate stabilised power units.

It is also of interest to note that varactor tuners (of the form discussed in Chapter 2) usually employ an integrated circuit voltage stabiliser to provide the voltage for the d.c. tuning control. This is necessary because of the very large tuning drift which can occur if this voltage changes—even in receivers fitted with a.f.c. A common type in use suitable for varactor tuners is the Mullard TAA550; with an input of about 40 V it shunt-stabilises the output to between 31 and 35 V (depending on which voltage group TAA550 is being used) with very little effect if the mains voltage or the ambient temperature changes.

SERVICING RECEIVERS WITH I.C.S

The most important thing to remember when servicing any receiver incorporating an integrated circuit, or a number of integrated circuits, is that the i.c. is inherently more reliable than the discrete components that surround it. This reliability stems from its encapsulated nature and the very complex and rigorous design work that goes into any i.c. We in fact expect an i.c., to perform a specific purpose at least as good as and usually better than its discrete component counterpart.

Unfortunately when an i.c. does go wrong we must, of course, replace the whole unit so that in terms of expense in maintenance it means 'more money but a great deal less often'.

When the first receivers came on the market fitted with i.c.s the manufacturers found themselves saturated with returns, with many dealers automatically blaming the i.c. whenever a circuit with which it was involved developed a fault. A great deal of this was probably due to a feeling of awe and fear of the new.

The business of fault location in a receiver fitted with i.c.s is no different to any other circuit fault-finding. But when the area of the fault has been located at an i.c.—such as one output being incorrect—

then the procedure differs. We are used in television to making quick substitutions of suspected components, valves and transistors; with the i.c. this is not done.

Not only is the i.c. more expensive than the average component, it is also usually more difficult to remove from the circuit board—because of the large number of connections—and it is a bit more fragile. A 'bit' more fragile because, provided ordinary transistor precautions are taken in soldering and desoldering, nothing should come amiss.

Well, how do we locate a faulty i.c. without removing it? The secret is in the types of circuit that the i.c. uses. They are, as we have seen, basically d.c. devices with few or no capacitors—in fact there are none in any of the chips we have looked briefly at in this Chapter, but, as an example, the Mullard TAA570 sound intercarrier i.c. has one capacitor fabricated in it.

If we test the d.c. conditions on the pins of the i.c., and if they are correct, then the i.c. is operational and we must look to the external circuit for our fault. If the d.c. voltages measured on each pin are outside the tolerances quoted by the receiver manufacturer in his handbook then it is *possible* that the s.i.c. is faulty.

A check procedure may be given by the receiver manufacturer to expedite servicing and a typical example from Bush is shown in Table 4.1. Although the circuit references do not mean anything to us without having the complete receiver circuit in front of us we can see that a lot of alternatives are given in each case before the S.I.C. IS FAULTY box appears.

To remove a known, faulty i.c. from a circuit board, the procedures detailed in Chapter 7 (for unsoldering multi-tag components) should be followed. In particular it must be emphasised that over-zealous mechanical pressure in the removal of an i.c. will either peel off the copper from the printed circuit board or break-off one of the i.c. leads. There is very rarely sufficient space under an i.c. to enable it to be removed by cutting off the device from its leads and removing each end separately.

When a replacement D.I.L. i.c. is fitted, the component should be pushed into the circuit board as far as the lead size allows it. With a TO-5 type header i.c. the component should be mounted at an identical height above the board as the original, and great care should be taken that the positioning of the marker-identification-tab is correct.

Manufacturers may quote d.c. voltages on the pins of an i.c. using a particular instrument—e.g. *Avo* 8 or *Philips Multitester*—but moves

161

Table 4.1 BUSH S.I.C. CHECK PROCEDURE

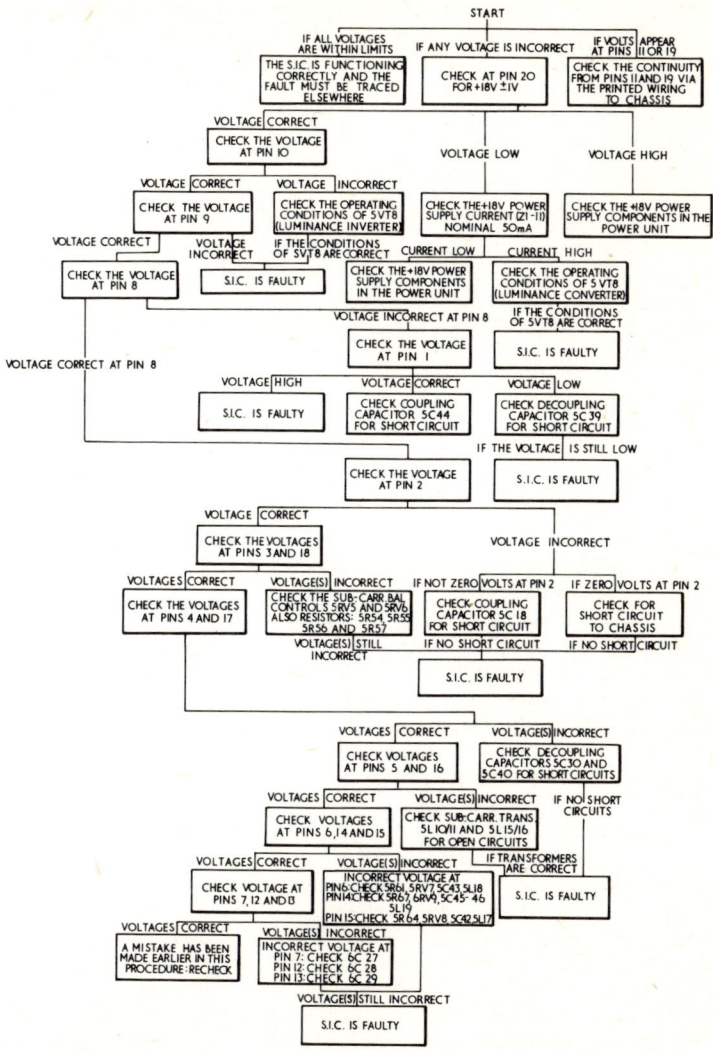

are being made towards the specification of much higher meter resistances, and voltages in the near future may well be quoted using an *impedance convertor* in association with a particular test meter.

5

COLOUR TELEVISION
by J. P. Hawker

The first public colour television service was inaugurated in the United States in 1951, although the first demonstrations of low-definition colour television had been made by Baird as early as 1928. The 1951 system, however, suffered from the major disadvantage that the transmissions were suitable only for reception on a colour receiver. It was soon appreciated that a successful colour system must be *compatible*, that is the transmission must be receivable as a satisfactory black-and-white picture on a conventional monochrome receiver. Similarly, another important requirement is *reverse compatibility*; that is the ability to receive monochrome signals on a colour receiver. In the 'thirties, the French engineer Valensi had shown that such a system could be devised by separating the colour information (chrominance) from the brightness characteristics (luminance).

Different colour systems

The requirements of compatibility and reverse compatibility were first met in a public colour system specified by the National Television System Committee in the United States, drawing on the work of R.C.A. and others, and transmissions using this system began in the United States on January 1, 1954. Later an identical system was adopted in Japan. A feature of this system (known as N.T.S.C. from the initials of the committee) and later systems is that the additional colour information is encoded into the transmissions *without occupying any additional bandwidth*.

The early spread of N.T.S.C. colour television in the United States proceeded for some years at a much slower rate than had been predicted, partly because instrumental shortcomings resulted in variable

quality of the colour. From about 1963, however, the number of receivers in use began to increase much more rapidly, and by 1966 sales
were limited primarily by the supply of the colour display tubes.

In Europe, some engineers considered that some of the reluctance
on the part of the public in the early days had stemmed from the
susceptibility of signals encoded by the N.T.S.C. technique to relatively
small changes of *differential gain* and *differential phase* anywhere
within the transmission path from camera to receiver display; and this
also presented broadcasters with problems in the field of videotape
recording, etc. Differential gain is a variation of chrominance gain as
the transmitted brightness signal varies; while any change of phase
of the subcarrier used to convey the colour information as a result of
changing brightness levels is termed differential phase; ideally variations
in brightness should produce no changes in the information carried
on the subcarrier either in phase or amplitude.

These problems encouraged the development of alternative techniques for encoding the colour information within the monochrome
transmission (though the systems still retained almost all the other
features of the N.T.S.C. system). Of the quite large number of different
systems so developed, a few have achieved importance: first of these
was SECAM ('sequential and memory'); and later PAL ('phase
alternation line').

A long controversy over the merits and demerits of these systems
relative to N.T.S.C., and also a number of variations within the systems,
has remained unresolved. A system which combined features of all
three systems, SECAM IV or NIIR was developed by French and
Russian engineers (and a similar technique had earlier been proposed
in the U.K. but not investigated). The result is that different countries
have adopted different encoding systems, the British choice being PAL.

PAL receivers can be made relatively immune to the effects of differential phase, although strictly this statement is true only of the PAL
receivers which incorporate a *delay line* (a technique which also forms
an inherent part of SECAM systems). Reception can still be affected
by differential gain but this tends to result only in slight changes in the
colourfulness (saturation) of the picture and is regarded as less objectionable than the variations of hue produced by differential phase on
N.T.S.C. transmissions. Because of this susceptibility to hue variations, N.T.S.C. receivers incorporate a hue control to allow the user
to adjust his picture; such a control is not necessary in PAL or SECAM
receivers, although user and pre-set controls labelled 'tint' may be
included for varying the colour temperature (see later).

In the receivers, the differences between the various systems are confined to the *decoding* sections. Those parts of the receiver concerned with receiving the sound and vision signals, the time-base generators, the display tube and associated circuits are basically similar for all systems. For the service technician, one of the most important operations is the setting up or readjustment of the display tube, and for a given type of display device this process is the same regardless of the encoding system used.

PRINCIPLES OF COLOUR TELEVISION

Colour television systems are based on the fact that almost any colour can be defined in terms of the amount of red, green and blue light present in the original. The sensation of colour derives from our eyes responding (as frequency-conscious devices) to electromagnetic radiation within a specific band of frequencies. Colour can thus be considered as the sensation conveyed to the brain when we look at light of a predominant wavelength: see *Figure 5.1.*

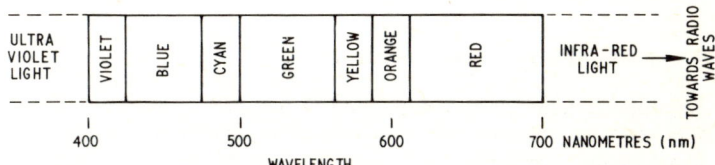

Figure 5.1   The electromagnetic spectrum of visible light. Short wavelengths such as these may be quoted in c.g.s. Ångstrom units ($10^{-10}m$)

As might be expected, the eye is more sensitive to radiation in the middle portion of the visible spectrum than to radiation towards the upper and lower edges of the spectrum. The brain associates this optimum sensitivity with the colour green. As can be demonstrated by means of a prism, white light is the sensation obtained from a combination of radiation throughout the visible spectrum; if an object reflects light uniformly it appears white; but if in the process the level of brightness is reduced, it will appear grey; and if little or no radiation is reflected, it will appear black. Note that colour may be registered by looking directly at a source of radiation, or alternatively at the light reflected from an object.

It should also be noted that the eye is much less sensitive to fine detail in colour than it is to detail in black and white; this factor is utilised in colour television since it means that much less information need be transmitted to provide colour detail than to provide the basic black-and-white detail.

Almost any colour can be obtained by adding together (called 'additive mixing') specific proportions of red, green and blue light (in printing and painting, however, where colour mixing is a sub-tractive process, the corresponding three primary colours are red, yellow and blue).

To define completely any colour we require three characteristics: its brightness or *luminance*; its *hue*, that is its dominant wavelength; and its *saturation*, which may be considered as the intensity (or 'colour-fulness') of the colour.

Also important as a standard for whiteness is the *colour temperature*, which for practical purposes is determined by the amount of blue or green in peak white. The present standard adopted in colour-television engineering is known as *Illuminant D,* which has a colour temperature of 6 500 K.

Luminance and chrominance information

Luminance is the characteristic used in monochrome television, and provides us with a picture in which detail is conveyed by varying levels of brightness. Saturation can be considered as the intensity of a colour. A desaturated colour is one in which a proportion of white is mixed with the colour. Thus 0 per cent saturation is entirely grey reproduction with no colour in it, 100 per cent saturation on the other hand has no white in the colour. Recognition of the saturation of a colour allows us to distinguish, for instance, between a pink and red light even though they may have the same dominant wavelength (i.e. hue): it is, so to speak, the amplitude of the colour.

In monochrome television only the luminance signal is trans-mitted; in all modern colour television systems this same information is radiated to provide the important characteristic of compatibility, but in addition information is also transmitted about the hue and saturation characteristics of the picture. This information is termed the *chrominance* (often abbreviated to *chroma*) signal. The trans-mission of colour television thus involves providing a luminance signal plus chrominance, and the encoding of the chroma signals in such a manner that they can be transmitted within the normal vision band-width (*in-band* systems): basically it is only in the encoding techniques that the N.T.S.C., SECAM and PAL systems differ.

We should note that a true chrominance signal is one which defines the hue and saturation only of the picture, and is not concerned at all with its luminance: this would mean that any distortion of the chroma information would result only in changes of hue or saturation, and not at all with the detail of the picture. Receivers designed to fulfil this condition are termed *constant luminance*, although in practice some departure from this ideal condition is used. Luminance is clearly that property of light which remains when all chrominance information is removed, as in monochrome television.

To obtain the luminance and chroma information, a colour tele-vision camera can analyse the light from the scene into three com-ponents by means of optical filters: these are R (red), G (green) and B (blue). R, G and B signals completely define a colour picture (that is provide information on luminance, hue and saturation) and in closed-circuit operation can be sent over three separate channels to the display device. In broadcast applications however such an arrange-ment would be too demanding in bandwidth, and furthermore would not involve the transmission of a luminance signal to which a mono-chrome-only receiver could respond.

Thus in many colour television cameras there are three similar pick-up tubes (usually Plumbicons*) each analysing the scene in terms of one of the colour components. Later, by processing the signal, separate luminance (Y) and chroma signals can be derived. In fact, in some cameras a fourth *separate luminance* pick-up tube is used, which provides the Y signal directly.

Other cameras use three Plumbicons* picking up the R, B and Y components of the scene; the G component can then be derived by later processing. The amount of gamma correction applied has to be a compromise because the transfer characteristics of a shadowmask and monochrome picture tube are not identical. Because of this, a camera which has a luminance (Y) tube may not use this directly when deriving a Y' signal for transmission.

In order that the R, G and B signals should reproduce in the correct proportions, it is necessary to take into account the differences between the response characteristics of the pick-up tube and those of the display

*Note. Plumbicon is a registered trade mark of NV Philips' Gloeilampen-fabrieken of Eindhoven, Holland.

device at the receiver, a process called *gamma correction*. Subsequent to this operation, the resulting signals are usually referred to as R′, G′ and B′.

The luminance signal (Y′) can be derived from R′, G′ and B′ signals by combining them in the correct proportions. Since the eye is most sensitive to optical wavelengths in the green region the luminance signal does not consist of equal proportions of the R′, G′, B′ signals, and by convention consists of approximately six parts green, three parts red and one part blue: or more precisely $Y′ = 0·3R′ + 0·59G′ + 0·11B′$. These proportions of the three signals are usually obtained by feeding the signals through resistive networks, a process called *matrixing*. This Y′ signal closely approximates the scene representation transmitted in monochrome television, and by modulating the vision carrier with it in the normal way the pictures displayed on a monochrome receiver will be indistinguishable from those obtained with a purely monochrome transmission system.

By slight variation of the proportions of the R′, G′, B′ signals, the colour temperature of the reproduction can be altered.

Chroma signals

Since we now have four different signals, R′, G′, B′ and Y′ it might be supposed that four separate channels would be needed to transmit them all; but since, as has been indicated, the four are related mathematically by a simple equation, the transmission of any three of them is sufficient to allow the fourth to be derived within the receiver.

In practice, since we have to transmit Y′ for the monochrome receivers, we fashion our other signals by taking the red and blue signals and subtracting from them the luminance signal: that is, we obtain R′–Y′ and B′–Y′. As already indicated, there is no need to transmit a G′–Y′ signal, since the magnitudes of the three signals Y′, R′–Y′ and B′–Y′ represent a series of simple equations (recalling the luminance equation given above) that give us all the information needed to reconstruct a G′–Y′ signal. The subtraction of the luminance signal from the colour signals means that the colour difference signals now represent solely the chromaticity (hue and saturation) characteristics with theoretically no luminance information involved.

It would of course be possible to transmit any three signals and derive the fourth, such as transmitting Y′, G′–Y′ and B′–Y′ and deriving R′–Y′ in the receiver. However, the G′–Y′ signal is the one chosen not to be transmitted because G′ is, on average, the largest amplitude colour signal and G′–Y′ is therefore, on average, the smallest colour difference signal and would be the one most prone to distortion by noise during transmission and in the receiver processing.

The important point to note from this rather involved series of operations is that a colour transmission consists of three packets of information: Y′ luminance signal and R′–Y′ and B′–Y′ chrominance signals. In this there is no difference between the N.T.S.C., SECAM and PAL systems; but from here onwards a fundamental difference occurs between N.T.S.C./PAL and SECAM. Both N.T.S.C. and PAL are designed to transmit the luminance and the two chroma signals simultaneously, whereas in SECAM there is continuous radiation of luminance but only one chroma signal is radiated at a time during the active line period, with a change over to the other chroma signal between each line. This means that the amount of vertical colour information transmitted on SECAM is only half that of the other systems, but rather surprisingly this is of far less practical importance than might be supposed (this is because of the inability of the eye to detect fine detail in colour).

In-band transmission

Although a monochrome television signal requires a wide bandwidth for transmission (amounting to some 6·75 MHz for a vestigial side-band 625-line signal), it has long been appreciated that relatively inefficient use is made of this bandwidth. In colour operation, this characteristic of television transmission is used to allow the chroma information to be transmitted *inband*, that is within the normal channel bandwidth, rather than, as in the case of stereo radio broadcasting, by multiplexing by means of a subcarrier outside the usual information bandwidth.

A monochrome signal has its information distributed in a series of packets spaced apart at a frequency equal to the line frequency, with relatively little spectrum energy in the gaps between. The relative amplitudes of the line harmonics and their side frequencies (at field frequency multiples) decrease rapidly at the higher video frequencies. By similarly modulating the chroma information on to a subcarrier accurately placed between two multiples of the line frequency, then similar packets of chroma signal energy will tend exactly to interleave the basic energy spectra of the luminance signal. (This characteristic has long been used to reduce co-channel interference in black-and-white television in the 'carrier off-set' technique.) It is this need to relate very precisely the chroma signal subcarrier frequency with the harmonics of the line frequency which results in the specification of the subcarrier frequency for PAL and N.T.S.C. to such accuracy (4 433 618·75 Hz for 625-line PAL). Unfortunately, this technique of *band-sharing* has some defects, and tends to result in a certain (fortunately very slight) amount of patterning interference on the screen of a monochrome receiver during colour transmission. The PAL sub-carrier—often abbreviated to 4·4 MHz for simplicity—is chosen to minimise the interference effects between the luminance and chroma information. The principal interference, most noticeable on monochrome receivers, is the movement of the sub-carrier 'dots' on pictures. Particularly annoying are slow 'crawls' across the screen; the PAL sub-carrier, based on $f_h(284 - \frac{1}{4}) + f_{v/2}$ (where, f_h = line frequency and f_v = field frequency), gives a pattern of very low visibility.

For both N.T.S.C. and PAL, we wish to transmit the two chroma or colour-difference signals on a single subcarrier. This is possible by a technique analogous to that used for stereo reproduction from the

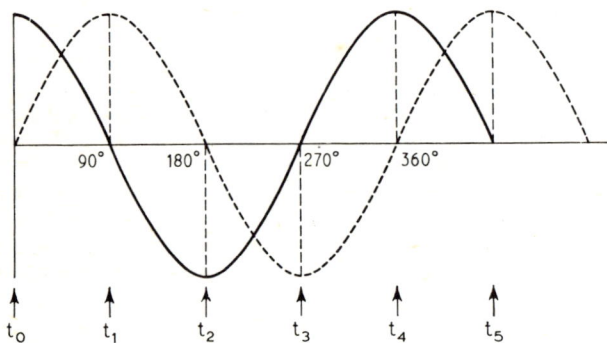

Figure 5.2   Principle of synchronous demodulation of quadrature modulation
Two carrier waves are shown of the same frequency but in quadrature (i.e. with 90° phase difference between them). If each is modulated by a different information channel and two demodulators, one synchronised with each carrier, sample the waves only at the times when one is at a peak and the other at null, then the two information channels can be retrieved in the receiver. The solid-line waveform is sampled at times t_0, t_2, t_4, and the broken-line waveform at times t_1, t_3, t_5.

single groove of disc records; it depends essentially upon the fact that two information channels can be carried simultaneously by shifting one of them through 90° of phase: so that the peaks of signal amplitude of one occur at the zero cross-over points of the other, see *Figure 5.2*.

This amounts to modulating the subcarrier in both amplitude and phase, and means that the complex chroma signal may be represented by a single phasor (vector) in which the length (amplitude) of the phasor represents the degree of saturation, and its angular direction (between 0 to 360°) represents the hue; see *Figure 5.3*. This phasor

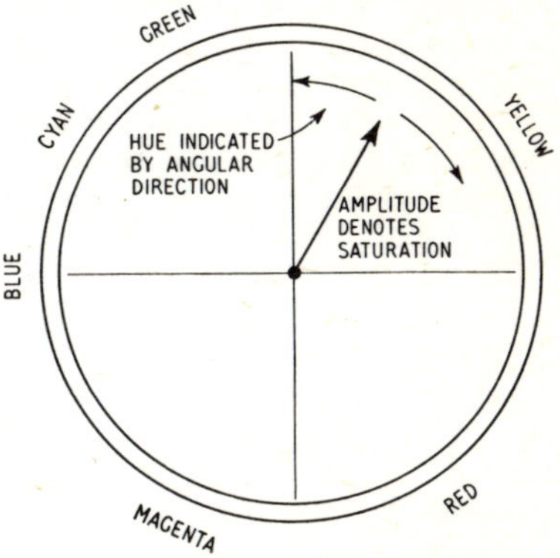

Figure 5.3 In the N.T.S.C. and PAL systems, any hue can be denoted by a phasor having a specific phase angle and an amplitude representing the degree of saturation

representation, it should be noted, involves the full chroma information, and can be resolved only by sampling the two colour-difference signals in time (phase).

At the encoder, as in stereo broadcast practice, the basic subcarrier is suppressed (except for short synchronising bursts) and only the chroma sidebands are transmitted. Since these are amplitude modulated, it means that only very small amounts of chroma sideband energy are present during the scanning of low saturation areas of the colour picture, reducing to a minimum the unwanted patterning of monochrome pictures in these more susceptible areas.

The signals leaving the studio are already coded so that the savings achieved in transmission bandwidth are also achieved in the circuits to the transmitters. This being so, the link and transmitter specifications must be held much tighter at the higher frequencies (chroma frequencies) particularly in terms of phase distortions, differential gain and phase and the products of interference between luminance and chrominance.

Reception of suppressed carrier transmissions

However, since the subcarrier is suppressed during transmission, it is necessary for the receiver to regenerate a local carrier which is locked in frequency and phase to the original (this is, similarly, a requirement in pilot-tone stereo broadcasting). This locally generated signal then forms a time reference for the *synchronous demodulators.*

To synchronise a local oscillator in both frequency and phase implies extremely accurate control of the oscillator, and to make this possible some ten cycles of the original subcarrier frequency are radiated during the back porch period of the television waveform; this is known as the *colour burst*. This signal provides an accurate timing reference to one (or, in the case of PAL, to two) of the axes of the colour-difference signals, from which any other axes can be derived by their phase angles. The differences in timing or phase-angle can be introduced by passing the regenerated carrier through a phase-shifting network.

Because the information represented by the phasors such as that of *Figure 5.3* comes from the combination of the separate chroma channels, it can be recovered only by two synchronous demodulators operating along different phase angles (i.e. with slightly different timing references from the regenerated subcarrier). The phase differences for the two axes need not be 90° and, in N.T.S.C. practice, it is common to use what are termed the X, Z axes, representing a phase difference of 62·1°.

Effect of differential phase

Consideration of *Figure 5.4* will show that information relating to the hue is conveyed by the phase angle or exact timing information, and that any disturbance of the synchronism between the studio 'clock' (as represented by the subcarrier) and that of the receiver (as represented by the local oscillator) will result in the display of incorrect hues. Unfortunately, there are a number of ways in which such errors can be introduced during transmission even though the local oscillator may be accurately locked in phase to the colour burst; such errors arise generally from differential phase in studio equipment, along the transmission links or due to multipath reception, or from mistuning or inaccurately aligned i.f. stages in the receiver. Such errors result in the phasor being shifted through a fixed or variable number of degrees and

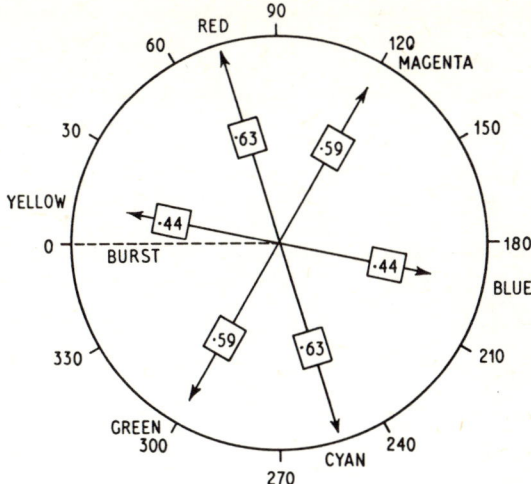

Figure 5.4   Phasor diagram of American N.T.S.C. colour signal, indicating fully saturated values

this causes the receiver to display the wrong hues. For this reason, an N.T.S.C. receiver contains a hue control by which the user adjusts the timing of his synchronous demodulators until the hue appears correct; and this control may need readjusting for instance when the colour pictures come from a different studio centre, etc. It was largely this feature of N.T.S.C. which caused engineers to develop alternative techniques such as PAL and SECAM, and led to the long controversy over the choice of a system for Europe.

The synchronous demodulation of the chroma signals can be compared with the manner in which a stroboscopic light 'fixes' a rotary motion by accurate timing of its flashes.

PAL transmission

One technique for overcoming susceptibility to differential phase is to reverse the phase of one of the chroma signals on alternate line periods; this apparently minor variation means that during one active line period we shall be working with a colour circle identical with that of the basic N.T.S.C. system, but in the next line the colour circle becomes that of *Figure 5.5(b)*. We must of course arrange for the receiver to recognise which colour circle is being transmitted and to alternate the signals by means of an *electronic switch* before displaying them.

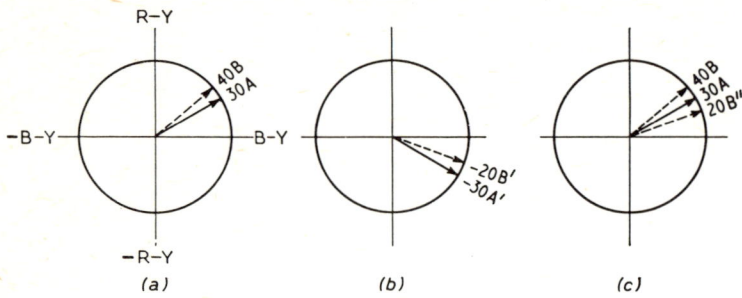

Figure 5.5   Principle of PAL automatic line correction by averaging
(*a*) Hue A transmitted with positive R–Y component; received as B due to spurious phase shift. (*b*) Same hue transmitted with negative R–Y (A′) component; received as B′ due to an equivalent phase error to (*a*). (*c*) After reversing polarity of (*b*), the two received signals B and B′ represent B and B″, which when averaged give the correct hue A.

Since the errors introduced by differential phase remain relatively constant, they will have opposite effects according to which colour circle is being used. For instance if a blue area is being transmitted, it may be shifted towards cyan during one line, but towards magenta in the succeeding line. If we display these signals on a picture tube, the eye will tend to integrate and average out the two hues and see the correct 'blue' hue, which will clearly be the 'average' of the two hues actually displayed. A receiver designed to make use of this effect is referred to as 'simple PAL' (PAL$_S$).

However, this averaging effect of the eye cannot cope really effectively when the phase errors are large; and in fact degradation in the form of a venetian blind patterning of alternate lighter and darker hues begins to become visible when the phase error exceeds about $\pm 5°$: this effect is sometimes referred to as *Hanover Blinds*.

But the same information which allows the eye to average the hues can more effectively be processed electronically to allow the correct hue to be displayed upon the screen. This can be done by means of a decoder containing a *delay-line* (a device which delays the incoming

signal by a precise time, in this case a complete active line period of
63·943 μs for 625-line systems). The incorporation of a delay line and
associated circuits enables two lines transmitted sequentially to be
available simultaneously, and these can be averaged electronically. A
receiver incorporating a decoder using this technique is sometimes
referred to as delay-line PAL or PAL$_D$. Such a receiver can tolerate
differential phase of as much as 40°. The averaging of phasors either
by the eye or electronically does however produce a resultant which is
shorter in length than the original. This reduction in saturation
increases as the differential phase increases but is far less objectionable
than errors in hue.

A PAL receiver must be able to recognise which line is transmitted
with which polarity, and thus identify whether the R′–Y′ signal is
being transmitted positive or negative. To do this a *colour ident* (colour
identification) signal has to be transmitted; in practice this is done by
transmitting the colour burst cycles alternatively at 135° and 225° from
the B′–Y′ axis.

To give a balanced view of the unquestioned advantages of the phase
tolerance of the PAL system, mention must be made of its problems:
the decoders are more complex than for N.T.S.C.; the picture may
suffer when there are horizontal colour edges (two adjacent lines in
different colours). In the latter case, the receiver interprets this as an
error in phase, and falsely averages the colours to produce a wrong
colour for the first line of the colour change. However, in practice this
is usually only noticeable on the horizontal edges of captions.

THE SHADOWMASK TUBE

The vast majority of colour receivers have been designed around the
shadowmask colour picture tube, originally developed by the Radio
Corporation of America, although various other forms of colour
display are beginning to emerge as serious rivals.

The shadowmask

The principle of the shadowmask is shown in *Figure 5.6*. Three electron
guns, one for each primary colour, are mounted round the axis of the
tube at 120 degree intervals and are inclined towards the axis so that
the undeflected beams converge to a common point on the screen. To

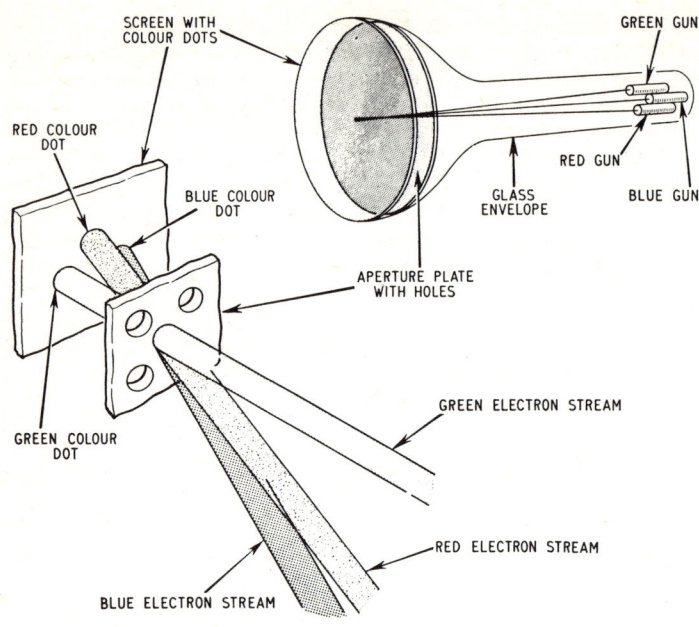

Figure 5.6   Basic action of a three-gun shadowmask tube showing how the electron beams from each of the guns strike only the appropriate colour dots

reach the screen the electrons must pass through the holes in a perforated steel sheet, the *shadowmask*, mounted about $\frac{3}{8}$ in behind the screen. Wherever the blue beam reaches the screen a blue phosphor dot is placed. These blue dots are in the *shadow* of the mask as far as the other two beams are concerned. Red and green phosphor dots are similarly placed wherever their respective beams hit the screen. There is then a *triad* of red, green and blue dots for each hole in the shadowmask.

The spacing between adjacent triads is approximately equal to the distance between adjacent scanning lines of the interlaced 625-line raster. Each beam energises only the phosphor dots of one primary colour and individual control of the colour components of each picture point is obtained. This principle is clearly illustrated in the simplified diagram of *Figure 5.6* but it should be borne in mind that in practice the beam diameter is sufficient to cover a group of adjacent triads. Despite continuing detail improvement of design, the holes in the shadowmask occupy only 17·5 per cent of the scanned area and thus 82·5 per cent of the electron beam energy is wasted in heating the mask. About 20 W may be dissipated in this way and the mountings of the shadowmask are designed to minimise the effect of thermal expansion.

The shadowmask is formed from 0·006 in cold rolled mild steel and is etched with the pattern of holes derived from a photographic master. The steel sheet is first coated with a photosensitive 'resist', i.e. a material which is hardened by ultra-violet light. This is exposed to the pattern produced by the master negative so that the intended position of the holes corresponds to areas of unhardened resist. This is washed away by a solvent and the plate is etched in an acid bath.

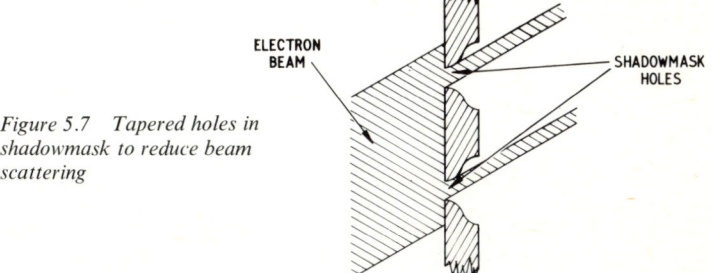

Figure 5.7 Tapered holes in shadowmask to reduce beam scattering

As a further refinement the holes are given a tapered cross-section by etching from both sides of the plate simultaneously (see *Figure 5.7*). This reduces the electron scatter that would otherwise occur at the sides of a cylindrical hole.

Phosphor dot screen manufacture

Correct positioning of the phosphor dots is even more critical than the location of holes in the shadowmask, and the manufacturing technique is of particular interest as it is related to the setting up procedure that will eventually be used in the receiver. Each screen is manufactured in conjunction with the actual shadowmask that is to be paired with it. The faceplate, which is a separate component, is

coated on its inner surface with green phosphor and photosensitive
resist. The shadowmask is accurately positioned relative to the faceplate; then an ultra-violet light is positioned so that the light originates,
in effect, from the intended position of the centre of deflection of the
green electron beam. (Because the tube has three electron guns there
are three centres of deflection within the deflection field).

Shining through the shadowmask, the light hardens a dot of green
phosphor in position for each hole in the mask. The remaining phosphor is washed away and a coating of blue phosphor with photosensitive resist applied; the ultra-violet light source is moved to the blue
centre of deflection and the blue dots are hardened in position. The
red dots are similarly accurately hardened in position relative to the
shadowmask (*Figure 5.8*).

In this way any errors in the pattern of the shadowmask are reproduced in the phosphor dot pattern and the operation of the tube is
unaffected. When the phosphor dot pattern is complete an aluminising
layer is added and the faceplate, with shadowmask fixed in position,
is sealed to the glass cone. A metal rimband provides protection against
implosion.

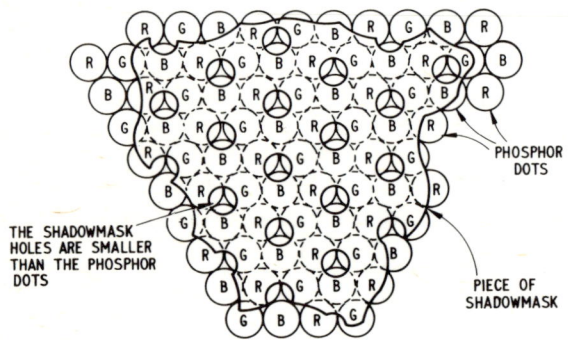

*Figure 5.8 How the shadowmask is aligned with the R, G and B triads of
phosphor dots on the screen of a shadowmask colour display tube*

SETTING UP SHADOWMASK TUBES

Rather paradoxically the adjustment of a shadowmask tube is directed
towards obtaining a perfect black-and-white picture. Any error in

setting up is seen as a colour *fringe* or tinted area in part of the picture. Two of the adjustments, purity and convergence, involve ancillary magnetic deflection fields acting on the three electron beams to ensure that they follow the intended paths. The field strength is small and comparable to the earth's magnetic field. Most receivers have a built-in demagnetising coil (*degaussing coil*) surrounding the part of the tube occupied by the shadowmask and this automatically compensates for the effect of the earth's field each time the receiver is switched on. However, if other metalwork in the receiver or in the vicinity of the receiver should become magnetised it may become impossible to achieve purity or convergence and an external demagnetising coil must be used. It is important that this should not be used to produce a stronger field than necessary or permanent damage to the shadowmask may result. A 50 Hz current is usually fed to the coil, and the field at the appropriate point is steadily reduced over a period of a few seconds either by reducing the current (as is done with the receiver's internal coil) or by progressively separating the coil from the metalwork to be demagnetised. It is hardly necessary to add that a colour receiver must not be used close to any apparatus producing a stray magnetic field.

Purity

The purpose of this adjustment is to ensure that the electron beams approach the shadowmask at the correct angle so that they must then hit phosphor dots of the correct colour. To do this the deflection centres of the three beams must be moved to the points in space successively occupied by the ultra-violet light source when the phosphor dot screen was laid down. Two adjustments are required: one field, produced by the purity magnets, shifts the three beams relative to the axis of the tube in the same way that the centring magnets of a mono-chrome receiver operate.

The second adjustment moves the three centres of deflection towards or away from the screen by moving the scan deflection coils along the neck of the tube. Since the position of the scan coils can only affect the deflected beams the effect is separated by using the following procedure:

(1) Switch off the blue and green guns and select the signal source and control positions which will give the nearest approach to a plain red raster while keeping the timebases correctly synchronised. Any

lack of uniformity in the red colour indicates that some electrons are reaching the green and/or the blue dots; and this can be confirmed directly by careful inspection with a magnifying glass.

(2) Move the scan coils forwards or backwards from their original position along the neck of the tube; about 50 mm will be sufficient to show a severe impurity around the edges of the raster.

(3) Adjust the purity magnets (*Figure 5.9*) for a pure red in the centre area of the raster. Ideally the red dots should be examined with a low-power microscope to check that the electrons are all landing within the area of the phosphor dot, but this may be considered a counsel of perfection.

TABS INDICATE DIRECTION
OF MAGNETISATION

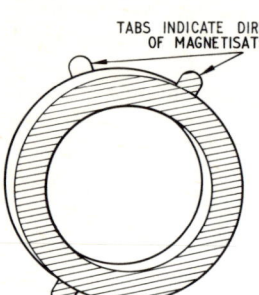

Figure 5.9   Purity magnets: one magnet is turned relative to the other to vary the strength of the resultant field. The magnets are turned together to vary the direction of the field. The design of purity magnets differs between manufacturers

(4) Move the scan coils along the neck of the tube until the optimum position is found when the pure red at the centre of the raster will spread to the edges.

(5) Switch off the red gun and separately check the purity of green and blue rasters. If these are not also correct it may indicate a large error in the convergence adjustment which is described below.

Convergence

Although the electrons should now be reaching the correct colour phosphor dots they may not converge to a common point. In this case the images of the three primary colours will not be accurately superimposed and coloured fringes will be seen on a monochrome picture. To correct this, additional magnetic fields are used which deflect each beam separately.

Figure 5.10 shows the convergence assembly mounted on the neck of the tube, operating in conjunction with internal pole pieces inside the tube which guide the magnetic flux to one beam while shielding it from the other two. The steady field applied to each beam is varied by rotating the permanent magnet at the top of each section of the assembly. The flux is guided by the U-shaped ferrite pole pieces to the internal pole pieces. These are shaped so that the flux crosses the path of the electron in the direction shown by dotted lines in the diagram. The resultant shift of the electron beam must be perpendicular to this direction and to the movement of the electrons, i.e. along a radius.

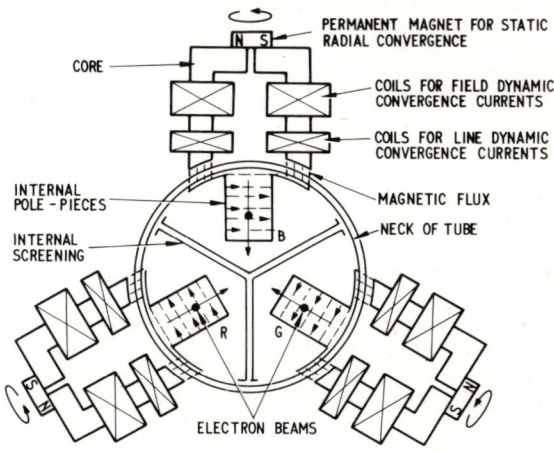

Figure 5.10   Radial convergence assembly

Any two beams may be converged by radial shifts, but the third beam will, in general, need a further shift perpendicular to its radial shift to superimpose it on the other two. This is provided on the blue beam as the two shifts then give vertical and horizontal movement of the blue raster which is convenient for setting up. The separate assembly to make this adjustment is called the *blue lateral shift* and is shown in *Figure 5.11*.

It will be seen that the flux paths, again guided by internal and external pole pieces, in fact cross all three beams. However, the strength and direction of the fields are such as to give equal horizontal shifts to the red and green beams in one direction and a horizontal shift to the

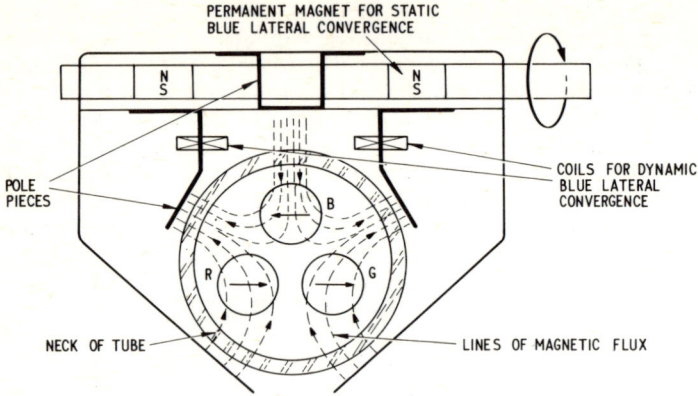

Figure 5.11   Blue lateral shift assembly

blue beam in the opposite direction. Looking at the raster one is conscious only of the blue image moving horizontally with respect to the other two.

The relative positions of all the magnetic deflection systems are shown in *Figure 5.12(a)* and *(b)*.

The adjustments so far described will give perfect convergence at the centre of the picture. This is called *static convergence*. Unfortunately it does not guarantee convergence when the three beams are deflected to form a raster.

Dynamic convergence

When the three beams have been correctly converged at the centre of the screen it is still found necessary to modify the convergence fields if the beams are to stay converged when deflected to form a raster. This is partly because each of the beams produces a slightly different pattern of pin-cushion distortion, and partly because the three electron guns are tilted slightly towards the axis of the tube.

Coils mounted on the external pole pieces of the four convergence assemblies are fed with current waveforms derived from the line and field timebases; thus the convergence fields are continually varying as the raster is scanned, and suitable choice of waveform gives correct convergence over the whole raster. This is called *dynamic convergence*.

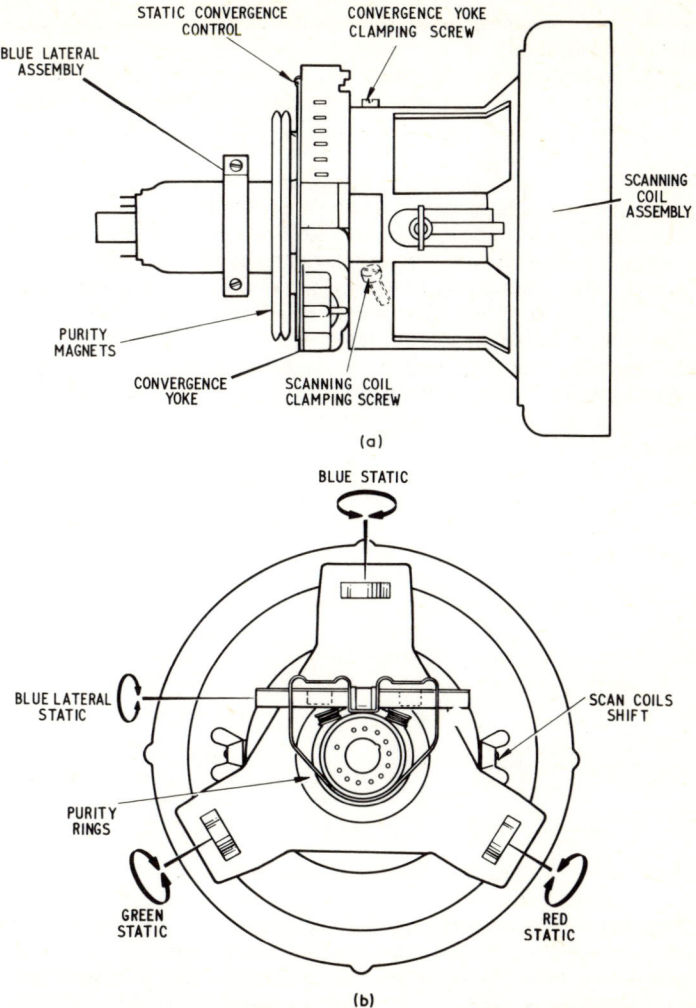

STATIC CONVERGENCE CONTROL

CONVERGENCE YOKE CLAMPING SCREW

BLUE LATERAL ASSEMBLY

SCANNING COIL ASSEMBLY

PURITY MAGNETS

CONVERGENCE YOKE

SCANNING COIL CLAMPING SCREW

(a)

BLUE STATIC

BLUE LATERAL STATIC

SCAN COILS SHIFT

PURITY RINGS

GREEN STATIC

RED STATIC

(b)

Figure 5.12 Relative positions of magnetic assemblies on shadowmask tube
(a) General cluster (from side of tube), *(b)* purity rings and static adjustment magnets (from base of tube).

A few basic principles can easily be established to indicate the type of waveform required.

A cross-hatch pattern of thin white horizontal and vertical lines on a black background is displayed on the tube. The blue gun is switched off and the static convergence magnets have been adjusted to converge the red and green patterns at the centre of the tube.

The first point to note is that any dynamic convergence current waveforms must have zero amplitude at the centre of the line or field period if the static convergence is to remain unaffected. Suppose there is a convergence error on the right and left hand sides of the picture which can be corrected by increasing the red convergence field in both cases. Then the required waveform in the red convergence coils will be a line frequency parabola as shown in *Figure 5.13(a)*. Had the error been asymmetrical requiring an increase of convergence field at the beginning of the line and a decrease at the end (or *vice versa*) then the required waveform would be a line frequency sawtooth (*Figure 5.13(b)*).

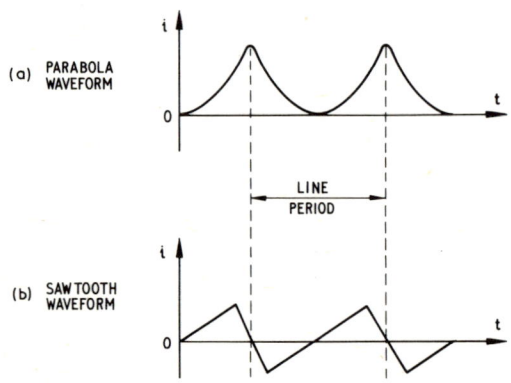

Figure 5.13   Convergence waveforms

Errors at the top and bottom of the picture are similarly corrected by sawtooth and parabola waveforms at field frequency. In practice each of the three rasters has its own distortion pattern shown (*Figure 5.14*) with a cross-hatch pattern with static convergence but no dynamic convergence correction.

Each radial convergence assembly must be fed with a line frequency current waveform in which the sawtooth and parabola waveform

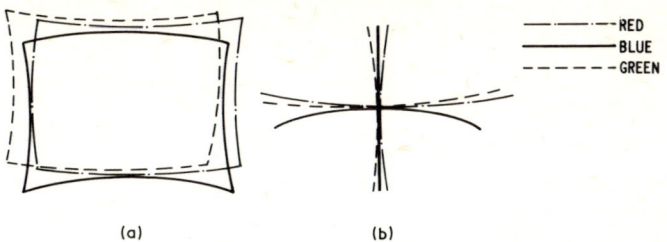

Figure 5.14   Convergence errors with static correction applied
(*a*) Individual pincushion distortions of each raster, and (*b*) the effect shown on a single 'cross' of a crosshatch pattern.

components may be separately adjusted and a field frequency current waveform with similarly adjusted sawtooth and parabola components. In addition the blue lateral shift field should have a dynamic component although a line frequency sawtooth is usually sufficient. Since the convergence waveforms are derived from timebase waveforms (see later), the picture size and centring adjustments should be made before adjusting convergence.

Matrixed convergence

A radial shift of the red or green beam moves the spot along a line making an angle of 30 degrees with the horizontal. Using the conventional cross-hatch pattern, adjustment is much easier if the beam shifts can be separated into vertical and horizontal components and this is achieved, in effect, by the technique of matrixed convergence which has been universally adopted. The red and green convergence controls in this case operate on both red and green waveforms, either to change both in the same sense or to change both in the opposite sense. *Figure 5.15(a)* shows the principle applied to one intersection of a cross-hatch pattern.

An increase of current in both the red and green coils moves point r along its radius to r'; the same increase moves g inwards along its radius to g'. The verticals of the cross-hatch have moved closer together but the relative *separation* of the horizontals is unchanged.

In *Figure 5.15(b)* the current in the red coil has been increased as before, but the current in the green coil has been reduced by the same amount, producing a radial shift in the opposite direction. In this case

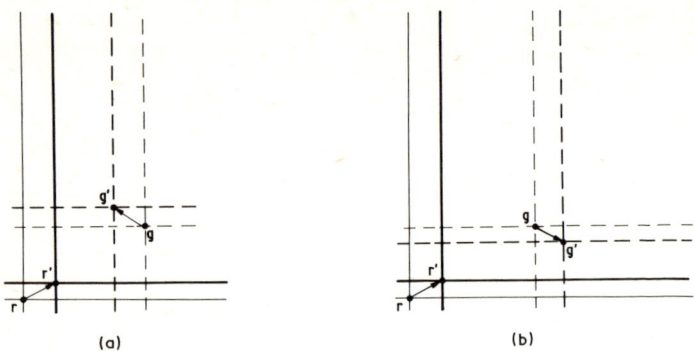

Figure 5.15   Principle of matrixed convergence

the visual effect is a shift of the horizontals towards each other without change of the verticals—the actual shift of the verticals is of no account since this adjustment is completed before the blue beam is switched on and blue has both vertical and horizontal shifts available.

The convergence controls affecting both the red and green rasters now become *R/G Amplitude* controls which affect only the *verticals* of the cross-hatch because they increase (or decrease) the red and green convergence currents simultaneously and *R/G Difference* controls which affect only the *horizontals* of the cross-hatch since they increase the red convergence waveform while decreasing green by the same amount (or *vice versa*).

In most modern receivers the function of each convergence control is indicated by a diagram on the convergence panel and this can be a great help. Even so there may still be occasions when it is desirable to relate the circuit function of a control to its effect on the raster and the following table summarises and completes the information given above:

Typical Title	Waveform	Part of Cross-hatch affected
R/G Amp	parabola amplitude to red and green coils changes similarly	red and green verticals on left and right hand side are similarly affected

R/G Diff	parabola amplitude increases in red, decreases in green (or *vice versa*)	red and green horizontals on left and right hand side are similarly affected
R/G Tilt '	sawtooth amplitude to red and green coils changes similarly	red and green verticals on left and right hand side are oppositely affected
R/G Tilt Diff	sawtooth amplitude increases in red, decreases in green coil (or *vice versa*)	red and green horizontals on left and right hand side are oppositely affected
B Amp	parabola amplitude to blue radial coil	blue horizontals on left and right move in same direction
B Shape	parabola waveform shape for blue radial coil adjusted in resonant circuit	flatness of blue horizontals
B Tilt	sawtooth amplitude to blue radial coil	blue horizontals on left and right move in opposite directions
B Width	sawtooth amplitude to blue lateral coil	blue verticals on left and right move in opposite directions

The above controls are all for line frequency waveforms. The table applies equally to field frequency controls if *left and right* are replaced by *top and bottom*. There is no equivalent to *B Shape* and *B Width* at field frequency.

Two remaining controls affecting convergence are connected to the scan coils and are used to balance the current in each half of the line and field scan coils. Any unbalance in these two currents shows itself as a differential effect on the red and green rasters. Thus *R/G Symmetry* (line) controls the parallelism of the red and green horizontals of the crosshatch (which have been straightened by the other convergence controls). *R/G Symmetry* (field) controls the separation of the red and green horizontals at top and bottom of the picture.

Grey scale tracking

The final adjustment to obtain a good black-and-white picture is to adjust the video drive amplitudes and the cut-off potentials of the three guns to give a white of the required colour temperature at all brightness levels from peak white to black.

The grid cut-off potential is adjusted by varying the screen (or first anode) potential of the electron gun. Some receivers make special provision for this. For example, some manufacturers provide a switch to remove the luminance signal or recommend unplugging the i.f. input and the three screen controls are then adjusted for a neutral grey. Others provide a switch which removes the field scan; the resulting horizontal lines can be adjusted to be just visible (or just extinguished) with some precision. Receivers fitted with a switch for this set up usually have it labelled as SET WHITE.

If no special provision is made, the screen controls may be adjusted for neutral dark tones in a monochrome picture and the video drive controls provided for two of the guns are then adjusted together with the contrast control for neutral peak white.

'Neutrality' should preferably be adjusted in comparison with an Illuminant D (6500) source. For best brightness one of the screen potentials should be at maximum or set to the voltage specified by the receiver manufacturer. It may be necessary to alternate adjustment of drives and screen potentials a few times to achieve the best overall effect.

The following table shows the adjustments that may be made to correct the tint shown in the left hand column. R.v.d. refers to Red video drive, R scr. to Red screen control etc.

Tint	Highlights	Lowlights
Red	Reduce R v.d.	Reduce R scr.
Green	Reduce G v.d.	Reduce G scr.
Blue	Reduce B v.d.	Reduce B scr.
Cyan	Increase R v.d.	Increase R scr.
Yellow or Sepia	Increase B v.d.	Increase B scr.
Magenta	Increase G v.d.	Increase G scr.

Since there will usually be only two video drive controls provided for the individual guns, the third is obtained by adjusting the contrast control in one direction and the two video drive controls in the opposite direction. It will be appreciated that where the table says, for example, 'increase green video drive', the same effect on the tint will be achieved by reducing red and blue video drives. The overall contrast will be slightly reduced instead of slightly increased. There are corresponding alternatives to each of the other adjustments.

Summary of setting up procedure

(1) Scan amplitudes and centring.
(2) Static convergence and a rough check on dynamic convergence (since there is some interdependence between purity and convergence).
(3) Purity.
(4) Convergence of red and green rasters followed by blue.
(5) Grey scale tracking.
(6) On initial setting up of a receiver repeat steps 2–5, if necessary, to obtain satisfactory results.

THE TRINITRON TUBE

The conventional shadowmask tube is by far the most widely used tube in colour television receivers but it suffers from a number of disadvantages:

(1) The shadowmask absorbs over 80 per cent of the electron beam power.
(2) Convergence is not a simple process and requires some skill and experience by the service technician.
(3) The focus and convergence are a compromise with one another; the two planes cannot be coincident when using 120 degree spaced electron guns so correct convergence cannot give perfect focus over the whole screen area.

The Trinitron tube, developed by the Japanese Sony Corporation, was released in 1968 as a challenger to the shadowmask tube of R.C.A. It is at present only available in small screen sizes* so it is difficult to conjecture its position relative to an overall use instead of the shadowmask.

*Author's Note: It is believed that an 18-in wide angle version of the Trinitron will be available shortly.

In the Trinitron the gun assembly is a single unit which emits three beams simultaneously. Using three separate cathodes which can be driven by —R, —G and —B video signals, each beam can be individually controlled. They all occupy the same electron lens position, all three beams passing through the central part of the main focus electron lens (*Figure 5.16*). The two outer electron beams (red and blue) diverge from the crossing point and are deflected back by the convergence electrodes (acting as electron-optical *prisms*) so that all three beams would strike a common point on the phosphor coated screen.

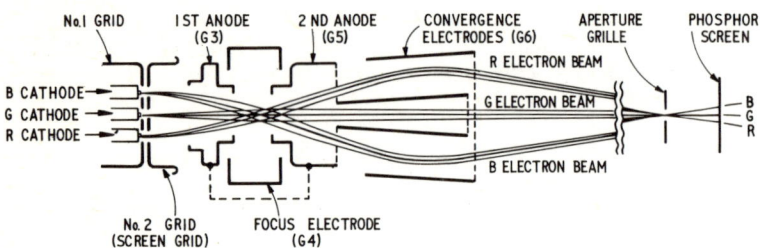

Figure 5.16   Electrode positions and beam trajectories in a 13-in Trinitron tube

All three beams should be in focus because they all suffer the same action under the main focus electrode. Prefocusing to the centre point of the main electrode is achieved by the lens formed by the screen (no. 2) grid and the first anode.

The colour restriction system in the tube (*cf.* shadowmask in the R.C.A. tube) is formed by an *aperture grille*. This involves a large number of vertical slits, instead of a large number of holes, etched into a metal sheet (*Figure 5.17*). This is no more difficult to construct than the shadowmask, but the larger aperture of the sheet gives a greater transmission of beam power—in fact about 1·33 times the transmission of the shadowmask. Because, too, the electron gun assembly does not restrict the beam current flow due to physical size— all three beams passing through the same lens system—there can be about 1·5 times the amount of beam current flowing anyway. This gives an overall improvement in picture brightness of 1·5 × 1·33— that is just double the brightness.

Just as phosphor dots are laid in a triad for the shadowmask tube, phosphor dots must be laid on the Trinitron screen to give the correct

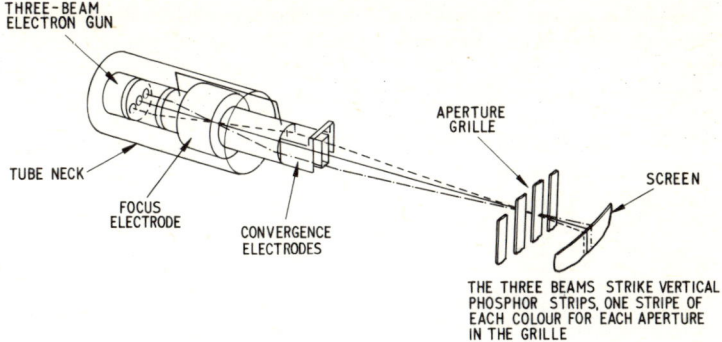

THREE-BEAM
ELECTRON GUN

APERTURE
GRILLE

SCREEN

TUBE NECK

FOCUS
ELECTRODE

CONVERGENCE
ELECTRODES

THE THREE BEAMS STRIKE VERTICAL
PHOSPHOR STRIPS, ONE STRIPE OF
EACH COLOUR FOR EACH APERTURE
IN THE GRILLE

Figure 5.17   Basic principles of Trinitron gun and aperture grille

colour rendition in front of the aperture grille. The pictures produced
by the Trinitron therefore have, on close examination, a vertically
divided appearance whereas the shadowmask tube gives a 'dotted'
picture appearance. Neither effect is unpleasant although many
viewers state an immediate preference for the pictures obtained from
the Trinitron.

Part of this preference may be due to the basically improved vertical
resolution of the tube, the resolution only being limited by the number
of scanning lines used by the transmission system; this is because of
the vertical slits adopted in the grille. Horizontal resolution in the tube
is limited in the same way as in the shadowmask—by the number of
vertical apertures across a line.

Convergence in the Trinitron

Horizontal alignment of the electron beams in the Trinitron is a
design feature of its construction: the mis-convergence on a single
horizontal plane is very small. Any misconvergence will, in fact, be
due to errors in manufacture of the convergence electrodes or in the
deflection yoke.

The pattern of misconvergence is indicated in *Figure 5.18*. Apart
from the lack of divergence on horizontals, the most noticeable thing
is the symmetrical misconvergence of the verticals from the centre.
This is because of the exactly symmetrical construction of the gun and

convergence electrode assembly. Only static and parabolic line convergence correction is therefore required.

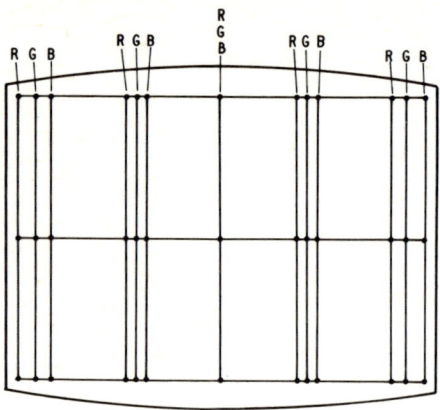

Figure 5.18   Misconvergence on Trinitron screen for various places relative to centre-beam spot

This arrangement makes convergence adjustment on the Trinitron a very much simpler operation than on the shadowmask tube, the typical number of adjustments being only six compared to an average of fifteen on the shadowmask tube receiver.

COLOUR RECEIVERS

In this section the basic requirements of a PAL receiver are outlined. Actual circuits of course vary from one manufacturer to another. There are, however, certain basic requirements that must be met for the production of colour pictures on a three-gun shadowmask tube. These requirements are fulfilled by the arrangement shown in block schematic form in *Figure 5.19*. The circuits up to and including the vision detector do not differ significantly from those of a monochrome receiver, since the total bandwidth remains the same, so that these sections are not shown, but it should be noted that some receivers use separate detectors for sound and chroma.

The signals used to modulate the shadowmask tube comprise the luminance information, which may be fed to the three cathodes, and the three (red, green and blue) colour difference signals then applied to the three grids. To control the scanning of the three beams to obtain accurate convergence, it is necessary to obtain from the timebases waveforms for the convergence unit around the neck of the tube. This unit comprises three groups of correction coils and magnets spaced 120 degrees apart around the tube neck, i.e. one group for each beam, giving radial beam displacement, together with a further unit giving

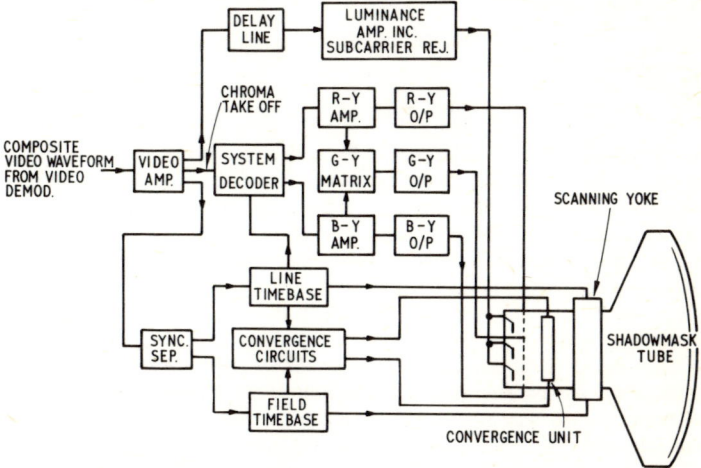

Figure 5.19   Block diagram of stages required to control a three-gun shadow-mask with colour-difference drive

N.T.S.C., PAL and SECAM systems differ mainly in the system decoder. RGB drive receivers differ in that the Y signal is matrixed with R–Y, G–Y and B–Y signals *external* to the tube.

horizontal beam displacement. The timebases also supply waveforms for pin-cushion distortion correction purposes and include provision for picture centring.

In the arrangement shown, a stage of video frequency amplification follows the vision detector. This stage feeds the luminance channel, the colour decoder (which recovers the separate colour difference signals from the composite chroma signal) and, as in monochrome receiver practice, the sync. separator.

Luminance channel

The luminance channel follows normal video frequency amplification practice except for the need for two additional features: first a delay line to keep the luminance signals in step with the chroma signals, which are effectively delayed slightly by being passed through frequency restricted circuits in the decoder and by luminance/chrominance delay in the i.f. amplifier and secondly a notch filter to remove the subcarrier which would otherwise appear on the screen as an interference signal.

The decoder

The decoder comprises a number of stages concerned with the recovery of the colour signals, which are radiated in the form of colour-difference or chrominance signals upon the subcarrier, their amplification and application to the shadowmask display tube; this entire section clearly has no equivalent in the monochrome receiver. The form of the decoder will depend upon the encoding system used at the transmitter and will differ for the PAL, N.T.S.C., SECAM III and SECAM IV (NIIR) systems.

The colour decoder for a PAL receiver will include stages for recovering the colour burst and colour ident signals and for using them to control the subcarrier regenerator and the electronic switch for reversing the polarity of the chroma signal on alternate lines; in addition two *synchronous demodulators* are required, with a matrixing network to derive the G′-Y′ colour-difference signal, and amplifiers to provide sufficient chroma drive to the picture tube.

Also within the decoder will be a *colour killer*, the function of which is to switch out the chroma circuits automatically when the set is tuned to a monochrome transmission (this can be achieved by arranging to detect the presence or absence of the colour bursts on the transmission waveform). For PAL_D receivers there will also be the *delay line* and its associated *adder* and *subtractor* circuits. To use the *colour ident* signals (sent in the form of a swinging colour burst) and the colour burst there will be a phase discriminator and reactance valve or equivalent arrangement to control the *crystal oscillator* regenerator.

A more detailed account of the operation of the decoder is given later in this section.

Timebases

For a shadowmask display tube, the timebases are called upon to
supply appreciably more scan power than for a black-and-white tube
of equivalent size, because of the larger tube neck required with the
three electron guns. The timebases must also supply various waveforms for the convergence circuits and other purposes. And the line
output stage has to be capable of providing a well regulated e.h.t.
supply of the order of 25 kV at beam currents which in total amount to
more than 1 mA.

Because of the wide range of beam currents needed for the three guns
good e.h.t. regulation is essential. In addition to the normal line
output stage stabilisation feedback circuit it was always usual practice
to include a triode e.h.t. voltage regulator in shunt with the picture
tube e.h.t. supply so that the e.h.t. power drawn from the supply
remains substantially constant. A simplified line output stage of that
form for a colour receiver is shown in *Figure 5.20*, in which the voltage

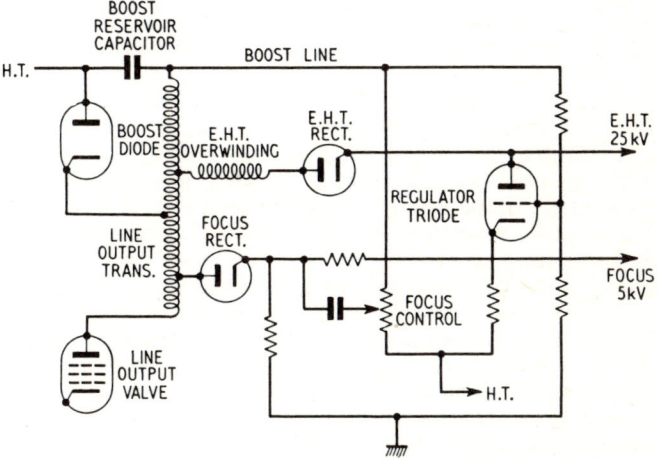

*Figure 5.20   Basic arrangement for producing a stabilised e.h.t. supply in a
hybrid colour receiver*

regulator circuit is indicated. The voltage regulator tube operates to
maintain a constant total load current on the h.t. supply regardless of
variations in the picture-tube beam currents. Its grid is taken, for this

purpose, to the boost supply line, since the boost voltage varies in proportion to the current drawn by the stage from the h.t. supply. As can be seen, the line output stage also provides the picture tube focus potential, which for a shadowmask tube is about 5 kV.

This kind of circuit can be found in the original generation of colour receivers. More modern receivers have made use of e.h.t. triplers and fifth-harmonic tuning of the line output transformer in the manner indicated in Chapter 3.

Further waveforms are derived from the timebases to provide pin-cushion distortion correction and convergence control, and in addition the line output stage provides pulses for the decoder circuits. Provision must also be made in the timebases for d.c. shift (picture centring).

The timebases are therefore more complex than in monochrome receivers, and operate at higher powers.

Convergence circuits

Three sets of radial convergence coils, one for each beam, are mounted 120 degrees apart around the neck of the tube as in *Figure 5.10*. In addition, a permanent magnet is provided with each set of coils to enable static convergence adjustment to be made as indicated earlier. The currents fed to the coils are obtained from the timebases, and result in radial movements of the beams as shown to provide good convergence over the entire picture. Extremely complex and precise waveforms would be required to give perfect overall convergence, so that practical circuits are to some extent a compromise. Nevertheless the sawtooth and parabolic waveforms obtained from the timebases provide adequate results. A further assembly is mounted behind the radial convergence assembly and is fed with a waveform from the line timebase to give lateral displacement of, mainly, the blue beam.

Figures 5.21 and *5.22* show in simplified form methods of obtaining the convergence waveforms from the line and field timebases, and the associated adjustments. The red and green sets of line convergence coils (*Figure 5.21*) are connected in parallel, and the blue set in series with them, the entire network being connected in series with the deflection coil winding of the line output transformer. Thus in this arrangement the deflection current waveform, suitably shaped, is used to control line convergence. The amplitude of the currents in the convergence coils is adjustable by means of the 20 and 5 Ω variable resistors in parallel with them. To obtain the required waveshape for the blue convergence coils, the correction network in parallel with the blue amplitude control is included—incorporated in this network is the blue tilt control. To provide red–green tilt adjustments, a negative-going 15 V pulse from a separate winding (of 1 turn) on the line output transformer is fed to the red and green coils via a variable resistor (*R–G tilt*) and the red–green difference control.

To provide blue lateral control, a 200 V pulse is tapped off the line output transformer and applied to the blue lateral convergence coils, which are series connected. Parallel connection may be used with a lower amplitude pulse. The current applied to the lateral convergence coils is controlled by the amplitude blue lateral adjustment coil shown.

In practice, the line-convergence system in earlier receivers may be complicated by dual-standard line timebase operation, so that extra switching and duplication of controls and correction networks are required. A clamp diode is generally connected in parallel with each set of line convergence coils to maintain the required d.c. conditions.

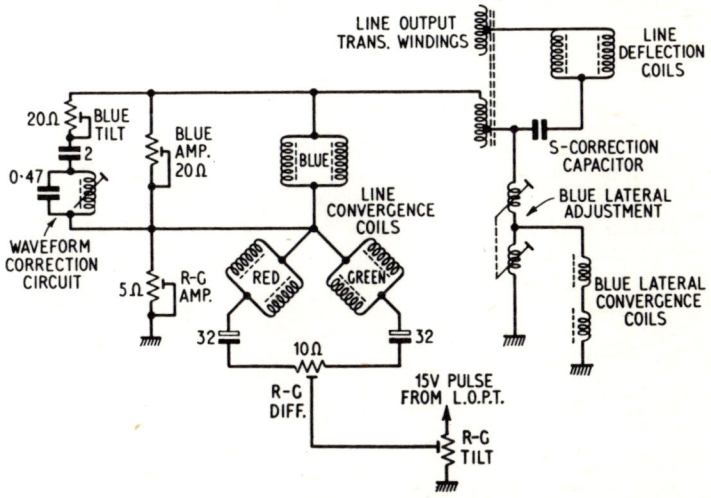

Figure 5.21   Simplified circuit of a line convergence arrangement

Among the complications found in practice would be the duplication of certain sections, with associated switching, required for dual-standard operation.

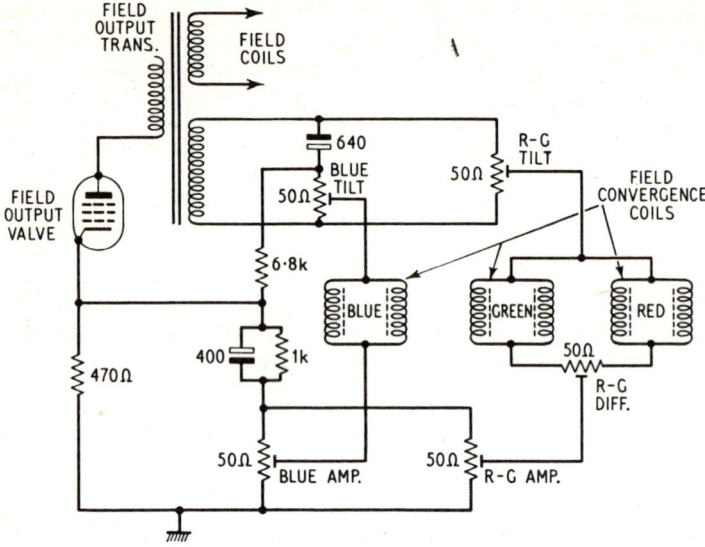

Figure 5.22   Simplified circuit of a field convergence arrangement

In the field convergence circuit shown in *Figure 5.22* a waveform of combined sawtooth-parabola shape is taken from the cathode of the field output valve and an opposing sawtooth waveform is derived from a separate winding (32 turns) on the field output transformer or out of phase waveforms are taken from convenient points in a transistor/transformerless output stage. The amplitude controls enable the overall amplitude of the current waveform fed to the convergence coils to be adjusted, while the tilt controls adjust the degree of sawtooth cancellation between the two waveforms. For ease of adjustment, the R and G controls are combined as described earlier.

It should be noted that the arrangements indicated are only one version of a multiplicity of possible arrangements which can involve different sources of convergence waveform at different amplitudes with and without subsequent amplification.

Picture centring and pin-cushion distortion

In monochrome receivers, centring and pin-cushion distortion correction (necessary with wide-angle deflection tubes) is achieved by means of adjustable permanent magnets around the neck of the tube. This technique is not practicable with shadowmask tubes since such magnets would interference with the purity and convergence of the picture.

Pin-cushion distortion at the sides of the screen may be corrected in high-quality monitors by taking the line deflection signal, modulating it with a signal from the field timebase, and applying the resultant signal to the line output valve. To correct pin-cushion distortion at the top and bottom of the screen, a waveform from the line timebase may be applied to the field deflection coils to vary the vertical deflection. An alternative approach that has been almost universal in commercial receivers is to use the signals from the line and field timebases to control a transductor linked to the line and field deflection coils. The principle is to modulate the line waveform at field frequency and the field waveform at line frequency, using the resultant waveforms to effect the necessary correction.

B.R.C. colour receivers all have the shadowmask tube mounted with the blue gun upwards. This reduces the visual effect of pin-cushion distortion when the screen is viewed slightly from above. Pin-cushion correction is only applied in the larger 26 in tube versions of the B.R.C. receivers.

Shift control is achieved by adjusting the d.c. component of the currents flowing in the line and field deflection coils. Small value potentiometers enable the amount of shift to be varied.

PAL DECODING

As shown in *Figure 5.19*, the composite video waveform is fed to the system decoder which derives from the composite chroma signal R–Y and B–Y colour difference signals. The G–Y signal is derived from a matrix circuit, to which the R–Y and B–Y signals are fed, the principle involved having been previously explained (see section headed *Chroma Signals* p. 168). The matrix and chrominance amplifiers are often regarded as a part of the system decoder. In this section, however, the decoder will be taken to mean the circuits handling the chroma signals between the original chroma take-off point and the R–Y and B–Y outputs.

The essentials of a PAL$_D$ decoder are shown in outline form in *Figure 5.23*, from which it will be seen that a number of different operations must be carried out in order to obtain the separate R–Y and B–Y outputs. These may be briefly summarised as follows:

(1) A *chroma amplifier* is required with bandpass filtering to separate the chroma signal from the composite waveform and provide initial amplification.

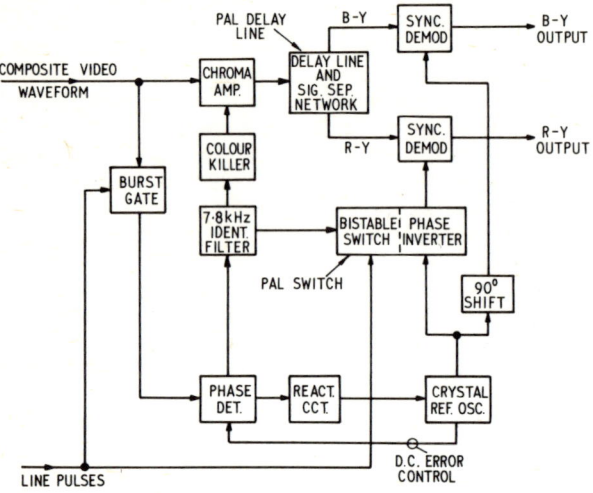

Figure 5.23   Simplified block diagram of a PAL$_D$ decoder

(2) The output from the chrominance amplifier is fed to the *delay line and signal separating network* which separates the R–Y and B–Y signals (by a process of adding and subtracting the direct signals and the signals which pass through the delay line) and carries out the averaging process between the alternate lines.

(3) The output from the signal separation circuit is applied to the two *synchronous demodulators*, which demodulate and pass only signals that are in phase with the reference signals fed to them from the *reference oscillator*. As there is a 90 degree phase difference between the transmitted B–Y and R–Y signals a 90 degree phase shift is introduced in the feed from the reference oscillator to one of the synchronous demodulators. In this way one synchronous demodulator passes on the R–Y signal and the other the B–Y signal.

(4) As the burst signal is used to synchronise the reference oscillator, a *burst gate* is needed to separate the colour bursts from the composite waveform; this can be achieved by gating the stage with a pulse from the line timebase so that it is switched on only during the appropriate portion of the line flyback period when the burst signal is present. The burst gate can be regarded as analogous to the sync.-separator stage, though its mode of operation is rather different.

(5) As the subcarrier is suppressed in transmission it is necessary to have a local oscillator to provide reference signals for the synchronous demodulators. This *reference oscillator* must operate at the same frequency and with the same phase relationship to the colour signals as the subcarrier. For this reason a crystal-controlled oscillator is used, with a phase detector and reactance circuit which operate on similar principles to flywheel sync. circuits.

(6) The burst signal varies in phase by ± 45 degrees on alternate lines, resulting in a signal at 7.8 kHz (half line frequency) appearing in the output of the phase detector. This, the *ident signal*, is taken off by the 7.8 kHz acceptor filter and used to control the colour killer and PAL switch phase-inverter circuits.

(7) The colour killer circuit, controlled by the ident signal, is used to cut off the chroma signals during monochrome reception. This is highly desirable in a colour receiver since otherwise spurious signals entering the decoder may result in cross-colour (parc) patterns of colour on the screen during black and white reception. When the ident signal is not present, the colour killer cuts off the chroma amplifier.

(8) As in the PAL system the R–Y set of chroma signals is phase-reversed on alternate lines, it is necessary to incorporate a phase inverter (180 degree phase shift) controlled by a bistable switch which brings it into operation on alternate lines. The switch is triggered by pulses from the line timebase, and these are locked in phase by the 7.8 kHz ident signal. The PAL switch phase-inverter may, as shown, be included in the feed from the reference oscillator to the R–Y synchronous demodulator.

As in all electronic systems, alternative approaches to decoder design are possible. For example, in some designs the PAL switch inverter circuits are included in the R–Y signal path following the delay-line circuit instead of being in one of the reference oscillator outputs as in *Figure 5.23*, and considerable variety is possible in the design of colour killer circuits. A.C.C. (automatic chrominance control) is also found in practice. The basic requirements, however, are as outlined above.

A more detailed block schematic diagram of a PAL_D decoder is shown in *Figure 5.24*, in which the technique used in the delay line-signal separator network is more clearly indicated. As shown, the output of the final chroma amplifier stage is fed to adder and subtractor circuits and also via the delay line. In this way the B–Y and R–Y signals are separated and the averaging process carried out before the

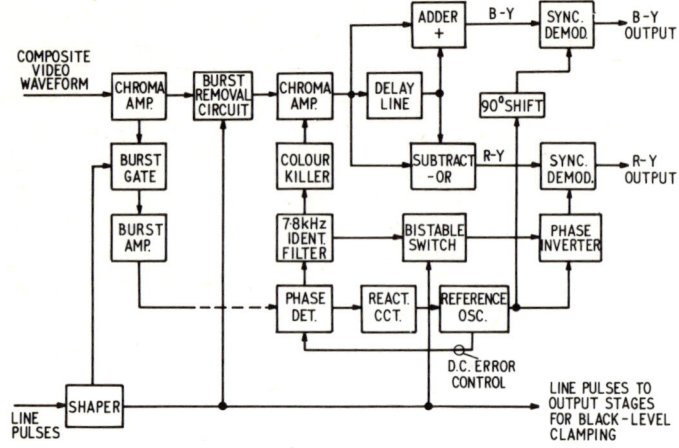

Figure 5.24 PAL_D decoder shown in more detail

B–Y and R–Y signals are applied to the synchronous demodulators. A burst signal removal circuit to remove the burst signal from the chroma channel is also required since the burst signal will otherwise interfere with the colour reproduction; this circuit is controlled by line flyback pulses which, as shown, are also used for black-level clamping in the colour output stages.

Delay line and R–Y, B–Y signal separation

The heart of the decoder in a PAL_D receiver is in the separation of the chroma signals impressed on the subcarrier and so processing them that the electrical average between each successive two lines is applied to the synchronous demodulators. This is achieved through the use of a delay line whose purpose is to make available at any given instant a

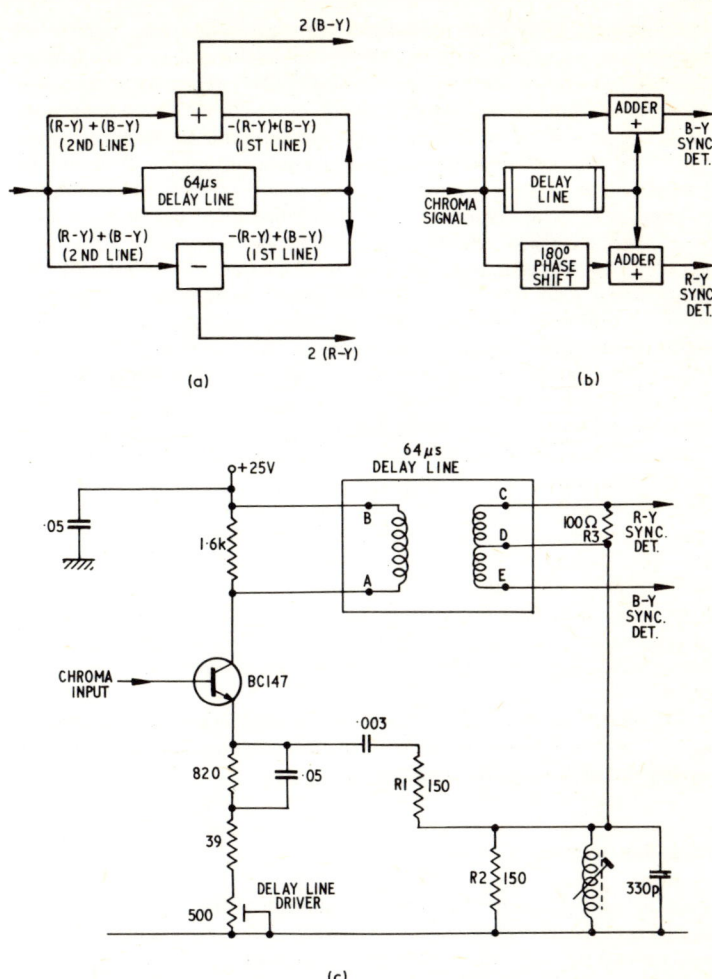

Figure 5.25 (a) Basic technique for using a delay line to provide electrical averaging. (b) Practical delay line demodulator technique. (c) Typical practical circuit using DL20 version of delay line

signal based on the line actually being transmitted and that of its predecessor. This involves delaying the signals by an exact line period, which for 625-line operation is 63·943 μs (63 943 ns), to an accuracy of the order of about ± 3 ns. If this can be achieved, we can recover both the $R - Y$ and $B - Y$ components by addition and subtraction of the direct and delayed signals, i.e. those passing around or through the delay line: see *Figure 5.25(a)*; we can also carry out the averaging process between successive lines. It should be noted that it is possible to subtract electrical signals by reversing one of them in phase (180 degree phase change) and then adding them (see *Figure 5.25(b)*). An arrangement such as that of *Figure 5.25(b)* can be achieved in a number of ways, and the early circuits used only passive components (e.g. phase-shifting transformers). A present method is indicated in *Figure 5.25(c)*. The important aspect of this arrangement is to remember that the chroma signal on the emitter of the transistor, TR1, is 180 degrees out of phase with that on the collector.

The signal enters the delay line at A with point B effectively at chassis (decoupled by 0·05 μF). At the output a 180 degree out of phase signal from the emitter of TR1 is effectively added to the delayed signal to produce R-Y and put through a further 180 degree phase shift at the output transformer (between connections D and E) to produce B-Y. Matrixing is determined by the values of R_1, R_2 and R_3, but as the exact balance of direct signal must be obtained against delayed signal in order to get pure R-Y and B-Y at the outputs, the gain of the driver transistor is varied by the 500 Ω emitter potentiometer.

This *delay line driver* control will hardly affect the direct signal level but will allow sufficient variation of the transistor gain for it to balance exactly the signal loss through the delay line.

Delay-line techniques

Since the delay line involves some attenuation of the signal, it is usual to feed the chroma signals into it via a *delay-line driver* amplifier, which may comprise an emitter-follower transistor to provide a low-impedance match. The delay line itself is conventionally of glass (steel has been used for this purpose in SECAM decoders where the tolerances are greater), with input and output ceramic transducers. In effect the input signals are converted into ultrasonic acoustic waves which pass through the glass, taking a finite time, and are then converted back into electrical signals. Typically, the special type of glass used for this application provides a path for the acoustic signals approximately 8 in long, but this is folded in order to make the component more compact. The transducers are usually barium titanate converters akin to those used in ceramic pick-ups. The delay line for a PAL_D receiver must be capable of passing the chroma signals to the necessary timing accuracy (within about 3 ns of the line period) over the chroma bandwidth throughout the temperature range likely to be encountered. This implies a bandwidth of roughly 2 MHz, a temperature range of 0–70°C and an insertion loss of some 10 dB into a 150 Ω impedance load.

Manufacturing tolerances can achieve this precision with little difficulty by accurate grinding of the folded length of glass.

Synchronous demodulators

Every N.T.S.C. or PAL decoder contains synchronous demodulators which are basically devices for obtaining time-sampled signals from the amplitude-modulated chroma sidebands. By feeding both the chroma sidebands and the local reference signal to the synchronous demodulator, the output with reference to the particular phase of the reference signal may be obtained. By feeding two synchronous demodulators with reference signals from the same carrier re-insertion oscillator but delaying one of the reference signals by a fixed number of degrees of phase (90 degrees), demodulation at the required axes can be obtained.

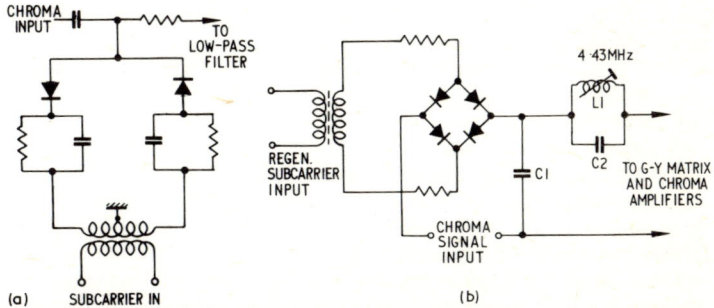

*Figure 5.26   Synchronous demodulators: (a) two diode sampler type,
(b) diode bridge*

Figure 5.26 shows several possible forms of synchronous de-modulator. A two diode sampler type is shown in (*a*), and a four diode balanced bridge type in (*b*).

90 degree phase shift networks

There are many possible ways in which a phase-shift of 90 degrees can be achieved in order to obtain the two frequencies with the necessary quadrature phase relationship to form the reference signals for the two synchronous demodulators. Typically, these may make use of the principle that in two loosely coupled resonant circuits the voltage induced in the secondary will lag 90 degrees behind that of the driven primary; alternatively, that if an r.f. signal is applied to resistance and reactance in series, the voltage at the junction will lead or lag in phase the applied voltage depending on whether the reactance is inductive or capacitive (a combination of these lags may be used to obtain a phase difference of 90 degrees in a low Q network).

This point is explained further a little later in consideration of the PAL switch requirements in this part of the circuit.

Crystal oscillator

The crystal oscillator provides an accurately controlled and stable sinewave output as a reference timing signal for the synchronous demodulators The crystal is equivalent to a resonant circuit of extremely high-Q characteristics, its resonant frequency being deter-mined by the physical constants of a small slice of piezoelectric quartz crystal vibrating between metallic plates. It will normally be supplied as a sealed unit at a frequency as close as possible to that of the subcarrier.

A crystal oscillator, although appreciably more stable than one using conventional inductor/capacitor combinations, could not by itself provide a reference signal of the correct phase and frequency, and is therefore 'pulled' into precise synchronism by means of the short bursts of sub-carrier radiated from the transmitter. This is normally done by means of a signal derived from the colour bursts acting on a phase discriminator in which the phase of the colour burst signals is compared with that of the crystal oscillator, and the difference used to obtain an 'error' signal which is amplified (by means of a d.c. amplifier) and used to reactance tune the oscillator: this may be done by utilising the voltage variable capacitance characteristics of a semiconductor diode (as in some automatic frequency control and tuning circuits in u.h.f. tuner units). A representative crystal oscillator circuit for a PAL decoder, controlled in this way by means of a variable capacitance diode (varactor), is shown in *Figure 5.27*.

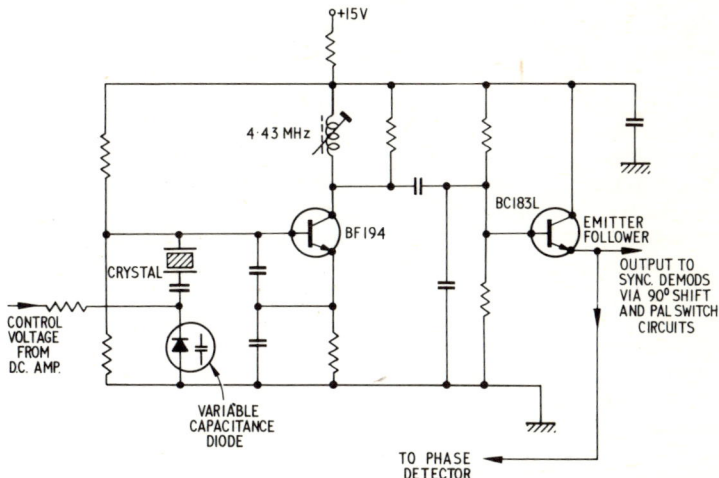

Figure 5.27   Reactance controller crystal reference oscillator ('subcarrier regenerator') with emitter follower buffer stage

An emitter-follower buffer stage will usually be connected between the crystal oscillator and its load to provide more constant loading on the oscillator.

Phase detector and ident filter

The phase detector obtains by comparing the colour burst signals with the reference oscillator output waveform a d.c. signal which after amplification is used to pull the crystal oscillator into precise synchronism (techniques have also been developed but not so far used, to allow the crystal to be replaced by an *LC* oscillator). Also, because of the phase 'swing' of the colour burst signal, a ripple at half-line frequency (about 7·8 kHz) can be recovered from this d.c. signal, and is fed to a high-Q coil so that this oscillates at half-line frequency. This sinusoidal waveform then has added to it pulses at line frequency derived from the line timebase; these pulses add to the peak-amplitude of the positive going 7·8 kHz signal, but are suppressed when the sinewave signal is negative going, providing us with an asymmetrical waveform having sharp positive going peaks at half line frequency and with its sense controlled by the swinging burst colour ident signals.

PAL switch

This waveform can then be used to control a bistable circuit, which is the basis of one type of PAL switch. A bistable circuit has two stable conditions, either conducting (on) or non-conducting (off), and can be switched from one to the other by means of trigger pulses. The basic bistable circuit is the Eccles-Jordan form of multivibrator, shown in *Figure 5.28*. The two transistors are triggered on and off repeatedly by means of a train of pulses derived from the line timebase, so that each transistor provides a square-wave pulse output. As in other forms of

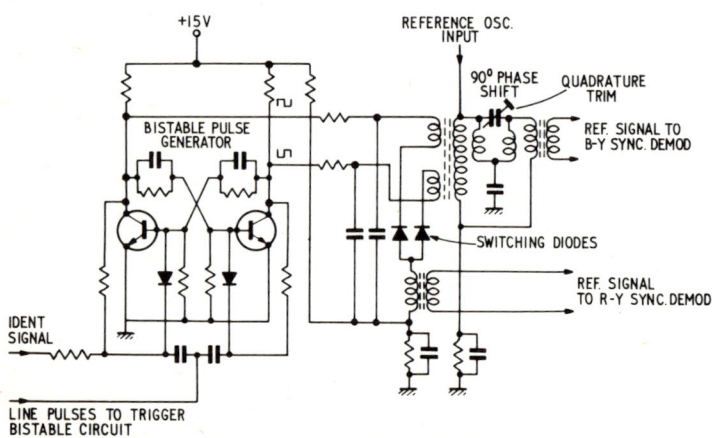

Figure 5.28   Simplified circuit of a bistable generator and electronic phase inverter switch for a PAL$_D$ decoder

multivibrator, one transistor is on when the other is off, and vice versa. Therefore the output waveforms, as shown, are of opposite polarity. The ident signal ensures that the triggering is synchronised with the alternate line phase reversal of the transmitted signal.

In the PAL switch the outputs of the bistable circuit are used in turn to control a pair of diode switches, which operate in conjunction with inductive networks to provide the required polarity switching. For those unfamiliar with the concept of using diodes as switches—a technique widely used in many branches of electronics—the following brief explanation is given. A diode can be used as a switch by making use of the low forward conductive resistance and the high resistance when the diode is reverse-biased; by suitably changing the bias by means of a switching voltage, the diode can be turned 'on' (i.e. presents very low resistance to the signal path) or by reversing its polarity the switching voltage turns the diode 'off' (i.e. presents very high resistance). A basic diode switching arrangement is shown in *Figure 5.29(b)*, and has the advantage over mechanical switches in that the switching is almost instantaneous and can thus be carried out at high speed; there is no tarnishing or other form of dirt; and no contacts to arc. The switching voltage is normally isolated from the circuit being switched by chokes or resistors.

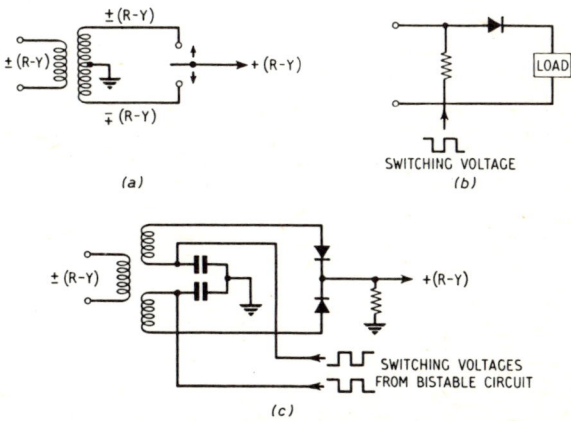

Figure 5.29   Principles of PAL electronic switching
(a) Switching principle, (b) simple diode switch, (c) two diode switches to provide a PAL switch.

In representative PAL decoders, two switching diodes are used, controlled by the bistable pulse generator, to form the electronic switch which inverts the polarity of the R–Y signal on the alternate lines so that it is always demodulated as though it were transmitted continuously with the same polarity, as in the N.T.S.C. system.

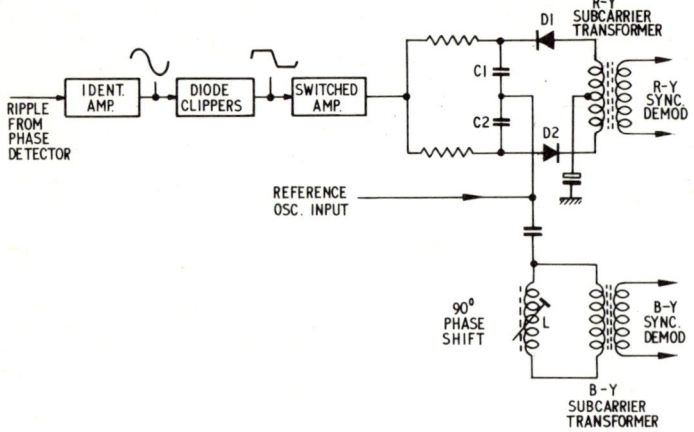

Figure 5.30   Alternative 180 degree PAL switch arrangement

A second type of PAL switch is indicated in *Figure 5.30*. This uses ident signal (7·8 kHz) which is amplified and the top and bottom peaks of the sine wave are clipped off using diodes to form *quasi*-square-waves.

Alternate half cycles of the clipped waveform will, after amplification, drive diodes D1 and D2 into conduction; so that on one half cycle reference subcarrier passes through C_1 and D1, and on the next half cycle reference subcarrier passes through C_2 and D2. Alternate ends of the R–Y subcarrier transformer are therefore used on alternate half-cycles of ident, creating a 180 degree phase shift—the PAL switch.

The B–Y subcarrier is fed at the constant 90 degree phase shift required, set up by L.

COLOUR DRIVE AND BLACK LEVEL CIRCUITS

Each of the three electron beams of the shadowmask tube must be modulated as a representation of the red, green and blue video signals.

This can be achieved with either $-R'$, $-G'$ and $-B'$ on the cathodes or, R', G' and B' on the grids of the tube. These are equivalent to one another. In the receiver we have demodulated $R'-Y'$, $B'-Y'$ and Y' signals directly available from the decoder.

We can therefore use these two difference signals to derive the third difference signal $G'-Y'$, and then add Y' to all three to produce R', G' and B' signals to directly drive the tube—RGB drive.

In colour-difference drive, the RGB signals are not produced external to the tube but in the tube itself. With $R'-Y'$, $G'-Y'$ and $B'-Y'$ on the grids, $-Y'$ is introduced at the cathodes. This $-Y'$ represents $+Y'$ at the grids so acting with the colour-difference signals there to produce the required R', G' and B'. Only one matrixing operation is then needed—that of deriving the $G'-Y'$ signal from the other two colour-difference signals.

RGB versus—colour difference drive

It is often thought that the essential differences between the two sorts of drive are external to the tube itself and that colour-difference drive simply replaces the external matrixing by electronic addition due to the tube action. This is unfortunately not the case. There is a large basic difference when driving the cathode or grid of any cathode-ray-tube and this applies equally well to the shadowmask tube.

There is first of all a difference in sensitivity in the tube: simple grid drive requires about 30 per cent more voltage than cathode drive to produce the same amount of light from the screen. This is due to the so-called first anode/cathode effect that is present in any thermionic device: a positive-going signal on the grid will not affect the first anode/cathode voltage directly—it will only act as a control on the flow of electrons leaving the cathode for the anode.

With a negative-going signal on the cathode, however, the first anode/cathode voltage *increases* as the video goes towards peak level. This effective increase gives a higher accelerating potential in the tube as the video level increases—giving not only modulation of the electron beam but greater sensitivity; the 30 per cent already mentioned. In fact, because of the standardisation that has taken place over the years with cathode drive in monochrome receivers, we normally only talk about cathode sensitivity.

A secondary symptom of the first anode/cathode effect is naturally that the effective transfer characteristics—i.e. the gammas—of grid and cathode drive are different. The degree of difference is not terribly important from the visual aspect but it does mean that in colour-difference drive receivers the matrixing in the picture tube can never be perfect; for example, $R'-Y'$ on the grid is subjected to a different gamma change compared to $-Y'$ on the cathode. The matrixing is therefore inexact, particularly in the mid-range of video signals.

White balance is not affected because this can be set up using peak drive amplitudes. Because the sensitivities of the three phosphors are not equal, different drive conditions are required on each gun. The degree of gamma distortion on each electron beam will therefore be slightly different and this means that the colour fidelity can never be perfect.

The distortions do not affect grey-scale tracking because on mono-chrome signals the colour-difference signals are zero and there is effectively only a cathode drive of luminance.

Additionally, the inter-electrode capacitances of the picture tube— between grid and cathode—change with different drive levels contributing to the matrix inaccuracy. This inaccuracy means that there is either additional luminance signal surplus to the matrix or chrominance surplus. The result is a form of channel crosstalk between the components. In practice, this is usually most visible on picture content with very low luminance levels—the appearance being of loss of the remaining luminance to produce just a colour-difference display.

These distortions can all be summarised by saying that the dynamic g_m of the picture tube with cathode drive is greater.

No such problems exist with RGB drive, but there are a rather different set of difficulties. As already pointed out, the colour-difference distortions do not upset the display grey-scale which to most viewers would be the most noticeable picture impairment. With RGB drive, any change in relative channel outputs will directly affect the white balance and the grey scale.

Also, the RGB channels must have the full video bandwidth (5·5 MHz) compared to the 1 MHz chroma passband required of colour-difference signals, and linearity of the three video channels must be accurately maintained to give reasonable grey-scale tracking. The tolerances in frequency response, gain and linearity must be extremely tight otherwise relatively large picture impairments result. The drive requirements are, however, rather less than those for colour-difference drives.

For example, for saturated blue, the drive voltage on the blue cathode might be 100 V.

Luminance: $Y' = 0.3R' + 0.59G' + 0.11B'$
$ = 0.3 (0) + 0.59 (0) + 0.11 (100)$
$Y' = 11 \text{ V}$

Therefore the colour-difference drive required will be:

$B'-Y' = 100 - 11$
$ = 89 \text{ V}$

Similarly for the complementary colour of blue (i.e. yellow) the $B'-Y'$ signal required would be -89 V. For grid drive the figures would have to be about 30 per cent greater, as we have already mentioned, so that the $B'-Y'$ output stage would be required to develop about ± 116 V; that is, a peak-to-peak range of some 232 V. We therefore require more than twice the amount of cathode drive.

For the other two channels the figures are not quite so excessive (182 V for $R'-Y'$ and 107 V for $G'-Y'$), but the requirements are always set by the worst. Simply from the drive level aspect it would therefore be valuable to be able to use RGB on the cathode; but with the original range of colour receivers this was not possible. At that time economy dictated that this kind of output level should be provided by valves. But, using valves the stability of channel performance cannot be readily maintained. Colour-difference drive was therefore a necessity.

Latterly, the availability of economic semiconductors makes it possible to transistorise the output stages in an RGB receiver. It is also easier to arrange duplication of performance between channels and this has been eased further by the development of more stable matrixing circuits in the form of integrated circuits (see Chapter 4)—leaving virtually only the output stages to be matched.

Similarly, of course, it would be possible to use $+R'$, $+G'$ and $+B'$ on the grids of a shadowmask tube. There would be no advantage in this over $-R'$, $-G'$ and $-B'$ driving the cathodes and there would be the disadvantage of requiring the extra 30 per cent drive level already mentioned, and the errors in gamma. Grid RGB drive has not been employed in any British commercial receiver design.

One additional advantage in not having any video circuits driving the grids is that any flashover in the tube is less likely to strike the cathodes rather than the grids, which will be at chassis or near chassis potential. Output device protection is therefore less necessary on transistors and any clamp diodes; sufficient protection may be assumed by the inclusion of a relatively small resistance in series with the amplifier feeds to the cathodes.

D.C. requirements of colour drives

One of the most repeated complaints about monochrome receiver quality over recent years has been the absence of any effective black-level control—preventing the increase and decrease in picture 'lift' that occurs with changing picture content when the signal passes through a.c. coupling circuits. The nature of this process and the basic details of correction were detailed in Chapter 2.

The effect of loss of black level on monochrome pictures can be very annoying, but in colour television it is a great deal worse. What might happen without any form of d.c. control would be that the average levels of all the signals, R', G', B' and Y', could all vary.

It is therefore quite probable that the amount of picture lift or sit on each channel would be different; in other words white light would be effectively added to or subtracted from each colour channel in the shadowmask tube. Varying the white content of colour information changes the saturation. Each channel will therefore have either greater or less saturation than was intended in the original scene.

There would therefore be a loss of colour fidelity; for example, let us imagine that the display is intended to be yellow (i.e. red + green) and that the red channel is more saturated than it should be, and that the green channel is less saturated than intended, both because of lack of d.c. control. The display is then in error due to the changes and in this case there will be a hue shift towards red—a reddish yellow.

Similar errors, which may be gross or only small depending on picture content, would occur on any displayed colour.

Colour-difference stages

The colour-difference feeds in a receiver are separate from any circuit involvement with the luminance feed to the cathodes, and we will therefore consider the problems separately.

First we must establish some idea of the range of voltages necessary for grid colour-difference drive; we have already noted these as $R'-Y' = 182$ V, $G'-Y' = 107$ V, $B'-Y' = 232$ V. But these figures assume equal sensitivity of each colour phosphor. In practice, although blue and green sensitivities are comparable, red sensitivity is lower and where 100 V might be required for red, the relevant figures for green and blue would be about 96 and 94 V.

Modification of our original figures with these balances, gives figures for the red, green and blue colour-difference signals of 182, 103 and 218 V respectively; these figures must not be considered absolute because there will always be some manufacturing spread in the picture tube.

These final voltages must be derived from the fairly low outputs of the standard bridge-type synchronous demodulators providing about 500 mV for R′–Y′ and 300 mV for B′–Y′. The ratio of these signals is, however, incorrect because the two signals are *weighted* at transmission to limit the overmodulation of fully saturated chrominance signals that would otherwise occur. Rather than restrict the amplitude of both colour-difference signals each is restricted only as necessary.

Because of the lower demodulator output on B′–Y′ and the higher drive required from it at the output, the gain of the B′–Y′ channel must be about 800 compared to about 400 in the R′–Y′ channel. Neither of these gains can be accomodated from a single stage with any stability, and it is usual to use a valve output amplifier with a single transistor preamplifier.

The G′–Y′ signal has to be derived from the two other colour-difference signals. From the luminance equation given earlier in the Chapter it can be shown that:

$$G'-Y' = -0 \cdot 51 \ (R'-Y') - 0 \cdot 186 (B'-Y')$$

So that in the G′–Y′ matrix we need about 37 units of B′–Y′ for every 100 units of R′–Y′. Taking into account the reduced level of demodulator output on B′–Y′ we will need a matrix where 66 units of B′–Y′ are taken for every 100 units of R′–Y′. This could mean a resistive matrix where the feed from R′–Y′ was, say, 330 Ω and that from B′–Y′ would be 470 Ω with some small balancing between them.

Practical circuits

These resistance values are used in the practical circuit for the G′–Y′ matrix shown in *Figure 5.31**. This circuit is to be found in the emitters of the two preamplifiers, and the matrix is followed by its own preamplifier to gain sufficient signal level to drive the G′–Y′ amplifier. A similar network would be employed in an RGB drive system but will normally be contained in the processing integrated circuit.

The derivation of the feed to the G′–Y′ matrix can be seen in the circuit of *Figure 5.32*. The whole of this B′–Y′ circuit would be repeated for the R′–Y′ channel and from point X for the G′–Y′ output.

The output colour-difference amplifier stages are usually clamped during line flyback period to provide correct zero value. The arrangement

*On the figures, primes have been omitted from R, G and B for simplicity.

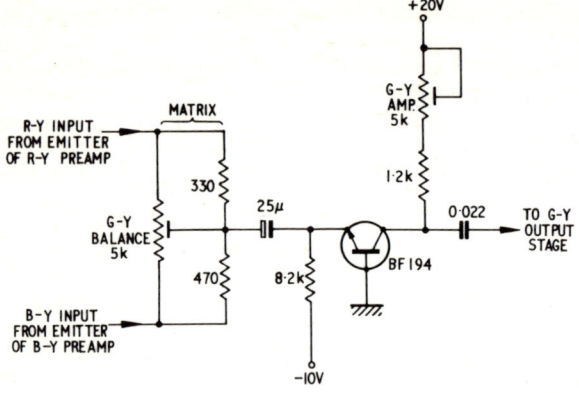

Figure 5.31 G'–Y' matrix and amplifier stage

The G'–Y' pre-amplifier operates in the common-base mode to maintain correct phase relationship between the three colour-difference signals.

shown is used in some decoders employing a triode–pentode in each colour difference output stage; the pentode is used as a cathode-compensated amplifier and the triode section as a driven clamp.

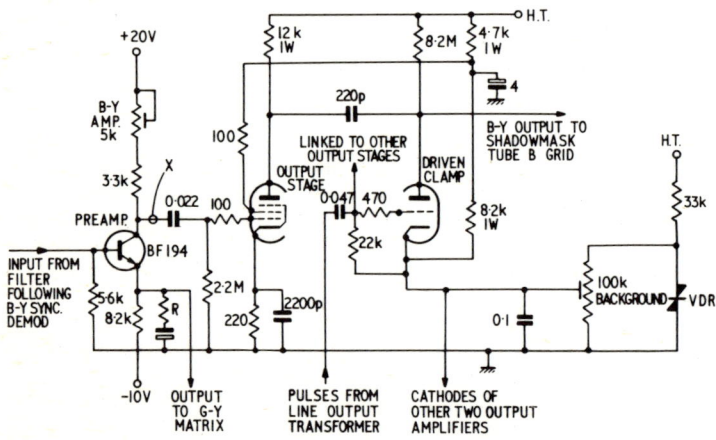

Figure 5.32 Representative colour-difference amplifier with driven clamp

The value of resistor R in the pre-amplifier circuit is selected to set the gain of the stage independently of the G′–Y′ matrix setting. The R′–Y′, B′–Y′ and G′–Y′ outputs must—as has already been noted—be preset to different levels because of the different efficiencies of the three types of phosphor used for the shadowmask tube screen.

The background control, effectively setting the output clamping level, sets the *lift* of the three chroma drives and is used during the set

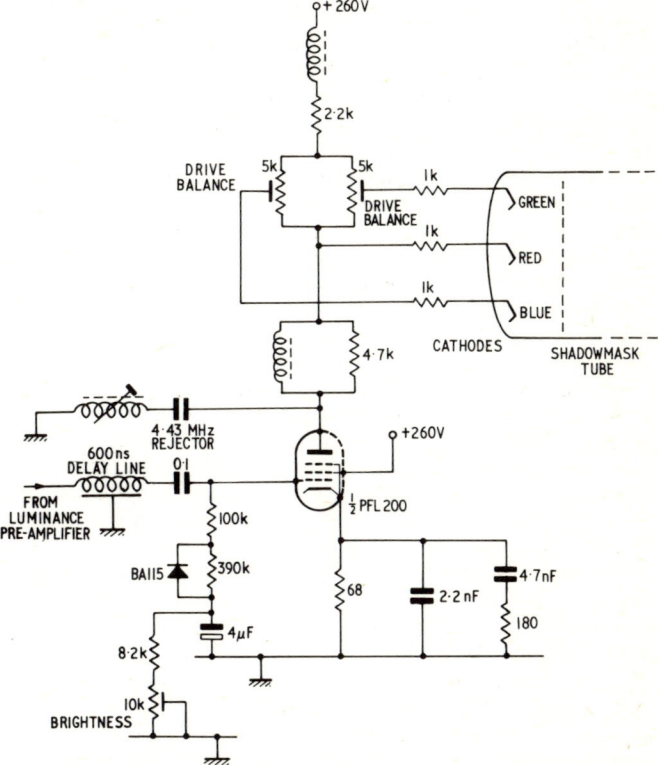

Figure 5.33 Representative luminance output stage for colour-difference drive receiver. (A flying-lead arrangement is normally provided so that any of the three cathodes can, if necessary, be driven by the maximum luminance signal. Red is normally the most insensitive)

up of the receiver to establish the correct tube cut-off point while keeping at least one of the screen potentials at maximum.

Simple resistive loading can be used in the colour-difference amplifiers because the stages need have only 1 MHz of bandwidth. The luminance output amplifier must, however, carry the full signal bandwidth and requires inductive peaking (*Figure 5.33*).

The drive from a preamplifier, which may also provide the feed to the sync. separator, passes through the luminance delay line which in the majority of receivers is 600 ns. This small delay, to balance the effect of the bandwidth restricted chroma channels, is created by a wire-wound coil around a longish former with a number of capacitive elements formed with respect to chassis.

In the example shown, the d.c. conditions of the luminance drive are maintained by d.c. restoration before amplification by the elements in the grid circuit. The reference potential is made variable so creating a brightness control for the receiver. The anode of the valve has a 4·43 MHz rejector connected to remove any possible interference effects that might be created on the pictures; there may also be a rejector in the preceding preamplifier stage.

The difference sensitivities of the phosphors in the tube demand that the luminance level fed to each cathode is slightly different. One cathode is fed direct and this will be the most insensitive—normally the red. The other two cathodes are fed from the two drive balance controls, a setting of which represents a smaller anode load and there-fore a lower output. The 1 k resistors in the three feeds to the cathodes are to limit voltage on the output valve should there be a tube flashover.

RGB driven receivers usually derive their R′, G′ and B′ signals inside a silicon integrated circuit (s.i.c.) as indicated in Chapter 4. There are, however, a few receivers which use RGB drive but whose design shortly pre-empted this s.i.c. development. These use discrete components for the matrixing of G′–Y′ and the addition of Y′ to all the signals, and one particular design performs both operations in a single resistive circuit.

A fairly representative design for a chroma output amplifier is shown in *Figure 5.34*. Only one channel is given, the other two channels being identical. The input, assumed to be from an s.i.c., passes through the emitter follower transistor TR1 to directly couple into the output transistor TR2. By careful layout of the stage and control of the driving impedance (by use of the emitter follower) no frequency correction circuitry is necessary to gain the required 5·5 MHz band-width.

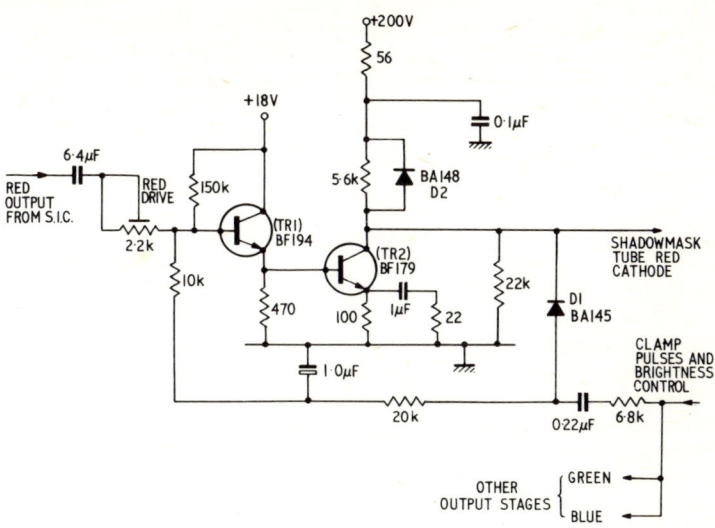

Figure 5.34   Representative RGB output circuit: only red channel shown, other two channels identical

D.C. clamping in this circuit is by means of a feedback arrangement. The clamp pulses derived from the line output stage and varied in level by the brightness control, are passed to the clamping diode D1. This compares the pulse amplitude with the reference period amplitude on the signal and any error is fed, as a d.c. error, back to the base of the emitter follower transistor, TR1. This system controls the d.c. level through the amplifier without directly placing any unwelcome, low impedance across the output load of the transistor amplifier, TR2. With an alteration in the clamp pulse amplitude caused by the *brightness* potentiometer the d.c. reference is changed and the amplitude of the d.c. error is different. As all three stages use the same reference, the overall lift or brightness of the picture is altered.

The diode D2 in the collector circuit of TR2 is a protection diode in the event of a flashover to the c.r.t. cathode.

6

VALVE TIMEBASE FAULT FINDING
by Gordon J. King

This chapter is intended to offer general guidance on the difficult fault finding area of valve timebases in 405-line and dual-standard receivers. This material should be used in conjunction with the circuit details of Chapter 3 and the general fault finding material of Chapter 9 for work on transistor timebases.

The purpose of the timebases is to deflect the electron beam in the picture tube, and hence the scanning spot on the screen, from top to bottom at a rate of 50 times per second and from left to right at 10 125 times per second for 405 lines and 15 625 times per second for 625 lines. The timebases thus have to generate signals of suitable form and frequency to drive amplifiers the anode circuits of which are loaded to the scanning coils which produce the magnetic scanning fields. There are two basic stages to a timebase, therefore, the generator which produces the driving signal and the amplifier which translates this signal to scanning-coil current.

The timebase for vertical deflection is called the *field* and that for horizontal deflection the *line*. The current in the scanning coils is of sawtooth nature, so that the rising part of the waveform deflects the beam linearly from top to bottom and from left to right, giving the scanning stroke, while the falling part gives the retrace or flyback stroke, which occurs in a very much shorter period of time than the scanning stroke, as shown in *Figure 6.1*.

The action of the two timebases creates the raster on the screen which is composed of the number of lines determined by the line repetition frequency. The number of lines is fixed by the nature of the transmitted signal, since the timebases are locked frequency-wise to the sync. pulses of the signal. The 50 Hz field frequency has been used in Great Britain since this corresponds to the frequency of the mains system and originally eased problems of hum in the vision circuits of the

220

receiver. This is no longer the case because the tight tolerances required of colour subcarrier generation mean that field frequency must be derived from this source and not the mains.

Thus, on both 405- and 625-lines, the same field frequency is used, the number of lines being altered by a change in frequency of the line timebase. This means that at least one section of the changeover switch in dual-standard sets is concerned with changing the line repetition frequency to the other.

Although the field repetition frequency is 50 Hz, only half this number of complete pictures is produced each second. This is because one complete picture is composed of two fields with their lines inter-laced. Thus, each field contains half the number of lines of a complete picture—$202\frac{1}{2}$ lines for 405-line pictures and $312\frac{1}{2}$ lines for 625-line pictures. Clearly, if the line frequency is 10 125 Hz and the field frequency 50 Hz, the number of lines produced by this combination is 10 125/50, or $202\frac{1}{2}$. With the same field frequency and a line frequency of 15 625 Hz, the number of lines is 15 625/50, or $312\frac{1}{2}$. Note also that because of the time required for the insertion of field-sync. pulses, in the 405-line system there are only 377 lines of picture information and in the 625-line system 575 lines of picture information.

The interlacing of the lines of one field with those of a subsequent field is controlled by the line- and field-sync. pulses of the composite video signal, and their timing relative to the picture parts of the signal.

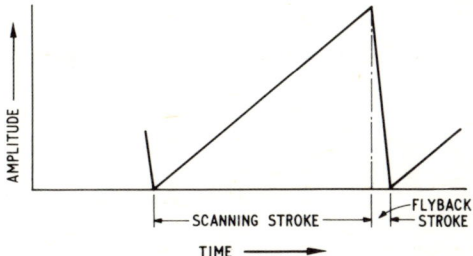

Figure 6.1   Sawtooth waveform providing the scanning and flyback strokes

These pulses also serve to time or lock the line and field generators. The pulses, in fact, instigate the retrace (flyback) strokes of the timebases.

There have been several circuits used in the types of receiver we are considering in this chapter for the generation of line and field signals. These include the blocking oscillator, multivibrator and sinewave generator with suitable networks for shaping the signals so that they represent the correct drive for the line or field amplifier. The blocking oscillator has been popular since the early days of television, and is still used extensively in today's sets. A typical blocking oscillator circuit is shown in *Figure 6.2*.

It is necessary, of course, to have some idea how this works before faults developing in it can be localised and cured. The fundamental operation of the circuit is explained below.

Blocking oscillator

The master component is the blocking-oscillator transformer T1. All blocking oscillators have this, which is a quick way of identifying them, as the transformer has the appearance of a miniature speaker transformer. The primary of T1 is connected in series with the valve anode, while the secondary connects to the grid circuit via C_1, R_1 and associated resistors.

Let us suppose that the valve is commencing to pass anode current. As this flows through the primary, an e.m.f. is induced in the secondary of polarity such that the grid is made positive with respect to cathode.

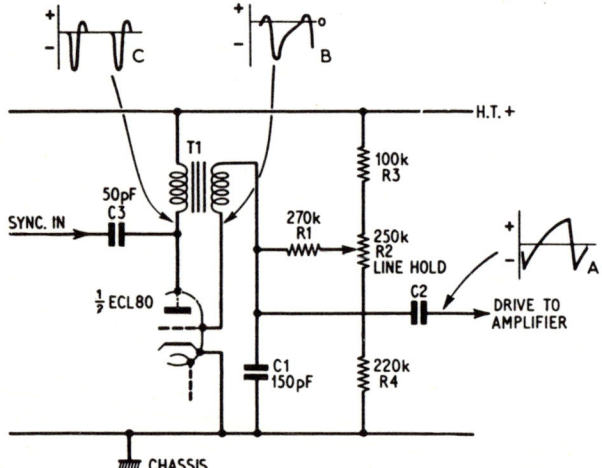

Figure 6.2   Practical blocking oscillator circuit showing waveforms

The grid/cathode circuit of the valve then acts like a diode and C_1 charges with its upper plate negative. The grid is still a little positive, however, due to the swing of voltage across T1 secondary.

Eventually, the rise in anode current ceases and the e.m.f. in the secondary of T1 collapses, which results in the grid going negative. This causes the anode current and the current in T1 primary to fall, an action which very rapidly makes the grid go further negative and blocks the valve by cutting off anode current completely.

At this stage, the grid is held negative and the valve at cut-off by the charge on C_1. This charge is not retained, but gradually leaks away through R_1, R_2, R_3 and R_4, and it is also countered by the positive potential at the slider of the line-hold control R_2. When sufficient charge has leaked away to exhaust the negative blocking voltage at the grid, anode current again commences to flow through the primary of T1, and the action described is repeated.

It will be understood that this action results initially in a fairly high charge across C_1, which then leaks away relatively slowly when the valve is blocked. A waveform similar to that shown at A on the circuit thus appears across C_1 and, in this particular circuit, this waveform represents the drive signal for the amplifier.

Indeed, in many generators the drive waveform is derived from the charging or dicharging of a capacitor. As waveform A shows, the discharging of a capacitor through a resistor gives a near sawtooth wave. A similar wave is produced by the charging of a capacitor through a resistor. Both methods are used, with the valve producing a very rapid charge or discharge for the retrace.

The charging or discharging period through a resistor, giving the scanning stroke, is not very linear, but is exponential, which is the way that a capacitor charges or discharges. By using only a small portion at the beginning of the charge or discharge curve, however, linearity is improved, and this is arranged by the choice of components and valve operating conditions. B and C in *Figure 6.2* show the waveforms at the grid and anode of the valve.

It will be understood that the repetition frequency of the blocking oscillator is governed by the time that the C_1 takes to discharge. The frequency can thus be adjusted either by varying the value of the discharging resistor or by varying the value of the countering positive voltage from the h.t. line. If the resistance is reduced or the positive voltage increased in value, then the frequency is increased. The value of C_1, of course, is also a prime timing element, since the time-constant is the product of C and R.

In practical circuits, the time-constant components are chosen to give the nominal generator frequency with the appropriate hold control at range centre. The frequency can thus be increased or decreased by turning the control either side of centre. The circuit in *Figure 6.2* is designed for producing a line signal at 10 125 Hz, and the line sync. pulses, for instigating the 'firing' of the oscillator, are applied to the anode through C_3.

Blocking oscillator faults

Complete failure of any generator will, of course, cut off drive to the output stage and destroy beam deflection. In the field timebase, this will give rise to a bright, horizontal line across the screen, and when testing the intensity of this display should be reduced as much as possible to avoid burning the screen.

Failure of the line timebase does not usually cause a vertical line to appear on the screen for the reason that this timebase plays a major role in the generation of the e.h.t. voltage for the picture tube. Thus, when the line timebase fails an accompanying symptom is nearly always that of e.h.t. failure. But more will be said about this later.

Complete failure of a blocking oscillator is not difficult to diagnose, but it is first necessary to make sure that failure is in the generator and not the amplifier or coupling components. The quickest way of discovering this is to employ an oscilloscope for displaying the generator waveforms. If these are present in the amplitudes expected or as given on the circuit or in the service manual or sheet, then the amplifier section is responsible for the symptom related to complete breakdown.

A fairly quick diagnosis of whether it is the generator or the output stage that has failed can often be made on the field side by a close examination of the collapsed field; if there is a *single line* across the screen then almost certainly the output stage is inoperative. If that stage is operating there is usually some 'noise' in the stage causing a *bounce* on the collapsed line. Caution must obviously be taken in interpreting such a diagnosis.

A common cause of failure in a circuit such as that in *Figure 6.2* is open-circuit of the primary of T1. This can quickly be checked, however, by measuring the voltage at the valve anode. Lack of voltage here, but normal voltage at the h.t. side of the winding, is a clear indication of the trouble.

Open-circuit of the secondary or grid winding may not be so obvious, and it is necessary here to check the winding resistance against that given in the service data. The normal secondary resistance is about 100 Ω or so, the primary being a little below this, sometimes about 70 Ω, depending on the circuit and whether the transformer is line or field.

The secondary sometimes goes high in value due to a bad connection of the termination to a tag or due to deterioration of the wire in the winding. In the latter case, the winding resistance can often be varied by applying pressure to the outside of the winding with the flat of a screwdriver while the ohmmeter is connected. Should this happen, the transformer must be replaced.

Complete failure can also be caused by valve trouble, of course. In many of these receivers the blocking oscillator is the triode section of a triode-pentode valve, with the pentode serving either as field amplifier or sync. separator. If possible, the best idea is to check the suspect valve by substituting with a like valve known to be in good order.

In field generators, the supply may be fed in through the *height* control, usually from the boost h.t. line rather than the ordinary h.t. supply line. Voltage stabilisation may also be found at the generator h.t. input, as shown in *Figure 6.3*. This means that trouble anywhere in

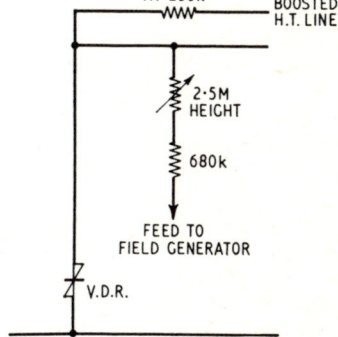

Figure 6.3   Method of stabilising the field timebase

the h.t. supply and associated components could mute the generator, but a few checks with a high-resistance voltmeter should soon bring to light a shorting capacitor or an open-circuit resistor or inductor.

Voltage stabilisation is provided in *Figure 6.3* by the voltage-dependent resistor (v.d.r.), the resistance of which decreases with increase in voltage across it. This forms a potential-divider with R_1, so if the voltage drops the resistance of the v.d.r. rises, and as this forms the bottom leg of the divider, the voltage to the generator, via the height control, remains reasonably stable.

An open-circuit height control will cut off the supply voltage and mute the generator, as also can an open-circuit vertical hold control. These are vulnerable components, especially the former, which, in modern versions of this circuit, is often valued in terms of megohms.

Rolling picture

Although the time-constant components in *Figure 6.2* relate to the line frequency, the generator as a whole differs little from that used in a field circuit, so alteration in value of C_1 or R_1 will affect the repetition frequency and sometimes put the correct frequency outside the range of the hold control, making it impossible to 'lock' the picture on the screen. Field-wise, therefore, the picture will roll up or down the screen, often slowing down with the hold control hard against one of its stops. The time-constant resistor is mostly to blame, this often increasing considerably from its nominal value. However, the capacitor itself should not be overlooked. Change in characteristic of the generator valve (or low emission) can also affect the frequency, and sometimes when a generator valve is replaced it is necessary also to 'pad' the time-constant resistor to secure lock towards range-centre of the control. Some circuits allow for this by having two series resistors in the time-constant circuit, one of which can be shorted out if necessary.

The picture will also fail to lock, of course, should the sync. pulses be missing, distorted or attenuated. In this case, though, there will be no tendency towards picture lock at all, and by very critically adjusting the appropriate hold control it may be possible to hold the picture on the screen for a second or two.

In *Figure 6.2* the sync. pulses are applied through C_3. These can be observed on an oscilloscope by disconnecting C_3 from the anode and connecting the Y-input to the free end of C_3. To avoid misleading displays, it is a good idea to mute the field generator when the field sync. pulses are being viewed and the line generator when the line pulses are under examination.

Line breakup

What has been said so far about generator frequency applies both to field and line circuits, but from the line point of view the picture will break up horizontally when the line frequency is too far removed from the line-sync. repetition frequency. If the line-hold control is at the end of its travel, attention should be given to the line-generator valve, the time-constant resistor (and any series resistors to the hold control) and the time-constant capacitor.

Multivibrator

Another popular generator is the multivibrator, as already mentioned. This is found in various configurations, but they all have one thing in common and that is they all use two valve sections (usually two triodes, and sometimes the two triodes in a single double-triode valve), the output of the second being coupled back to the input of the first and the output of the first being coupled to the input of the second. This

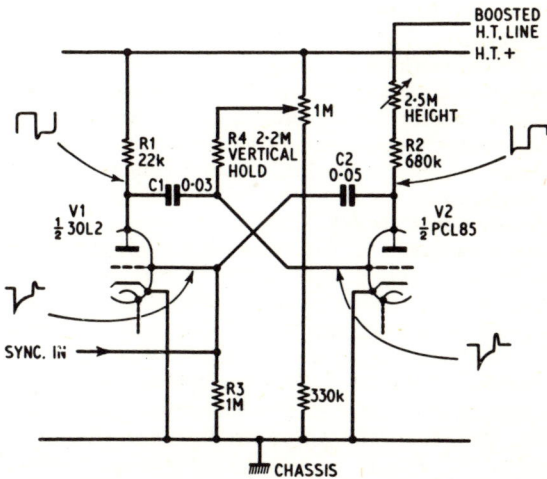

Figure 6.4   Practical multivibrator circuit, showing waveforms

cross-coupling introduces regenerative feedback and causes the stage to oscillate.

In practical circuits, one valve section may be a pentode, and the coupling may be by two capacitors connected between the grids and anodes or by a common cathode coupling arrangement. A cross-coupled field multivibrator-type generator is given in *Figure 6.4*, and the following explains briefly how it works.

When the valves V1 and V2 start to pass anode current, voltage is dropped across the anode resistors R_1 and R_2. It is impossible for absolute balance to be attained, so the voltage drops will be unequal, meaning that one grid, as coupled from the anodes to the grids, will go more negative than the other.

The grid that is more negative becomes quickly biased into cut-off due to the mutual amplification between the two valves as the result of their cross-coupling. This means that the grid of the other valve has a positive voltage communicated to it via the coupling capacitor. Thus, one valve goes into cut-off and the other into full conduction, and these conditions alternate between the two valves at a rate governed by the circuit time-constants.

Let us suppose that V2 is conducting and V1 cut-off. At this time, C_2 is charged and commences to discharge through R_2 and the height control and through the grid resistor R_3. Eventually, the negative charge on C_2 leaks away sufficiently to cause V1 to pass anode current and this is communicated to V2 grid as a negative pulse, via C_1, owing to the resulting volts drop across R_1 (the anode load of V1). This negative pulse is quickly amplified by the cumulative action of the two coupled valves and V2 is pushed hard into cut-off. This reflects a positive pulse, via C_2 to V1 grid, rapidly bringing this valve out of cut-off into conduction. This time C_1—which is now charged—discharges through R_4 and R_1.

The switching action is thus controlled in frequency by the discharging time of C_1 via R_1 and R_4, etc., and C_2 via R_3 and R_2, etc. The frequency of the generator is rendered variable by making the discharge time of C_1 adjustable. This is achieved in a similar manner to that described for the blocking oscillator by applying a countering positive potential from the slider of the hold control to the negatively charged plate of the capacitor. Thus, the more positive the potential reflected to this, the shorter will be the discharge time, and the higher the frequency.

The discharge time of C_2 comes under the control of the sync. pulses to some extent, and this is how the generator is 'locked' to the signal.

Multivibrator faults

Clearly, then, alteration in value of any component likely to influence the discharge time of the capacitors will upset the frequency of the generator and be likely to put the locking point outside the range of the hold control. R_4 and R_3 are the most 'sensitive' resistors in this respect, and also the capacitors, of course, but in service these seem to be less troubled by value change than the resistors. Change in the characteristics of the valves is another common cause of frequency error and, again, substitution tests are recommended.

Complete failure can only be caused by valve trouble or by serious value change (or open or short circuit) in an associated component. Voltage checks soon show up defects of this nature, but the voltmeter must have a fairly high resistance, not less than 20 000 Ω/V.

An oscilloscope can prove very useful for checking in multivibrator circuits, and waveforms expected at various points are shown on the circuit in *Figure 6.4*.

The drive signal is usually taken from the anode of the second valve and this is an approximate square-wave as the waveform shows. A sawtooth waveform, however, can easily be obtained by the use of an extra capacitor connected between the anode of the second valve and chassis, as shown in *Figure 6.5*.

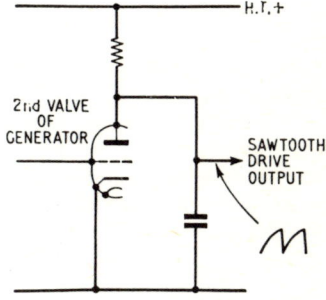

Figure 6.5   Obtaining a sawtooth output across a charging capacitor

This capacitor charges through the anode resistor when the valve is cut-off and thus gives a rising, approximate sawtooth voltage across it. The charge is rapidly expelled when the valve conducts, and this gives the flyback stroke (see *Figure 6.1*). This arrangement is also sometimes used with blocking oscillators to secure a greater amplitude signal than is possible from the grid-charging capacitor alone.

Ideally, the capacitor is allowed to charge only to a very small percentage of its full charge before discharge occurs. This ensures that the most linear portion of the charging cycle is used for the scanning stroke.

Figure 6.6 shows the basic cathode-coupled multivibrator with the charging capacitor to provide the sawtooth drive waveform. The action of this type of generator is similar to that of the symmetrical arrangement, one valve being cut-off while the other is fully conductive, the conditions alternating as before.

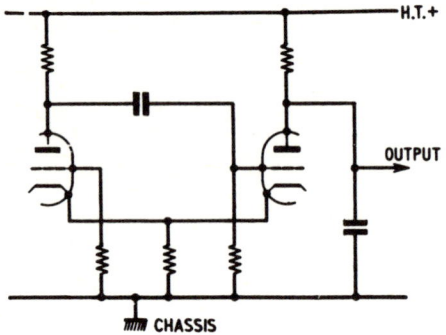

Figure 6.6   Basic cathode-coupled multivibrator

In some field timebases, the amplifier or output valve is sometimes arranged to act also as the second valve of the multivibrator pair, but it is more usual to find the triode of one triode–pentode as one section of the multivibrator with this valve's pentode as sync. separator (or some other service) and the triode of a further triode–pentode valve as the second multivibrator section, with this valve's pentode as the field amplifier.

Dual-standard line generator with flywheel control

In dual-standard receivers a blocking oscillator has often been employed as the line generator with switching for the two line speeds, as shown in *Figure 6.7*. Note here that the positive potential for timing the discharge of the grid capacitor is derived from the anode of a triode control valve, more about which will be said shortly; also that the output drive signal is obtained from across a capacitor which charges via R_1 from the h.t. line and discharges when the generator 'fires' to give the effect already explained relative to *Figure 6.5*.

Switch S1 is the line-speed-change section of the changeover switch, and in the 405-line position, the time-constant is made a longer (for the lower frequency) by R_2 in series with the 405-line hold control. Thus, this control can be set independently of the 625-line hold control, which comes into operation without a series resistor, thereby giving the higher frequency.

Now, the potential at the anode of the control valve is set by the current in the anode resistors and hence by the bias on the valve to provide the correct line frequency when the line-hold controls are set approximately to the middle of their ranges.

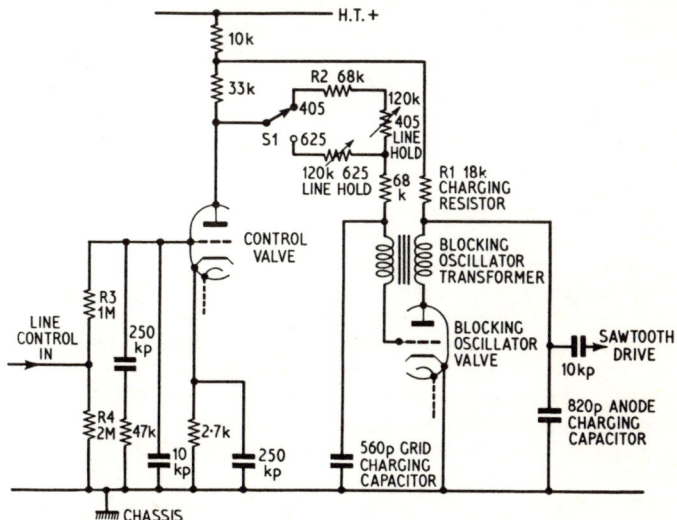

Figure 6.7   Practical dual-standard line generator using a controlled blocking oscillator circuit

If the line frequency is set correctly in that manner it can be altered up or down by varying the bias on the control valve, because this action will change the anode current and consequently the voltage at the anode, due to the corresponding change in volt drop across the anode resistors.

The control valve is statically biased by the cathode resistor and can also be biased about this static value by the application of a control voltage to the grid, this, in fact, being applied via the potential-divider R_3/R_4. This control voltage is derived from a discriminator circuit consisting of a pair of diodes which compares the line-sync. pulses with sample pulses obtained from the line-amplifier section. Of course, when the frequency and phase of these two sets of pulses coincide, the generator frequency will exactly match that of the sync. pulse repetition frequency. Under that condition, the picture would be locked correctly on the screen.

Thus, the output of the discriminator under that condition would be zero. However, should the sampled line pulses tend to drift relative to the sync. pulses, the discriminator delivers a plus or minus output voltage, depending on whether the generator is tending towards increasing or decreasing frequency. This output is the control voltage which is applied to the grid of the control valve in *Figure 6.7*. This influences the valve's conductivity, as already explained, and adjusts the generator frequency so that it is pulled back accurately in step with the sync. pulse repetition frequency.

Flywheel-sync. discriminator

A circuit of a typical line discriminator is given in *Figure 6.8*. The 47 k resistor in parallel with the 15 pF capacitor in the sample-pulse feed to the discriminator times the sample pulses so that they correspond to the edge of the sync. pulses as applied to the discriminator through the 73 pF capacitor.

To avoid the line generator falling out of sync. should the line-sync. pulses fail momentarily or should they be badly affected by interference, a long time-constant is given to the control voltage feed to the control valve. This time-constant is provided by the R and C elements on the grid of the control valve in *Figure 6.7*. The effect is that the control voltage remains at the value for correct lock for a period even though the control voltage from the discriminator is falling due to sync. pulse failure on the transmission.

This is often called the *flywheel effect*, and the circuit as a whole, including the discriminator and generator, is known as flywheel-controlled line sync., to distinguish it from direct sync., where the line-sync. pulses from the sync. separator are applied direct to the generator, via a pulse-shaping or differentiating circuit.

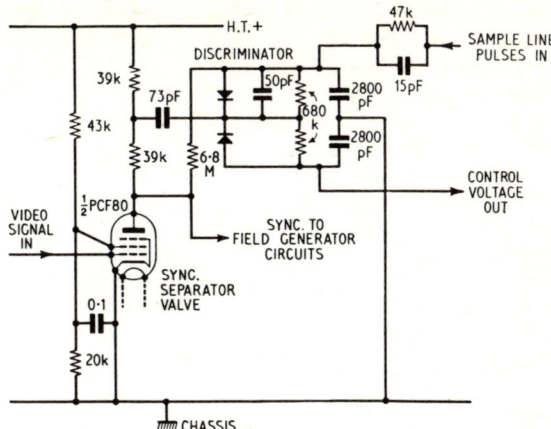

Figure 6.8   Sync. separator stage feeding discriminator circuit to provide a control voltage for a circuit such as that in Figure 6.7

Flywheel-controlled line circuits are nearly always featured in dual-standard sets, either on the 625-line standard alone or on both standards, and some 405-line-only sets of earlier vintage embody a similar circuit.

Effect of flywheel-controlled sync.

The great advantage of flywheel-controlled line sync. is that distortion or interference on the sync. pulses does not easily disturb the operation of the line generator. In severe cases of interference, receivers using this system exhibit only a slight horizontal waving of the picture, while receivers using direct sync. are often disturbed to the extent of horizontal tearing of the picture and very ragged edges to vertical picture content; these effects being caused by irregular 'firing' of the line generator due to interference pulses on the line-sync. pulses.

The trouble is aggravated in areas of weak signal and high local interference, for then the 'noise' of the tuner can itself influence the line generator synchronisation, which is the reason why some early sets designed for 'fringe area' operation incorporated flywheel-controlled line sync.

The reason for its revival on the 625-line standard is related to the higher scanning speed and energy of the line generator and to the fact that on the negative-going picture modulation of the 625-line standard interference pulses and noise signals rise in the same direction as the positive-going sync. pulses, making it that much more difficult to discriminate between the line sync. pulses proper and noise and interference signals in direct line sync. circuits.

The flywheel effect is sometimes given by the use of inductors in the anode circuit of the line generator. Some Pye models, for instance, used a pair of switched inductors in series with the anode circuit of the first triode of the line multivibrator. One inductor is tuned (with its stray and circuit capacitances) to 10 125 Hz on 405 lines and the other to 15 625 Hz on 625 lines. The oscillatory effect provided by these inductors tends to hold the generator in sync. for brief periods during

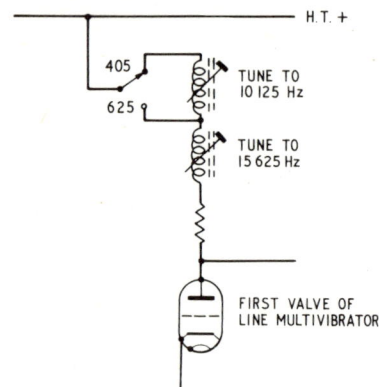

Figure 6.9 The 'flywheel' effect is sometimes achieved by loading tuned circuits in the anode of the line generator, as shown
On 405 lines the two inductors are in series, tuning to 10 125 Hz, while on 625 lines the bottom inductor only is used, tuning to 15 625 Hz.

bursts of interference. The switched inductors are shown in *Figure 6.9*, the switch section being in addition to that required for generator frequency change, and the generator—in spite of it being a multi-vibrator—is held in lock by a control voltage from a discriminator similar to that shown in *Figure 6.8*.

Sinewave line generator

A sinewave oscillator is sometimes used to provide the line signal, its waveform being shaped by an *RC* network in the coupling to the line amplifier so that it resembles more of a sawtooth than a sinewave. One advantage in the use of a sinewave oscillator is that its frequency can be controlled by an electronic reactance connected across the tuned circuit.

An electronic reactance is basically a triode valve with an *RC* or *RL* phasing circuit connected to it so that to an oscillator it appears either like a capacitance or inductance, depending on the nature of the circuit. The value of its reflected reactance can be varied by altering the grid bias, which means that the oscillator frequency can be controlled by the correction voltage delivered by a discriminator. A typical sinewave line generator, reactance valve and discriminator flywheel-controlled line circuit is given in *Figure 6.10*.

Here V2 is the sinewave oscillator designed to run nominally at 10 125 Hz. Feedback is between L_1 and L_2 (cathode and control grid) L_1 with its parallel capacitor is the main tuned circuit. Now, in parallel with this tuned circuit is the reactance valve V1. This can quickly be identified by the capacitor connected between control grid and anode. It is this capacitor which endows the stage with the effect of capacitive reactance.

The discriminator contains the two diode sections of V3, the oscillatory signal being coupled to their cathodes by L_3. Line sync. pulses from the sync. separator are fed to the centre tap of L_3, via the 50 pF coupling capacitor, and when the phase and frequency of the sync. pulses coincide with those of the oscillatory signal the discriminator gives zero output.

However, should the oscillatory signal drift in frequency, a positive or negative control voltage is delivered by the discriminator, and this is fed to the control grid of the reactance valve, via the time-constant (470 k in parallel with 0·02 μF). This control voltage changes the effective capacitance across L_1 and thus alters the oscillator frequency, bringing it in step with the sync. pulse repetition frequency. A manual control of frequency is provided by the line lock control, which adjusts the bias on the control grid of the reactance valve.

The time-constant in the control bias feed contributes towards the circuit's flywheel effect, and this is further assisted by the resistor and capacitor series combination in the control grid circuit of V1.

Phasing adjustment

The oscillator can be adjusted to the correct frequency by the core in L_1/L_2, and this would be done to lock the picture horizontally with the line lock control set to range-centre. However, it may now be found that the picture is displaced to the left or right in the *raster*. This means that the phasing of the two discriminator signals is incorrect. This can be corrected by adjusting the core in L_3 until the picture becomes central within the raster.

This misphasing symptom should not be mistaken for incorrect adjustment to the picture shift magnets on the tube neck. In this case, the whole raster is displaced, not just the picture within the raster; this can be checked by increasing the brightness sufficiently to be able to see the line blanking edges—with the picture width reduced.

Faults in flywheel-controlled circuits

Flywheel-controlled circuits are more fault-prone than the straight-forward direct-sync. circuits, and technicians in the early days often used to modify flywheel circuits to direct sync. to minimise the fault potential. Modern circuits, however, are far more reliable and such modification should never be necessary.

Circuits employing small metal rectifiers in the discriminator can cause trouble by one of the rectifiers deteriorating causing unbalance and consequent poor locking conditions. A typical symptom is the line swinging out of lock at the least provocation, such as on a burst of electrical interference or changes in picture content or scene. If there is any doubt about the condition of the metal rectifiers or semiconductor diodes, it can save much time and avoid later service calls to replace both components while the set is under attention: the diodes should always be replaced as a pair.

The discriminator diode load resistors should also be checked for balance, and leakage in a coupling capacitor can upset the rectifier operating conditions.

Unstable line lock

Another frequent cause of trouble is drift in the characteristics of the control valve or emission increase as the valve temperature rises (typical of an ageing valve). The best plan if the line lock is unstable is to

237

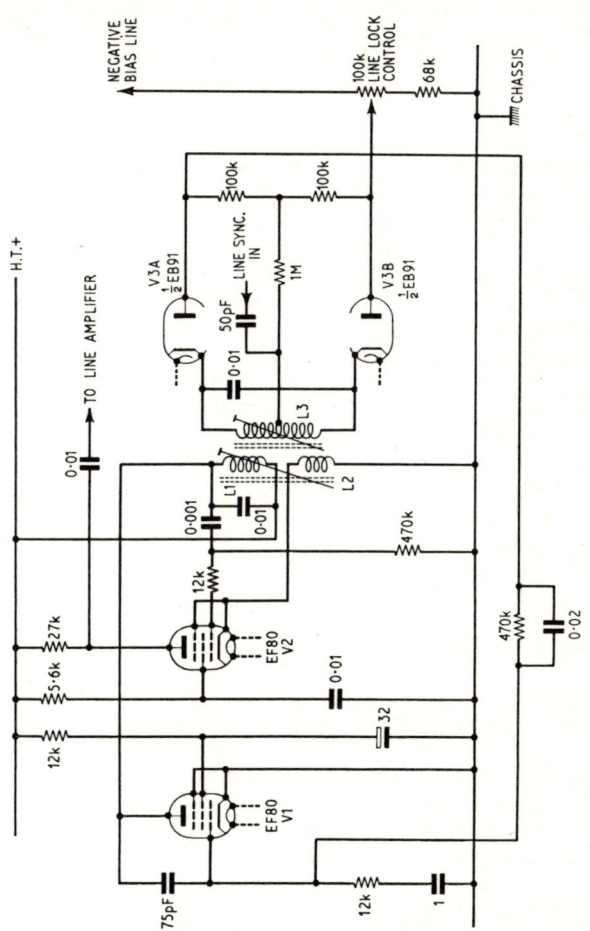

Figure 6.10 Sinewave line generator with reactance valve control, showing also the discriminator circuit

test the control valve by substitution before delving too deeply into the rest of the circuit.

Change in value of resistors, especially those on the control valve electrodes, will cause the control potential to the line generator to alter and thus unlock the timebase when the drift exceeds the stage locking range. The anode and cathode resistors on the control valve in *Figure 6.7* are vulnerable in this respect.

Alteration in value of a component in the time-constant circuit between the discriminator and the control valve can make the circuit abnormally sensitive to impulsive interference and can also cause the generator to swing out of lock on camera changes.

In the sinewave circuit in *Figure 6.10* locking instability mostly results from a noisy reactance or oscillator valve. This can often be checked by tapping the valve envelope with the handle of a screwdriver while the set is working. If this action causes the lock to fail, the valve responsible should be replaced. The discriminator diodes are less of a problem in this respect, but they should not be overlooked.

Another cause of frequency and/or phase drift in a circuit such as that in *Figure 6.10* is change in value of the capacitor across L_1 or that across L_3. These are often located inside the metal screening can of the oscillator/discriminator transformer (i.e. L_1, L_2 and L_3). Sometimes the tuning and phasing cores can shift in their formers, and it has been known for the cement holding the windings to the former to soften and for the winding to shift, again unlocking the generator.

Self-oscillating field timebase

Having now examined most generators, let us continue by investigating a composite field timebase, embracing both generator and amplifier. A typical stage of this kind is shown in *Figure 6.11*. An interesting feature about this circuit is the formation of a multivibrator between the triode and pentode, with the pentode acting also as the field amplifier. A number of dual-standard receivers adopted this technique but it has been used far less in the line timebase.

The height control, hold and the stabilisation of the anode voltage of the triode (first multivibrator section) are based upon the operating principles already described. The major difference is that when the pentode section switches on a relatively large rise in anode current results, and this flows through the primary winding of the field output transformer T1.

When the pentode switches on, there is a current build-up in the field coils connected across the secondary of the field output transformer, which is similar to the sawtooth scanning stroke shown in *Figure 6.1*. The electron beam in the picture tube is thus deflected vertically. When the pentode switches off, the primary current in T1

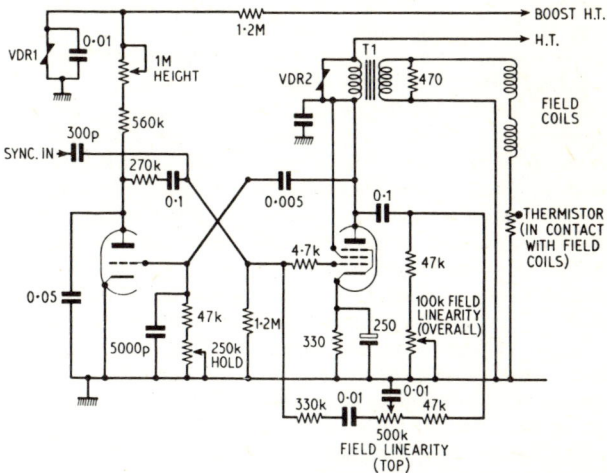

Figure 6.11   Practical field timebase using PCL85 in multivibrator circuit, with the pentode section acting also as the field amplifier

suddenly reverses, causing the retrace. The mark/space ratio of the multivibrator is arranged to provide optimum scanning and flyback operation.

The voltage-dependent resistor (VDR2) in parallel with T1 primary serves to suppress the large back e.m.f. which occurs across this winding on the retrace stroke. These pulses of high peak amplitude cause the resistance of the v.d.r. substantially to reduce, and the pulses are thereby damped. The considerably smaller rise in amplitude of voltage across the primary during the scanning stroke is insufficient significantly to reduce the value of the v.d.r., so the scanning operation is unaffected by the presence of this component.

Stabilising for temperature change

The height control, of course, is adjusted to provide the correct field drive for full scan amplitude. However, since scan amplitude is based on the rise in current through the scanning coils, increase in resistance of these coils will reduce the current in them and consequently the scan amplitude. The resistance of the coils can and does rise with increase in temperature, and to combat this a thermistor is connected in series with them. This component is generally arranged to be in thermal contact with the coils, so that any change in coil temperature is immediately reflected to the thermistor.

Since the resistance of the thermistor decreases with increase in temperature (it is said to have a *negative temperature coefficient*), the increases in resistance of the scanning coils (positive temperature coefficient) can thus be neutralised, the load 'seen' by the field amplifier thereby remaining reasonably constant over a wide temperature range, which is a requirement to prevent the height from reducing as the set warms up. This was a shortcoming of some early sets.

Because the electron beam is deflected over an arc and the screen of the modern picture tube is flat, geometric distortion of the picture can only be prevented by the incorporation of linearity correction. Two pre-set linearity controls are found in modern field circuits, one controlling the overall linearity and the other linearity mostly at the top of the picture.

These controls work on the basic principle of frequency-selective feedback in the field amplifier, and the two controls in *Figure 6.11* are seen to be connected in a network between the anode and control grid. Here the overall linearity control varies the magnitude of the anode-to-grid feedback, while the 'top' control varies a reactive (capacitor) element.

Apart from correcting for the tube geometry, these controls tailor the current waveform in the scanning coils so that it is of the most ideal shape to deflect the scanning spot at a linear rate over the surface of the screen, which is the requirement for a picture of optimum field linearity.

Time-constant faults and complete failure of the field generator have already been considered, but there are a host of other faults that can plague the field timebase as a whole. These will now be dealt with.

Insufficient height

A common fault is low scan amplitude, it being impossible to increase the height sufficiently even by fully advancing the height control. If the linearity is fairly good over the reduced scan, a likely cause is increase in value of the resistor from the height control to the boosted h.t. line (the 1·2 M resistor in *Figure 6.11*). This resistor is sometimes common to the first anode feed to the picture tube, so if it goes high in value an accompanying symptom could well be poor picture focus along, possibly, with reduced brightness.

Insufficient height can also result from low emission of the amplifier valve or open-circuit of the electrolytic capacitor across the cathode resistor of the amplifier valve (the 250 µF capacitor in *Figure 6.11*). Usually, however, defects other than increase in value of the height control series resistor incite a degree of linearity distortion as well. Low valve emission, for example, often causes compression at the bottom of the picture, this developing into a bright, horizontal line owing to the valve not being able to cater for full drive. This condition is aggravated by the height control being advanced in an endeavour to combat the valve weakness.

Receivers in which a thermistor is used in the scanning coil circuit can exhibit very much reduced scan due to a fracture developing in this component or the lead-out wires making poor connection to the element. This trouble can generally be brought to light by applying pressure to the element or the lead-out wires while the set is working, the scan then jumping up and down as the component is so disturbed.

A feed resistor on the h.t. side of T1 primary winding in *Figure 6.11* going high in value or an associated electrolytic decoupling capacitor reducing in value or developing poor insulation resistance can also produce the symptom.

In sets using a blocking oscillator or multivibrator independent of the field amplifier, which may be a PL84 or similar class of valve, reduction in the value of the capacitor coupling the drive from the generator to the control grid of the amplifier is another cause of the effect. Intermittent height symptoms are also produced by this coupling capacitor being itself some way intermittent. The best check for this component is by substitution or by applying pressure to its case and lead-outs while the set is working, if this is physically possible, and it usually is in receivers where the field timebase is built on a printed-circuit board.

Instead of being coupled direct from the secondary of a transformer, the field scanning coils may be capacitively coupled to an inductive anode load, or coupled through an auto-transformer. Lack of height can be caused here by a section of the autotransformer going open-circuit, the d.c. path being completed through the scanning coils themselves, or the capacitor, which is an electrolytic, going low in value or almost open-circuit.

The use of an oscilloscope for checking the overall performance of the field timebase can save a great deal of diagnosis time. Complete failure, for instance, can quickly be located either to the generator or

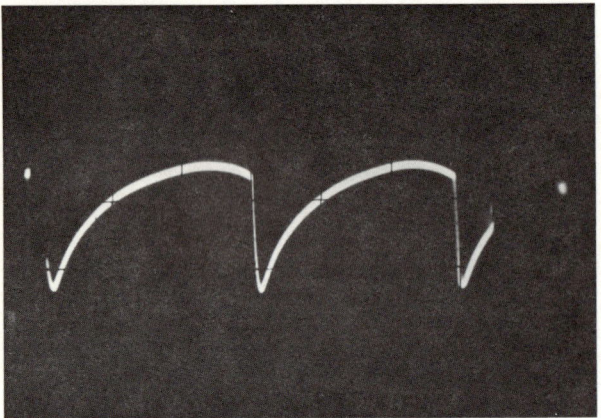

Figure 6.12   Waveform at the control grid of the field amplifier

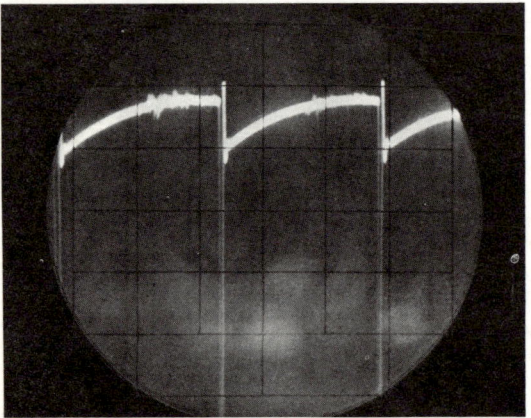

Figure 6.13   Waveform at the anode of the field amplifier

amplifier by picking up the Y-signal first from the generator output (i.e. control grid of the amplifier) and then from the amplifier anode circuit, following through to the scanning coils.

Figure 6.12 shows a waveform as monitored at the control grid of the field amplifier, while the waveform in *Figure 6.13* is that at the anode of the amplifier. In the latter, the large amplitude negative-going spike is caused by the flyback, which, in some receivers, is suppressed by a v.d.r., as in *Figure 6.11*, for instance. It is interesting to note, though, that this pulse, usually suppressed by a large amount, acts as a blanking signal to the tube during the field retrace. The effect is that it back-biases the tube so that the scanning spot is extinguished, thereby avoiding the display of the flyback lines.

Field linearity faults

Assuming that the height can be advanced sufficiently to over-scan the screen, yet the linearity is poor and outside correction range of the pre-set controls, attention should be given to the possibility of one or more of the resistors and capacitors associated with the linearity correcting network having altered in value. In open-circuit wiring, resistors tend to increase in value, while in printed circuits experience has shown a tendency towards the opposite effect, that of value decrease.

Other components that should be investigated are the valves (both generator and amplifier), the generator charging capacitors and resistors, the coupling capacitor for insulation resistance and the electrolytic across the amplifier's cathode resistor. However, open-circuit of the latter often tends to reduce the height, as already mentioned. A short-circuit, though, can severely impair the field linearity.

Delayed field flyback

A slow field flyback can cause the display of a number of horizontal lines at the top of the picture, often seen with the field unlocked, as shown in *Figure 6.14*. This trouble can be caused by increase in value of the resistor in the anode of the first valve of a multivibrator or a fault in a blocking oscillator transformer.

Some early receivers are prone to a slow field flyback, and without modifying the circuit, little can be done about it. On most transmissions, system test pulses are carried during the field interval of the

Figure 6.14   Effect of slow field retrace

signal, and these can give rise to one or two sets of horizontal lines at the top of the picture, which are graded in brightness along their length.

Field bounce

Another field timebase fault is so-called *field bounce*, where the whole of the picture tends to judder or bounce vertically over a small distance. The speed of the bounce is sometimes so rapid that the display appears as two vertically displaced pictures with, of course, very poor interlace. A lesser displacement simply impairs the interlace performance and thus reduces the vertical definition.

Field bounce is nearly always caused by a defective field generator valve or by trouble in the charging capacitor or resistor. In blocking oscillator generators, intermittency in the transformer windings can cause similar trouble.

Another cause is instability in the field amplifier due to failure of a decoupling element, such as an electrolytic capacitor, or even due to the development of spurious oscillations in the valve itself. Valve change soon clears the effect resulting from this cause.

A very rapid bounce over a small distance can be caused by the influence of the line pulses developed in the line amplifier (see later) on the field generator. A major contributory factor in this respect is poor screening of the line output stage and e.h.t. section. It is important that all the screens and shields round this part of the circuit are securely refitted after a servicing operation. The net result on the picture is *line pairing*, where the lines of one field fall directly on top of the lines of a subsequent field, completely destroying the interlace.

Misplacement of components in the field generator or line amplifier and e.h.t. sections can also cause the trouble, which is very real in practice. There have also been cases where the thermistor in series with the field coils is intermittent: this can be quickly checked by temporarily short-circuiting the thermistor.

Field-sync. troubles

The efficiency of the field hold and the interlace performance are governed basically by the field sync. In simple circuits, the field pulses from the sync. separator are fed to the generator through an integrating network, which consists of a resistor in series with a capacitor, the output being taken from across the capacitor.

The line sync. pulses are too fast recurring to influence the charge across the capacitor, so they produce virtually no output from the integrator. The longer period field pulses, however, which occur as a train of eight pulses or so at the finish of a field scan (405-line system), add up to a substantial charge across the capacitor and thus 'fire' the field generator (instigating the flyback) at the critical time. The time position changes relative to the picture signal after each field scan to provide the interlace.

Many modern receivers, however, use a semiconductor diode, to assist with the elimination of the line sync. pulses and to produce a field sync. pulse with a fast-rising edge, very accurately timed.

Lack of interlacing or poor field locking should first lead to a check of the diode in this circuit, along with the associated components.

The efficiency of interlace is usually appraised by the percentage of displacement of the lines of one scan between the lines of the partnering scan, making up the complete interlaced picture. If the interlaced lines fall accurately in the centre of the space between the lines of the partnering scan, then it is said that the interlace is 50:50. The figures change in accordance with the degree that the interlaced lines deviate from the centre space, and in fact, on the British 405-line system interlace cannot be better than 55:45 whereas the 625-line system, with its field equalising pulses, can offer a perfect 50:50 interlace.

Sync.-separator troubles

The basic sync. separator has been given in *Figure 6.8*, and if the operating conditions of this change substantially due, for instance, to a weak valve, alteration in the value of associated components and so forth, picture signal is likely to emerge from the stage and its presence at the sync. input of a generator can result in a poor lock or one that is affected by changes in picture signal.

Components which are particularly important here are the screen grid h.t. feed resistor and decoupling capacitor and the resistor and capacitor on the control grid, connected from the video amplifier output.

The sync. separator stage acts essentially as an amplitude limiter; it passes pulses of anode current only on the sync. pulses, while completely limiting the picture signal on the opposite polarity. *Figure 6.15* shows a waveform of the line sync. pulses as developed across the anode load of the sync. separator valve. *Figure 6.8* shows how these pulses are applied to the discriminator of a flywheel-controlled circuit, and later it will be seen how they are fed to a directly synchronised generator.

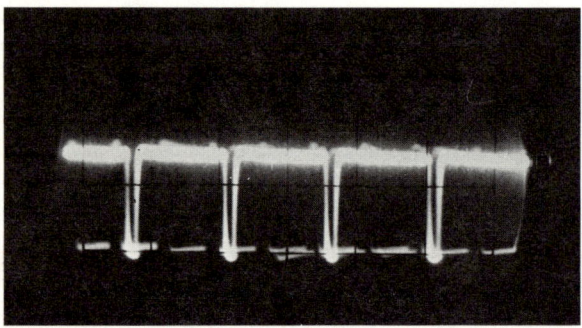

Figure 6.15   Series of line-sync. pulses at output of sync. separator with picture signal removed

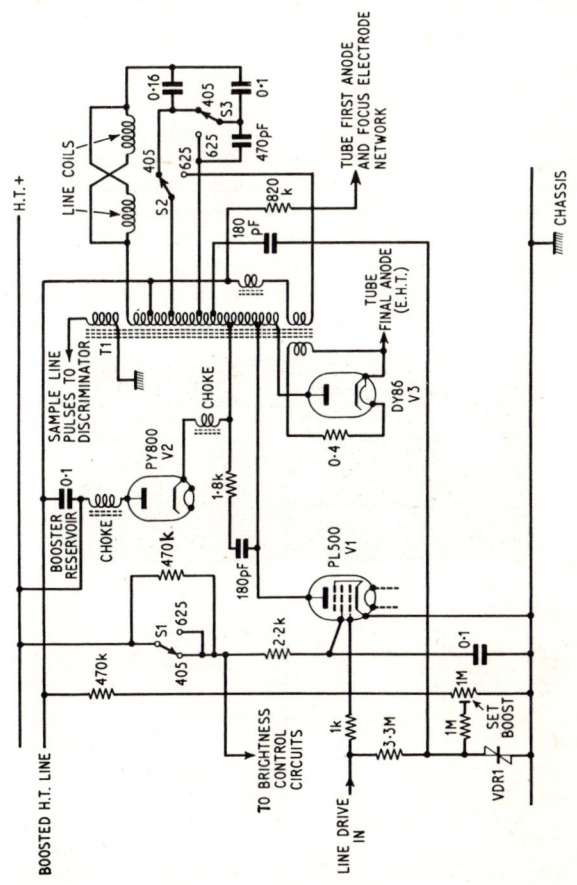

Figure 6.16   Circuit of dual-standard line output stage

Line amplifier or output stage

So much, then, for the field timebase. We now turn our attention to the line amplifier and associated circuits. The line amplifier of a dual-standard receiver is shown in *Figure 6.16*. Here V1 is the amplifier, V2 the boost (or efficiency) diode and V3 the e.h.t. rectifier. The line drive waveform at the grid of V1 is portrayed in *Figure 6.17*. This has a rising exponential scanning stroke and a very fast-falling retrace stroke.

The rising drive on the scanning stroke pushes a sawtooth current wave through the line coils and deflects the beam horizontally, while the rapid change of direction of the current instigates the flyback. Actually, the circuit operation is somewhat more complicated than this owing to the action of the boost diode and the need to obtain e.h.t. voltage from the amplifier to energise the final anode of the picture tube and of course, to provide the boosted h.t. voltage for the tube's first anode and for the field generator (see Chapter 3).

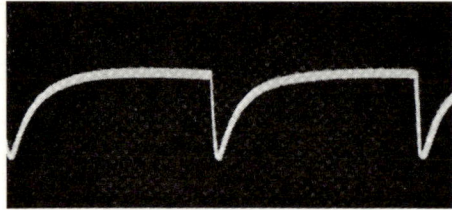

Figure 6.17   Typical line drive waveform (405-line)

Towards the top of the drive scanning the grid bias of the output valve is almost zero, and the valve is thus fully conductive. It remains in this condition for the remainder of the scanning stroke, and during this period current continues to flow through the inductance of the line output transformer (T1 on the circuit in *Figure 6.16*) between the anode of the line output valve and the cathode of the boost diode.

While the line output valve is passing this current, the tap on the transformer winding connected to the boost reservoir capacitor (marked on the circuit) is driven positively due to normal auto-transformer action. At this time the boost diode is also conductive, since it is through this that current is supplied to the anode of the line output valve.

On the flyback stroke of the drive waveform, the control grid of the line output valve is suddenly driven hard negative, quickly resulting in anode current cut-off. This causes a collapse in the magnetic fields of the line output transformer and line scanning coils, and as a consequence swings the anode of the line output valve violently positive (which is an important action of the system from the generation of e.h.t. voltage aspect). The connection from the line output transformer to the cathode of the boost diode also swings highly positive with respect to its anode and the diode ceases to conduct.

Directly following this disturbance, the voltage at the anode of the line output valve swings in a negative direction due to ringing effects, which are damped oscillations caused by resonance due to the inductive and capacitive elements of the circuit.

This negative swing-back is also reflected to the cathode of the boost diode, and when this makes the cathode more negative than its anode, the diode again commences to conduct, an action which suppresses further rings. The magnetic fields of the transformer and line scanning coils then start a relatively slow collapse due to the load of the conducting boost diode, and charging energy is fed to the boost reservoir capacitor.

This happens at the finish of the flyback or at the start of the scan, but at this time the line output valve is still non-conductive. Thus, the current in the line scanning coils for the first part of the scanning stroke is provided by this magnetic field energy collapse, and when the energy is all used up, the drive at the control grid of the line output valve has risen sufficiently to bring the valve back into conduction so that it can supply current for the final part of the scanning stroke.

Thus, collapsing magnetic field energy, derived when the boost diode is switched off, provides the first part of the line scanning stroke during the period when the line output valve is switched off and the boost diode switched on, while rising current in the line output valve and boost diode provides the final part of the stroke. The boost reservoir capacitor remains charged throughout the entire cycle, losing some of its charge during the second part and regaining it again during the first part.

To ensure optimum efficiency, the ringing frequency is arranged to be approximately three times the frequency of the retrace pulse (see Chapter 3).

The boost reservoir capacitor charges to some 500–1 000 V, while an over-wind on the transformer steps up the high peak positive pulses produced at the anode of the line output valve during flyback to some 18 000 V. These pulses are then applied to the anode of the e.h.t. rectifier, where they cause the rectifier to conduct and charge the e.h.t.

reservoir capacitor almost to their peak voltage. Thus, a d.c. e.h.t. voltage is available for the picture tube final anode.

The e.h.t. reservoir capacitor is nowadays formed entirely by conductive coatings on the inside and outside of the picture tube, with the glass between acting as the dielectric.

Standards switching

Switch sections S1, S2 and S3 on the circuit in *Figure 6.16* are coupled to the main standards change control. S2 switches in an extra inductance on 625-lines to correct the third-harmonic tuning of the line output transformer, while S3 changes the value of the capacitance in series with the line scanning coils. This series capacitance is for so-called *S-correction* (scan correction). Without this, the modern, flat-faced picture tube would display a picture with vertical compression in the centre and expansion at either side. This, which has already been considered in terms of field linearity, results from the fact that the screen is essentially flat while the beam deflection describes an arc.

S1 in *Figure 6.16* may appear on the face of it to do nothing! Actually, it has an important job to do. It is a break-before-make switch section, and it brings the 470 k resistor into circuit during changeover from one standard to the other. The additional voltage drop across it reduces the line output valve anode current and greatly minimises peak voltages that would otherwise cause arcing across the contacts of S2 and S3 and probably result in other damage as well.

An interesting feature is that the bottom end of the resistor feeds the brightness control circuit. Thus, when the resistor is passing the changeover current, the voltage to the tube grid is also made more negative (i.e. less positive so far as the feed point is concerned), muting the beam. Without this facility, the screen would display bad flashes during the changeover.

Line stabilisation

The voltage-dependent resistor (VDR1) in the control grid circuit of the line output valve serves to stabilise e.h.t. and line scan current with changes in e.h.t. load, resulting from changes in picture brightness. Since the resistor has a nonlinear characteristic, the pulse voltage fed to it through a capacitor from a tap on the line output transformer produces a potential across it, and this potential is adjustable by the *set boost* pre-set. This is adjusted to give a specified value of boost voltage. Now, should the load on the line output transformer rise due, for instance, to the picture going extra bright and calling for a large amount of e.h.t. current, the amplitude of the pulses applied to the v.d.r. will reduce, as also will the voltage developed across it.

The effect is then that the control grid goes less negative, thereby calling upon the line output valve to deliver more power to combat the increased load. The action is similar to that of automatic gain control (a.g.c.).

A winding on T1 (*Figure 6.16*) shows where the sample pulses for the flywheel-controlled line sync. are derived in some receivers.

To conclude this description of a dual-standard line output stage, it should be noted that several variations may be found in practice, but the basic principles described apply to the majority of circuits, and these must be known before fault-finding can be attempted.

Finding faults in the line output stage

Owing to the very high peak voltages created in the line timebase, the amplifier or output stage in particular is prone to insulation breakdown troubles. The line output transformer is vulnerable in this respect as may well be imagined, for the pulse voltage developed at the taps to the anode of the e.h.t. rectifier, to the anode of the line output valve and to the cathode of the boost diode is sufficient to cause an arc discharge of almost half an inch. Moreover, the relatively high-frequency nature of this voltage incites a violent discharge to any metal item in proximity, though not necessarily earthed or connected in circuit. For instance, a substantial discharge from the anode of the e.h.t. rectifier valve to the tip of a screwdriver blade, with the screwdriver held by an adequately insulated handle, can be expected from a correctly operating set. Discharges of slightly less violence can be obtained from the anode of the line output valve and the cathode of the boost diode, and this arc test is often used by technicians—to the tip of a screwdriver blade, never to earth or chassis—as a quick means of checking the operation of the line timebase. This type of test, however, must not be done with transistorised e.h.t. systems.

If there is no discharge or if the discharge is weak, then there is no doubt that the line timebase is in trouble. Of course, failure of the line timebase means that the picture tube is deprived of e.h.t. voltage, so there is no screen illumination to assist with the diagnosis. When the e.h.t. circuits are inactive, the boost voltage is also low, and when the line timebase is really defunct the boost line voltage will be of the same order as the h.t. line voltage, and this may be abnormally low owing to excessive loading reflected from the faulty line timebase.

For instance, lack of drive to the line amplifier will cause the output valve to pass an abnormally high anode current because then the valve will be without bias (which is normally caused by the drive, as we have seen). This high current will severely load the h.t. supply circuit and pull down the voltage, often from a nominal 220 V to 180 V or so. Moreover, the high current taken by the line amplifier will result in this valve (and possibly the boost diode) overheating badly.

Many people can hear the whistle of a correctly working line generator switched to 405 lines. The higher frequency whistle on 625 lines, however, is inaudible to many. Even on 405 lines it is best to disconnect the aerial and adjust the line hold control to a low-frequency setting. Clearly, if the whistle can be heard, the line generator must be working. The whistle, incidentally, emanates in large proportion from the line output transformer, due to electro-mechanical stresses to which the core is subjected.

If there is no line whistle, then the generator section should be investigated. This is where an oscilloscope again shows its worth. A quick check at the output of the generator or input of the amplifier will show a waveform such as in *Figure 6.17* with the generator working correctly. Lack of drive or generator output should lead to normal servicing techniques, as described for the field generator.

Another check for line drive consists of connecting a high-resistance voltmeter between the line output valve control grid and chassis, with the negative lead of the voltmeter to grid. The reading should be substantially negative owing to the bias effect created by the line drive. Lack of deflection or very small negative or positive deflection would indicate either lack of drive or a defective line output valve.

If the line whistle can be heard, or if it has been otherwise proved that the line generator is working, yet tests reveal lack of pulse potential at the taps of the line output transformer, a strong possibility is shorting turns in the line output transformer. This component, however, should never be replaced out of hand before performing other tests.

The d.c. conditions of the line output valve should first be established. If the valve is relatively cool, for example (bearing in mind that it normally runs hot), its emission may be impaired or a resistor in the screen grid feed may have open-circuited. These are frequent causes of the trouble. Remember, also, that the boost diode must have good emission to pass the d.c. to the line output valve.

A shorting boost reservoir capacitor (see *Figure 6.16*) is another cause of the trouble, and the stage, of course, will fail to work correctly if the capacitor is low in value or open-circuit.

If the line output valve and boost diode d.c. conditions check as they should and the line generator is delivering drive to the output valve, other less obvious factors must be investigated. For instance, an abnormally high load across the line output transformer will greatly impair its required high efficiency and almost certainly reduce the pulse potential for e.h.t. generation and scanning coil current.

Checks should thus be made of the components directly associated with the transformer, including width and linearity inductors, if used, and even the line scanning coils themselves. If these components appear to be in order, the line output transformer itself then becomes the chief suspect. The windings must possess d.c. continuity, of course, since it is assumed that the d.c. conditions are correct. Sometimes a clue can be obtained by measuring the resistance of the windings and comparing this with that written in the service manual or service chart, but a

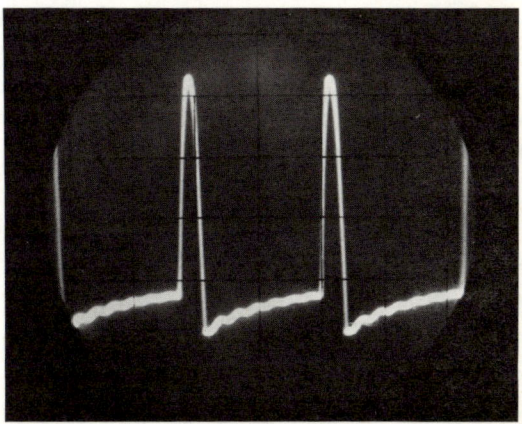

Figure 6.18   Waveform as picked up from the anode of the line output valve, via capacitive coupling through the lead insulation

A direct connection to meter or 'scope should never be made at this point because of the high peak voltages.

short between two or three adjacent turns will not change the overall resistance much, yet such a short can often be sufficient to mute the line output stage completely!

The symptoms are generally overheating line output valve, slight trace of line whistle, no e.h.t. voltage or pulse voltage at the transformer taps (or, in some cases, very small pulse voltage) and heater of the e.h.t. rectifier valve not alight. Assuming that all else is well, as far as can be tested, then the above symptoms represent a good case for checking the line output transformer by substitution which, incidentally, is about the only conclusive way that it can be checked outside the factory or laboratory.

An oscilloscope can reveal the line pulses generated in the output stage simply by clipping the Y-input to the insulation of the anode wire of the line output valve. The resulting waveform is shown in *Figure 6.18*. Note the slight ringing during the scanning stroke. This can cause vertical, dark and light bars on the left of the picture in some sets, but modern sets are less prone to this fault.

Poor insulation either in the transformer or windings of an associated inductive component can cause corona on the picture, and a severe case of this trouble is shown in *Figure 6.19*. Sometimes arcing can be seen from a high-voltage point, and extreme care must be taken to improve the insulation to prevent the discharge. Inside a component, of course, the discharge may not be visible, but it may pay to view the

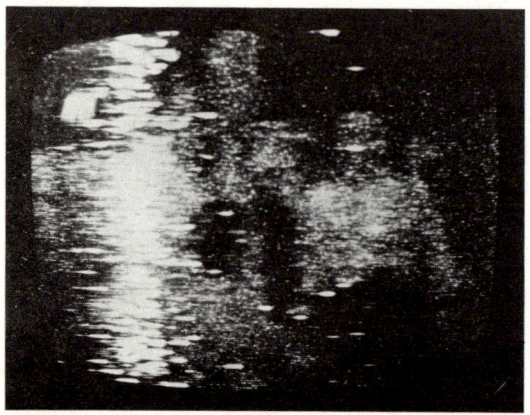

Figure 6.19   Display of severe corona

inside of the set in a darkened room. Suspect components can only be checked by substitution, including the line output transformer (which is a frequent culprit) and the line scanning coils. High-voltage points should be thoroughly cleared of dirt and dust but only, if these points stubbornly refuse to stop arcing, should they be treated with an anti-corona grease.

The corona symptom in *Figure 6.19* is due to pulse voltage discharges. Discharge from the rectified e.h.t. (i.e. at the tube final anode) produces random interference spots on the screen, rather than a distinct line effect as in *Figure 6.19*.

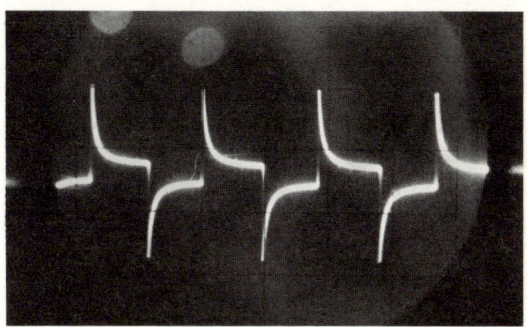

Figure 6.20   Series of differentiated line-sync. pulses

Lack of e.h.t. with plenty of pulse voltage at the anode of the e.h.t. rectifier is often caused by a worn e.h.t. rectifier valve, signified sometimes by a bluish glow within the valve. Trouble in this valve can also cause the picture to expand as the brightness is turned up, with a total fade-out at full brightness. This is so-called poor e.h.t. regulation. Similar effects can be caused by a low-emission line output or boost diode valve.

Failure of the line generator to lock should lead to a check of the feed from the sync. separator to the line generator if the line-hold point is within range of the control. If not, check the time-constant components, as explained earlier.

With direct line sync. the line sync. feed is generally through a low-value capacitor, the signal being developed across a resistor. This is the basic differentiator circuit which shapes the line pulses for accurate 'firing' of the line generator. A series of differentiated line pulses is shown in *Figure 6.20*. Again, an oscilloscope is useful for tracing these to the line generator, but, to avoid error, the line timebase should be muted by disconnecting the screen grid h.t. feed to the output valve and by shorting the output of the generator.

In conclusion it should also be remembered that the timebase circuits can be badly affected by low valve-heater current which would arise from the use of an incorrect mains voltage tapping on the receiver. The effect is that of a low-emission valve, causing foldover at the bottom of the picture on field and lack of full scan on the right of the picture on line. Lack of full scan on the left of the picture often indicates a low-emission boost diode, and this may be accompanied by a light, vertical bar towards the centre of the screen, at the point where the line-output valve commences to conduct on the scanning stroke. Presence of this bar by itself may however be an indication of poor balance in a fly-wheel-sync. discriminator.

Adequate ventilation of the receiver is also essential to ensure stable working of the field and line timebases.

7

INSTALLATION AND SERVICING TECHNIQUES

The mains supply socket must be rated at at least 2 A, and a television receiver must always be adjusted to suit the supply. The frequency of a.c. mains supplies should be determined, and the correct rating of the mains supply fuses is another important consideration. A.c./d.c. receivers will operate from d.c. supplies only when the plug is inserted into the supply socket the correct way round.

Earth leads should be of heavy gauge wire, and must be kept as short as possible. A copper earth tube or plate, with a large surface area and going deep into damp ground, is best. A main cold- or hot-water supply pipe, gas or telephone installation must on no account be used. Most television receiver chassis are at mains potential, and direct connection to the chassis should never be made.

Most receivers today are not provided with an earth socket, and no attempt should be made to use an earth with such a set.

The controls provided in a modern television receiver fall roughly into three categories: the main user controls, mounted at the front, side or top of the receiver; auxiliary controls, such as those for the timebases, which are pre-set but which can generally be readily adjusted either by the user or by the engineer; and finally, those which can be adjusted only with the protective cover removed and which may require the changing of soldered tappings, etc. With the development of circuits which are less subject to variations with changes in mains voltages, ageing of valves, etc., there is a tendency for this third category to increase.

Installation adjustments

Modern receivers are despatched with the pre-set controls factory adjusted for correct operation so that fewer installation adjustments are required today than in the past. The adjustments most likely to be required are: mains adjustment (not necessary in modern receivers which are rated at 240 V a.c. only); sensitivity, which is generally a 'local/distant' pre-set to adjust for the different signal strengths found in different areas and the different signal strengths that may be found on the various bands, and which generally takes the form of a system for varying the delay applied to the a.g.c. line; fine tuning and, on v.h.f. if a particular channel is found to be outside the range of fine tuner adjustment, oscillator coil tuning.

Pre-set contrast controls may also be provided to compensate for the different conditions on the 405- and 625-line systems on dual-standard receivers.

Timebase controls

Timebase controls include hold controls which vary the operating frequencies of the timebase oscillators so that the picture can be stabilised, amplitude controls to vary height and width, and linearity controls to provide adjustment of the scanning waveforms.

In dual-standard receivers, hold controls are provided for adjustment on the two line standards. Where these controls are in parallel they are independent; in some models, however, they are series connected so that one of them will then be effective on both standards. In addition, when flywheel-sync.-controlled sinewave oscillators are used, the tuned feedback coil provides further adjustment, and a variable tuning capacitor may be found in addition. With flywheel synchronisation there is generally a range of control operation over which correct synchronisation is obtainable; the points at which the picture can just be synchronised with clockwise and anti-clockwise rotation of the control should be found and noted, and the control set to the mid-position.

Two field linearity controls are usually provided, one affecting the top of the picture only. Adjust the overall linearity control first, then set the 'top' linearity control to compensate for any irregularity in the field scanning at the top of the picture.

With the 5:4 aspect ratio of many picture tubes (aspect ratio of the transmitted picture is 4:3) the width control is generally set for slight (not more than 8–9 per cent) overscan. On new receivers using 'squared-off' tubes and on much older receivers, the picture tube aspect ratio is 4:3 and no overscan is necessary. Width controls take various forms; for many years a variable inductor in series with the scan coils was common; more recently, some sets have relied upon a shorted-turn sleeve for the adjustment of both width and linearity; most modern receivers, however, include a pre-set potentiometer in the line scan/ e.h.t. stabilisation feedback network, and this is adjusted for a given value boost h.t. voltage. It is important to keep within about 5 per cent of the stated value, or damage to the line output stage may result. This type of control is often labelled a 'set boost' control. In addition it is common to provide pre-set height and width controls for picture equalisation on the two standards in a dual-standard receiver.

The line linearity control is generally of the shorted-turn sleeve type dealt with below, although the saturable reactor type of control with an adjustable magnet has been favoured by some manufacturers.

The picture is squared by adjusting the position of the deflection coils and centred by means of shuffle plate magnets on current models. The line-hold control may also affect centring.

In some models which include flywheel line synchronisation, a switch is incorporated in the line-hold control, and in such cases the knob of the control must be depressed before it can be rotated. Pre-set line-hold controls are also common on such models. The usual practice in adjusting these is to set the main hold control to mid-travel, short out the line sync. pulses (in some cases it may be easier to remove both the field and line-sync. pulses), and adjust the pre-set control so that the picture is synchronised (or slowly running through if the field-sync. pulses have been removed). The width and line linearity controls may need slight readjustment afterwards. In some models provision was made to short-circuit completely the flywheel circuit when the set is used in an area of good signal strength; this makes the setting of the line-hold control less critical.

In most modern receivers the line linearity control takes the form of a shorted-turn loop, which is fitted beneath the deflection coils, around the neck of the picture tube. This device may also be used to control the width. Over insertion may damage the coils; set makers usually give approximate settings; see Chapter 3 for further details.

Interference and sensitivity controls

Two other pre-set controls which may be found on some older receivers are the vision-interference limiter and the sensitivity control.

The vision-interference control may take the form of a pre-set potentiometer or a plug and socket adjustable in a number of fixed steps. Most forms of limiter, when set to give maximum clipping, will cause some flattening of the highlights of the picture. Vision-interference limiters should therefore always be adjusted to the minimum clipping position consistent with satisfactory reception. In some models the vision interference limiter circuit is an optional extra which can be removed completely if fitted.

A sound-interference limiter adjustment was provided in a few sets; as with vision interference limiters, the adjustment should be set to give the minimum limiting action consistent with freedom from interference.

The sensitivity control is fitted to vary the gain of the receiver so that it may be adjusted to suit the signal strength of the area in which it is installed. If the signal strength is too great, sound-on-vision and vision-on-sound interference may be caused, and the sensitivity control will in such conditions need to be reduced so that the interference is eliminated. A compromise setting may have to be found when signal strengths on different bands vary widely. The control may also be marked 'r.f. gain'.

Miscellaneous pre-set controls on older receivers

Pre-set controls fitted in some older models are picture quality, line anti-striation and line drive adjustment. The picture-quality control usually took the form of a compression trimmer or other arrangement for increasing or decreasing the capacitance in the cathode circuit of the video output valve. This control should be adjusted to give a sharp picture, free from smear, a compromise in some locations having to be sought between Bands I and III transmissions.

Line anti-striation or balancing trimmers were fitted on the deflection coils to balance the stray capacitances in the coils. The adjustment of these is normally carried out in the factory; they should only need readjustment if new deflection coils are fitted. The adjustment is for minimum waviness in the line scanning.

Line drive controls usually consist of a trimmer controlling the drive to the line output valve, and should need no adjustment unless valves in the line timebase are changed. The usual method is to adjust the control so that the e.h.t. voltage is a given value, as specified by the manufacturers. If out of adjustment, a faint white line may appear near the centre of the screen. Damage to the line output valve or transformer can be caused if this adjustment is not correctly carried out.

In fringe areas it is advisable to adjust the pre-set controls for average signal-strength conditions, as in these areas a certain amount of fading is to be expected, and unless some such allowance is made they will require constant attention.

Mechanical picture adjustments

Around the neck of the picture tube are situated the deflection coils and also usually several magnets for such purposes as centring and, in older models, for focusing the picture, and reducing ion burn.

The deflection coils are housed close up against the bulb of the picture tube, and may be rotated, after loosening the clamping screws, so as to level or square the picture. Correction magnets may also be found fitted against the bulb of the tube to correct for barrel and pin-cushion distortion, etc., and to remove corner shadowing. Behind the deflection coils are situated magnets controlling the focusing (if magnetic focusing is employed) and centring of the picture. These are adjustable by means of suitable levers or knobs.

Adjusting ion-trap magnets on older picture tubes

Ion-trap magnets are normally secured to the neck of the tube by means of a clamp, and to facilitate fitting there is usually an arrow stamped on the magnet, and a line along the neck of the tube. The magnet is normally fitted above the position of the electron gun with the arrow pointing in the direction of the screen, but alternatively may be fitted with the magnet underneath and the arrow pointing away from the screen. To fit and adjust the ion-trap magnet, the following procedure should be followed, preferably when a stationary test pattern is available.

With all power switched off, and reservoir capacitors discharged if necessary, the magnet is pushed over the base of the tube with the arrow pointing towards the screen, and placed immediately over the line marked on the tube neck. The cathode-ray-tube socket is then replaced, the receiver reconnected to the mains supply, ensuring that the chassis is not above earth potential, and the brightness control set to a position where the raster is just visible. To achieve this, it may be necessary to adjust the position of the magnet slightly.

Then with the arrow over the line, the magnet is moved towards the screen until the focused raster is at its brightest. The brightness control is then readjusted until the peak-white portions of the image are at a correct level, and, if necessary, the position of the magnet adjusted slightly to obtain maximum brilliance.

Where the picture cannot be centred by adjusting the position of the focus field, the ion-trap magnet may have to be rotated slightly around the neck; this operation, however, should not lead to any decrease in brilliance.

When the picture fulfils the above requirements, lock the magnet in position by tightening the thumbscrew, ensuring that the magnet does not change position while this is being done.

Should it not be possible to obtain a position of maximum brilliance, it may be necessary to substitute another magnet. The magnet should never be adjusted to remove a shadow if this involves reducing the brightness of the picture; this should be done by adjusting the focus coil and/or deflection coils.

Always handle an ion-trap magnet with care: it should not be subjected to strong magnetic fields or mechanical shocks. It should not be allowed to come into contact with metallic objects.

Steering magnets

Apart from picture correction magnets positioned close to the flare of the tube and centring magnets around the neck of the tube, a steering magnet (sometimes known as a 'beam centring magnet') in the form of a circlip has occasionally been fitted around the neck of the tube for the purpose of ensuring that the electron beam is central within the final aperture.

Whether or not a steering magnet is required is judged by the quality of the picture. A general smudginess or 'tails' on white spots may be an indication of the need for a magnet, though these faults can be caused by incorrect alignment of the receiver circuits. On some picture tubes where the steering magnet is not required it may nevertheless be fitted, usually at the extreme end of the tube neck, where it has no influence on the electron beam. The magnet should not be discarded, as it may be needed should the tube be changed. In practice, the steering magnet is used for much the same purpose as the final adjustment of an ion-trap magnet on a bent-gun tube, that is to say for general picture quality rather than for brightness. The setting is determined by rotational and axial movement along the tube neck to a position which gives optimum freedom from blurring of leading edges. The adjustment should be carried out on a transmitted test card (test card described later in this Chapter).

COLOUR-RECEIVER INSTALLATION

A colour receiver should always be adjusted in such a condition that primarily the receiver gives good monochrome pictures in terms of timebase settings, scan amplitudes, picture centring, brilliance, contrast, and general picture quality. Then the special chroma requirements of the receiver can be set up if this is necessary. A number of the primary installation details are similar to those for monochrome receivers, but as the methods of adjustment can be different the process is tabulated below:

Aerial   Details of suitable aerials for u.h.f. work are given in Chapter 10 together with general advice on their installation and positioning. For colour reception it is vitally important to have an efficient aerial system, and if possible this should be set up using a signal-strength meter. Make final adjustments after installation of the receiver, if necessary, to minimise the effects of ghosting.

Receiver location   A colour receiver should be located in a room position where there are no nearby metal objects of any size, where the ambient light is reasonably low during daylight and where the room lighting does not 'wash out' the pictures. After installation the receiver should not be moved because of the earth's magnetic effects (see Chapter 5) but if it is absolutely necessary (for cleaning of a room, for example) the receiver should be drawn out and pushed back along the axis of the tube: it should not be turned at an angle from the original position and it should certainly not be rotated.

Mains adjustments   Most modern receivers are a.c. only and many are for use only on 240 V mains supplies. Should the supply differ from this and if there are no voltage changing taps available, the manufacturer should be consulted to obtain details of suitable modifications. Some receivers are fitted with overload cut-outs instead of, or in addition to, the customary mains fuse; this is usually in the form of a little red button that must be depressed in order to reset the overload. Most receivers take under 2 A and the mains plug top should under no circumstances have a fuse rated higher than 5 A.

Tuning   After the receiver has been switched on for a few minutes the standard channels for the area can be tuned in (transmitter details for u.h.f. are given in Table 1.7). This may be achieved on many receivers by depressing the channel changeover button required and rotating the knob (clockwise for 'up' in frequency, anticlockwise for 'down' in frequency), or, on receivers fitted with varactor tuners the d.c. potentiometers controlling channel frequencies must be adjusted. This may also require setting a 'coarse' frequency control to bring the potentiometers into the required range of channels. Receivers fitted with a.f.c. can often be tuned by means of monitoring the a.f.c. voltage and on some the a.f.c. action must be negated by depressing the channel changeover button fully.

Local-distant control, or delay a.g.c. control   Most manufacturers advise leaving this control in the maximum position unless cross-modulation occurs in very strong signal areas. In this case, the control should be reduced until the cross-modulation effects just disappear. Cross-modulation will usually show itself as a 'buzz' from the loud-speaker varying with picture content (this should not be confused with the intercarrier sound 'buzz') or bars of interference on picture varying with the sound content.

Horizontal control   This should not normally require adjustments on a modern single-standard receiver. If it does, or if there are doubts about the lock being at the centre of the pull-in range of the oscillator, check as advised by the manufacturer for his particular circuit, or follow the general advice given for monochrome receivers fitted with flywheel synchronisation given earlier in this chapter.

Vertical hold control   This does not differ from that on a monochrome receiver (see earlier) although it is often not an external control.

Contrast and brightness (brilliance)   Adjust contrast for best repro-duction of the mid tones and highlights of the picture (with saturation control turned fully down). Maximum setting is not normally necessary or desirable and if such a requirement exists it may indicate a receiver or aerial fault.

Adjust brightness so that the black parts of the picture are just black in the normal viewing ambient light.

Focus   Adjust the focus control on the v.d.r. or tapping on the line output transformer to give optimum resolution at the *centre* of the picture. Note that on some v.d.r. focus controls an insulated screw-driver *must* be used.

Picture size   Modern receivers do not generally have width controls —this often tracks with the SET EHT (see below). The height control should be adjusted to give the best formation of the circle on test card.

Set e.h.t.   Two alternative methods exist for setting the e.h.t. on a modern colour receiver; a power supply adjustment or a tapping on the line output transformer. Manufacturers usually advise the use of an e.h.t. meter or e.h.t. adaptor plus multimeter to measure the 25 kV (or so) voltage. If the details are given in the manufacturer's manual an alternative method is to measure the current through the focus potential divider chain (often about 120 μA). Very great care should be taken in the making of all e.h.t. measurements: the e.h.t. on a colour receiver at 25 kV with a power capability of about 30 W *can kill*.

Vertical linearity   Adjust as necessary the two controls (usually) provided for this purpose. If necessary after correction height may have to be altered to bring the raster back to its correct amplitude.

Picture shift   Horizontal and vertical presets are usually provided on colour receivers to give the correct picture shifts. If there is insufficient range on the control, the polarity of the d.c. voltage being fed through each set of scan coils can usually be changed by moving a flying lead to an alternative tag.

Degaussing   All colour receivers are fitted with automatic degaussing circuits that operate when the receiver is switched on. The power in this circuit is limited to prevent possible damage and after moving a receiver or after a long period of operation this internal degaussing may be insufficient. In this case a separate degaussing coil may have to be used (see Chapter 5).

Static convergence
Purity
Dynamic convergence
Grey-scale tracking

These should be adjusted using the appropriate test signals from a cross-hatch generator in the manner detailed in the section: 'Setting Up Shadowmask Tubes' in Chapter 5.

Note: (1) If, after convergence is completed any of the scan controls are adjusted (height, shifts, focus, etc.) the convergence must be readjusted.

(2) In some of the earlier dual-standard colour receivers, 405-line convergence must be adjusted *before* 625-line convergence. Even if the

receiver is only now used on 625-line reception, the 405-line convergence should be set approximately correctly, otherwise disappointing
results may be obtained on 625-lines.

Saturation   In a correctly operating colour receiver this is the only
control that should have to be set for correct colour. It should be
adjusted on transmitted pictures to obtain the correct intensity of
colour. This is normally best indicated on a flesh-tone such as a face.

Channel balance   Although the same power of transmission is used,
the same transmission aerials are used at the same sites, and the care
taken in the amplitude of chroma signals is the same, it is unfortunately
sometimes the case that the saturation of pictures received on the
different programme services appears different. The best comparison
for this can be made when all services are radiating Test Card F. On
older receivers this can sometimes be caused by different accuracy of
tuning on the stations, but a receiver fitted with a.f.c. cannot have this
problem. Sometimes a small adjustment to the aerial can balance up
signals, but often a small compromise in the setting of the saturation
control has to be made so that all channels appear to be more or less
correct. If the errors in saturation are large, there is probably a fault
in the u.h.f. tuner making it rather insensitive to some channels.
'Aerial Boosters' have also been known to cause similar problems, but
there are also unfortunate receiving positions where reception of one
programme service is very poor. The B.B.C. and I.B.A. Engineering
Information services are always pleased to advise in these rare cases.

TEST CARDS

Test cards are transmitted on a regular basis by both the B.B.C.
and the I.B.A. to assist the television installer and repairer to assess
the performance of a receiver in the home. The latest version of test
card is Test Card F, details of which are given below. This will be
regularly seen on u.h.f. transmissions in colour and in monochrome
after being standards converted for 405-line/v.h.f. transmission. On
occasions, however, other test cards may be seen; an older Test Card E,
which is monochrome only, may be seen from some B.B.C. transmitting
stations when the station is radiating its own test card (used when the
network circuits are being used for inter-studio programme recordings
or when circuits are faulty) and the 405-line Test Card D may be seen

Figure 7.1   Test Card F as radiated on 405-line v.h.f. services

from v.h.f. transmitting stations of the I.B.A. All the features of these test cards are contained in Test Card F so the information will not be repeated. At the time of publication some experimental transmissions of electronic test cards may also be seen.

Test Card F designed jointly by the B.B.C., B.R.E.M.A., E.E.A and I.B.A.

The test card includes features which facilitate the testing of colour and black-and-white receivers.

The pattern shown in *Figure 7.1* includes the electronically generated colour bars which replace the first 24 lines of the test card proper for the reasons stated in paragraph 1 which follows. The test card includes coloured edge castellations to check colour synchronisation effects and picture size; a grid with corner diagonals and centre circle enabling picture geometry and focus to be assessed; a 'letter-box' pattern to test low-frequency effects, frequency gratings for assessment of bandwidth, and a six-step grey scale. Colour information is included in the edge castellations and within the circle.

The electronically generated colour bars transmitted with the test card enable colour decoder performance to be checked.

1. Checking decoder performance.   The top of the transmitted test card starts about 12 lines after the end of the field blanking period in each field on 625 lines (or about 8 lines on 405). Thus about 24 lines (on 625) or 16 lines (on 405) of each picture are made available for the electronic colour bar signal. An electronic signal is used to avoid the problems of generating a high-accuracy signal by optical means. The decoded colour-separation signals corresponding to the coloured bars in this part of the picture should take up values of 100% or 25% (according to the colour of the bar) of their value for peak white. This form of signal (sometimes referred to as '100% amplitude, 95% saturation') has been adopted in agreement with the British Radio Equipment Manufacturers' Association as being representative of the high-saturation and high-luminance signals obtainable from colour cameras and is regarded as a stringent but realistic test of colour television systems. On black-and-white receivers, and in any case on 405 lines, the colour bar signal will appear as a grey scale with the bright bars at the left and dark ones at the right.

Faults or adjustment errors in a colour receiver are easily recognised as asymmetry or changes in shape of the waveform and/or differences in amplitude from line to line in the waveforms of the red, green and blue components of the colour bars at the output of the decoder. If the colour bars are observed on the face of the tube, an assessment of the decoder performance can be obtained by switching off individual guns of the tube.

2. Colour-receiver reference generator faults.   The colour bars, being located at the top of the picture, also enable the recovery of the reference generator after the field-sync. period—during which the reference bursts are absent—to be clearly seen as variations in the saturation of the bars. The bottom castellations, being green, provide a means of assessing the reference generator performance at the end of the field as compared with the start.

The red and blue castellations on the left-hand side will give rise to the greatest disturbance of the displayed colour picture if the gating circuits permit picture information to pass to the reference generator. Thus, if this fault is present in the receiver, either bands of saturation changes or 'Venetian Blinds' will be visible in coloured areas across the picture depending on the decoding circuits employed.

3. Colour-receiver convergence checking.   Certain lines in the grid have been outlined in black to assist in checking display-tube convergence. Others have no outlining so as to avoid the confusion which might otherwise result from the low-frequency ringing seen when certain types of quadrature distortion are present in a receiver. The blackboard and white cross also provide a check on static convergence.

4. The colour picture.   The colour picture in the centre circle contains areas of flesh tones and of bright colours to facilitate overall picture quality assessment and the correct adjustment of saturation. A child was chosen as the model rather than an adult so as to avoid the effects of changes in make-up fashion.

The circle containing the colour picture is outlined in white to avoid any 'optical illusion' effects which might occur if the coloured areas abutted directly on to the background grey and white grid. It must be emphasised that Test Card 'F' is intended primarily for technical purposes and is only secondarily a demonstration picture. Further, although provision is made for the critical appraisal of convergence it is not intended that the colour test card should be regarded as a substitute for a grid pattern generator when setting convergence.

5. *Transmitted picture limits.*   The limits of the transmitted picture
are indicated by the points of the opposing arrowheads in the border.
For receivers having a display area with an aspect ratio of about 5 : 4,
—the transmitted picture aspect ratio is 4 : 3—it is usual to adjust the
receiver so that the visible picture height extends from the top line of
the inserted colour bars to the point of the bottom arrowhead; if the
picture width is then adjusted so that the side castellations of the test
card just appear in the display area of the receiver, the aspect ratio of
the picture will be correct.

6. *Contrast.*   To the left of the centre circle of the test card is a column
of six rectangles with an overall contrast range of about 30 : 1. The
difference in brightness between adjacent rectangles should be approximately constant on a correctly adjusted receiver. Within the top and
bottom rectangles are small lighter spots, both of which should be
visible; white or black crushing results in the merging of the top or
bottom spot respectively into its surrounding area.

7. *Resolution and bandwidth.*   At the right of the centre picture are
six gratings each consisting of vertical stripes corresponding to the
following fundamental frequencies (MHz):

625-line UHF services	1·5	2·5	3·5	4·0	4·5	5·25
405-line VHF services	1·0	1·6	2·25	2·6	2·9*	3·4*

The gratings, on a correctly adjusted receiver, will appear to extend in
value from white to black with their surrounding area white.

8. *Scanning linearity.*   The white lines in the background should
enclose equal squares and the central white ring should appear truly
circular.

9. *Line synchronisation.*   The left, right and bottom borders of the
test card are formed by a pattern of alternate rectangles in black and
colours with a high luminance value. On black-and-white receivers
these rectangles appear as black and various lighter tones ranging from
grey to white. The right side of this border serves as a test signal to
check the line synchronisation of receivers. Faulty line synchronisation
shows as horizontal displacement of those parts of the picture on the
same level as the lighter toned rectangles in this side; it will also give
the central ring the appearance of a 'cog-wheel'. The right-hand

*After standards conversion to 405 lines, these will not generally be resolved
on a receiver display.

castellations, being yellow and white, provide a check on sync.-separator performance in the presence and absence, in 625-line transmissions, of the colour sub-carrier. The spacing of the left- and right-hand castellations has been staggered to make it clear from which side any disturbance arises and to give the maximum possibility of phase change resulting from a gating error.

10. Low-frequency response. Low-frequency response can be checked from the appearance of the black rectangle within the white rectangle at the top centre of the test card. Poor low-frequency response shows as streaking at the right-hand edges of these rectangles and also of the border castellations.

11. Reflections. If there are reflections of the television signal, from hills or large buildings, these may result in displaced 'ghost' images of any significant feature of the picture. This effect will be most readily seen as displaced images of the white or black vertical lines, particularly where these are adjacent. The effects of short-term reflections are often most clearly shown by the noughts and crosses on the blackboard.

12. Uniformity of focus. In each corner of the test card there is a diagonally-disposed area of black and white stripes; the focus in these areas as well as in the central area of the test card should be good.

SERVICING PRECAUTIONS

In order to protect the public from the dangers of mains-connected chassis, a number of recommendations have been drawn up by the British Standards Institution (B.S. No. 415). The most important sections of the standard concern the measures to be adopted to prevent the user from having access to 'live' parts of the apparatus. Whether or not any particular part is 'accessible' is to be determined by consideration of a 'standard finger' consisting of a metal probe about the size and shape of the little finger of a human hand.

Points which are listed as normally requiring protection in receivers using a.c./d.c. technique include:

(*a*) *The Chassis.* The ventilation holes in the back-plate should be small enough to prevent access, and the back-plate itself should be removable only by means of a tool such as a screwdriver.

(*b*) *Control Spindles.* These should be either of insulating material or isolated from the chassis. Live spindles are considered to be acceptable only if the fixing holes for grub screws are subsequently filled with insulating material.

(*c*) *Fixing Screws*.   All screws securing such parts as the chassis, loudspeaker, etc., should be isolated from the user.

(*d*) *Chassis Outlets*.   All outlets from the chassis, such as aerial and earth terminals, should be isolated from the chassis.

It is also recommended that apparatus should be tested for insulation by the application of a test voltage between the live parts and the safety earth provided.

General precautions

(1) The chassis of almost all modern receivers are connected to one side of the mains supplies, and may therefore be 'live'.

(2) Never attempt to measure the voltage at the anode of a line output valve or the cathode of the boost diode directly, as the e.h.t. pulses will damage the meter.

(3) Before removing the picture tube make certain that the e.h.t. capacitor is discharged: remember that Aquadag coating may hold its charge for long periods.

(4) Removal of a scanning-coil connection plug while the receiver is operating may cause picture-tube screen burn.

(5) Never assume a receiver mains switch is operating correctly— disconnect at the mains plug.

(6) Never test for e.h.t. sparks around a transistorised line output stage.

(7) Never measure voltages around a thyristor or silicon controlled switch (s.c.s.) with a multimeter—use a high impedance valve or transistor voltmeter.

(8) The greatest care should be taken in handling picture tubes to avoid breakage—and hence implosion. Wear goggles and antislip gloves when handling a shadowmask tube.

(9) Never dispose of old picture tubes in public places.

V.H.F. TURRET REPAIRS

The repair of faulty turret tuners often requires delicate workmanship, and less skilled work is likely to impair rather than to improve results. It is for this reason that many manufacturers recommend that turrets should be returned to them for repair or adjustment.

For instance, the adjustment of the bandpass transformer coils between the r.f. amplifier and the mixer is generally considered most inadvisable, as this operation really requires laboratory-type alignment equipment. Even the adjustment of the oscillator and aerial coils should be tackled with care and as for all adjustments, the correct size and shape of alignment tool should be used.

The most frequent fault in turret tuners is oxidation of the stud contacts, resulting in noisy or intermittent reception on one or more channels. The studs are readily accessible by removing the turret cover. Conventional switch-cleaning fluids, such as carbon tetrachloride, will remove the oxidation but do not provide any protection against the fault reoccurring and, unless used extremely sparingly and carefully, may attack the turret materials. A safer method of removing oxidation is to polish the studs with a dry cloth. The channel-selector mechanism on turrets should be kept lubricated, and after cleaning it is advisable to apply a smear of petroleum jelly to each stator spring contact. There are also available protective lubricants, such as 'M.S.4' and 'Electro-lube', containing water-repellent substances, and these provide useful protection against oxidation.

It is important that no attempt be made to increase contact pressure indiscriminately by adjustment of the spring contacts, although some-times adjustment to the switch-locating device will change the point of contact slightly. The contact pressure should be adjusted only where there is clearly a loose spring.

A common cause of short-circuits is the wearing off of the insulating paint on the ends of the small ceramic capacitors, leaving an exposed wire which may rub against another component. Faulty components may be the result of overheating during soldering, either in manufacture or subsequently.

Where any attempt is made to replace a component, it is essential that the tuner should be handled with great care, and that only an exact replacement, physically as well as electrically, should be fitted in the precise position of the original component, with connecting wires of the same length and path.

Coils are often adjusted by spreading their ends; great care should be taken not to disturb them accidentally.

Noise in a turret tuner may be caused by poor rotor earthing. If a steel wire rotor retention spring is used for earthing purposes dirt or tarnish on this spring or loss of tension will affect earthing. Loss of tension in the spring holding the rotor in position for the various channels may also cause noise or loss of gain. Some tuners are fitted with a wafer switch at the aerial end of the tuner for selection of separate Band I and Band III sensitivity controls; dirt on this switch will also give rise to noisy performance, so that the switch should be cleaned if in doubt. Poor contact of valve pins is a further possible cause of noise; remove the valves and apply a light smear of silicone grease to the pins, rocking the valves gently after replacement.

Loss of gain in a tuner unit is often due to low emission in the r.f. amplifier valve or to a faulty r.f. amplifier transistor. A convenient check is as follows: advance the contrast and sensitivity controls to maximum, then reduce the brilliance control until the peak whites of the picture are just visible. Switch off and fit a substitute valve; if this restores gain, the rest of the picture will then show up. Other causes of loss of gain in tuners are: poor spring contacts, weak oscillator (check the anode load resistor in a valve tuner, which may have gone high-resistance), a faulty i.f. output screened lead, or incorrect biasing of a cascode r.f.-amplifier stage. A fault that can arise with cascode stages is incorrect d.c. bias of the 'top' section: increase in the value of the upper section of the potentiometer network supplying the grid, or an open-circuit grid decoupling capacitor, can lead to low gain in this type of circuit. Lack of gain may also be due to a fault in the a.g.c. circuit (check by shorting the a.g.c. feed to chassis).

Instability is more often caused by a fault in the i.f. sections of the receiver than by a fault in the tuner. However, faulty decoupling capacitors, poor earthing between the tuner and the main receiver chassis (check screened lead outer connections), or screening cans missing or not making good contact to chassis are possible causes.

U.H.F. TUNER REPAIRS

Unless very high quality alignment equipment is available, the average workshop should not attempt more than basic fault-finding on u.h.f. tuners. The majority of electrical faults are due to ageing transistors or valves. The older u.h.f. tuners can be particularly troublesome in their demands to have almost perfect valves in circuit—and usually the most critical position is the mixer/oscillator stage.

Valve screening cans should always be replaced and the external cover of the u.h.f. tuner should always be securely refitted together with its internal copper shielding. Transistor replacements, if made, should always be undertaken with very great care: lead lengths should be cut similarly to the original and the transistor should be supported in its 'screened hole' in the same way as the original.

If a fault is traced to a u.h.f. tuner, and the supply voltage, the heater supply (in a valve tuner), and the a.g.c. rail and output leads all appear to be in order, check any external dropper resistors mounted on the tags of the tuner. If no fault is obvious from these basic checks and if transistor or valve replacement fails to cure the problem, the tuner should be returned to the manufacturer for repair.

Mechanically there are few problems on push-button tuners but sometimes tuning threads may break or the locking plate may wear. These can be easily replaced as they are not part of the tuner itself. Continuously-tuned receivers come in two varieties—those with a small tuning knob and a separate 'display' of the channel being tuned, and those with a slow motion drive with the channel marked directly on the tuning knob. The latter rarely give problems but the separate 'display' variety can cause difficulties with broken cord drives and sheer mechanical inelegance. The same advice that might be given for radio receivers applies; always use a length of cord either as specified by the manufacturers or of a length sufficient to give some tension on the mounted cord system, but without too much tension (or else the cord will snap) or too little tension (when the cord will slip). Always make sure that the guides and rollers are completely free from grease and always use the specified number of turns of cord around spindles.

SERVICING PRINTED-CIRCUIT BOARDS

Tracing faults on printed-circuit boards calls for special techniques. It is not usually an easy matter to disconnect components, or even to trace out the connecting path of the printed wiring. Resistance and continuity checks are thus made difficult. Soldering requires extra attention, as application of too much direct heat tends to loosen the bond and raise the foil from the board. Great care must be taken not to allow solder to flow beyond the necessary connecting area and cause short-circuits. The construction of printed-circuit boards generally means that components will be closely packed, and often quite small; great care should be taken not to apply heat from the soldering iron where it may damage components.

Many manufacturers give some regard to serviceability by numbering components, or even printing a colour-coded 'wiring plan' on the surface of the board, and without doubt, this is a great help to the fault-finder.

As most boards and panels are translucent, a strong light held behind the board will aid circuit tracing, the print showing as a silhouette. Care must be taken, especially in confined spaces, not to damage components by close application of a hot lamp.

Perhaps the most elusive trouble is a hairline crack in the foil, often caused by flexing of the board. This can lead to intermittent symptoms of all kinds and is extremely difficult to trace—particularly if hit-or-miss methods are employed. A magnifying glass is a great help in detecting hairline cracks (*Figure 7.2*).

Where tests reveal the presence of hairline cracks, it is sometimes possible to scrape the adjacent foil and remove the solder resist or subsequent coating of varnish, then overlay the break with a thin skim of solder. Much more efficient, however, is a 'jumper' of tinned wire laid across the break and soldered firmly to the good portions of the print (*Figure 7.3*). These jumpers should not be too long, or loose, or they may themselves lead to more faults by straining the bond between foil and board. If a long lead has to be used, it is better to employ an insulated piece of wire and solder to an existing joint at each end of the portion of broken or raised foil.

After soldering, a clear lacquer can be applied to joints to provide the same protection generally given to the complete board. Before soldering, and sometimes before making tests, it may be necessary to remove this insulating coat of protective varnish, and acetone can be used as a solvent for most boards. Apply sparsely with a soft brush or

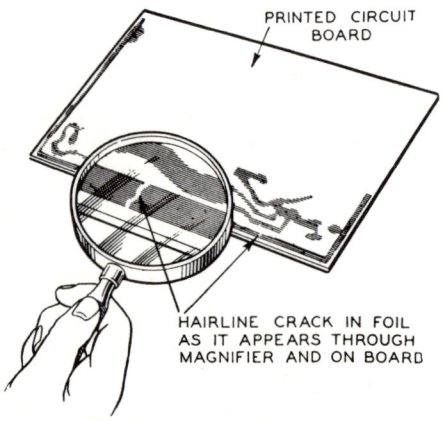

PRINTED CIRCUIT
BOARD

HAIRLINE CRACK IN FOIL
AS IT APPEARS THROUGH
MAGNIFIER AND ON BOARD

Figure 7.2   Using a magnifying glasss to examine printed circuit wiring for hair-line cracks

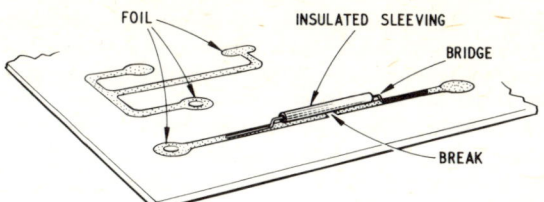

Figure 7.3   Bridging a break in a printed-circuit path
The technique is useful for mounting replacement components.

cloth and clean off carefully afterwards. Gentle treatment is the key to successful printed-circuit work.

Component replacement has its own peculiar problems on printed-circuit boards. In the first place, the method of mounting should be determined, and the method of removal will depend partly on this factor. Secondly, replacement with a component which may be electrically equivalent but physically different may require some modification to the mounting. To take a practical example: a miniature capacitor may need to be replaced with a larger component which does the same electrical job. Instead of a point-to-point lateral mounting, it may be necessary to employ the vertical type of mounting, and bend the lead wires of the component so that it is self-supporting. Care should then be taken that movement will not cause subsequent joint strain.

Another problem is the removal and replacement of the special types of variable controls used with printed circuitry. These may be soldered directly to the board, with the only support provided by the lead tags, or may have support brackets and tags all soldered into place. The same problem arises with multi-tagged components such as integrated circuits. The difficulty is to remove the solder from the tags either simultaneously, or with sufficient care and dexterity to be able to withdraw the control complete, and, what is equally important, avoid damaging the printed circuit. If the control is faulty and needs changing, it is often an advantage to cut off the tags, leaving only the stubs to be removed individually. The holes are then cleaned out and made ready for mounting the new control. A similar problem arises when valve bases have to be removed, and in this case the additional difficulty caused by the screen mounting makes it necessary to remove a fair amount of solder before the mounting is loose enough to be withdrawn. The risk of damaging foil bonds by excessive heat becomes very real. There are now special tools on the market, designed to suck surplus molten solder from joints, component mountings, etc. Some of these are quite sophisticated, with trigger-operated pumps and bench mountings, specially designed for production line use but a number of simple, cheap, spring loaded tools are available. With a little ingenuity and adaptation it is also possible to use a liquid detergent 'squeeze-bottle' for this purpose. The important point is that the molten solder should be removed cleanly from the joint without the application of too much heat and one of the methods suggested should always be used.

Useful for cleaning out fixing holes is a sharpened matchstick. As a component is removed the matchstick is inserted in the hole, while the solder is still molten. This keeps the hole clear, and the wooden sliver does not attract solder and is thus easily withdrawn. A simple trick, but remarkably effective.

Where component replacement is necessary, it is sometimes better to avoid disturbing the printed circuit by simply snipping the mounting leads of the existing component at a convenient length above the board, then looping the lead wires of the new component around these ready-made anchor points. The important thing to remember is to use a heat shunt between the new solder point and the board to avoid loosening the original joint. *Figure 7.4* shows the method. Where the end leads of the existing component are too short, and it is not convenient to mount the new component in the original holes, the end wires of the new component should be bent to provide a right-angled step and soldered to the approach portions of foil to give a good contact area. A similar method can be used when making a jumper strip to bridge broken sections of foil.

When making off solder joints, care must be taken to avoid blobs, to prevent solder running over unprotected portions of the board and

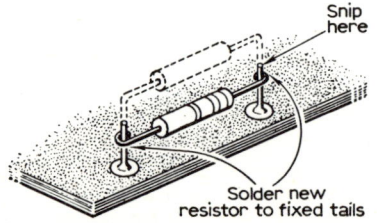

Figure 7.4   Method of mounting a new component using the connecting leads of the previous one

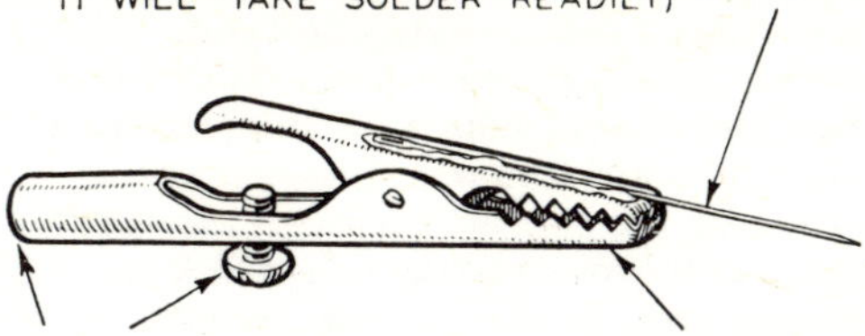

PIN OR NEEDLE SOLDERED TO CROCODILE
CLIP (NEEDLE SHOULD BE BRASS SO THAT
IT WILL "TAKE" SOLDER READILY)

SCREW OR END MADE   CROCODILE
ADAPTABLE FOR OSCILLOSCOPF   CLIP
OR TESTMETER PROBES

*Figure 7.5   Construction of a needle-point probe for testing
printed-circuit panels*

shorting to adjacent sections of foil, and also to reduce the risk of
raised blobs or spikes of solder contacting mounting brackets, etc.,
when the panel is refitted. The iron should apply just the right amount
of heat to make a good joint, should not be kept near the joint too long
and must be kept clear of other joints or components. Many small
irons have slim shanks for this purpose, and at least one popular range
employs a wire shank-guard that is very useful for avoiding accidental
damage to adjacent wiring or components in confined spaces.

Heat shunts are necessary to avoid conduction of the heat to the
board, and there are now a number of proprietary devices for clipping
on lead wires where a good deal of soldering has to be done. For the
occasional job, however, the end of a long-nosed pair of pliers, gripping
the wire between the heat sourse and the vulnerable part, serves the
purpose adequately. The vulnerable part, in this case, can be either the
foil or the component itself, many of which are small enough to be
damaged, even by radiated heat.

Reference to a 'sucker' device for removing solder has already been
made. Such a tool should always be used to take away the solder from
a multi-tag component after the solder has been melted and before any
attempt is made to remove the component from the printed-circuit
board. Methods of removal using shaking, brushing or similar techniques should not be used as they only encourage further damage and
can spread pieces of solder on to unwanted positions. One of the great
secrets of removing multi-lead components, and in fact in any solder
work, is to use the correct sized iron—both bit size and power—for the
particular job. This is something that is learnt with experience.

It becomes clear that random removal of components is unwise. Testing must be carried out according to a logical scheme, and a little longer spent studying a circuit will pay dividends.

Resistance tests need some care, where semiconductor devices are in circuit, as their presence will both modify the reading and restrict the voltage of the battery which powers the instrument.

To make any tests, it is often necessary to pierce a protective coating of lacquer put over the foil or solder points. A sharply pointed probe can be made and kept for this purpose, piercing the varnish where needed, rather than applying acetone solvent to test points (*Figure 7.5*).

In general, the common 60–40 resin-cored solder will prove quite adequate, except for special applications, such as chassis mountings.

When testing transistorised circuits with a meter, avoid using the wrong range. On many meters, a 'high ohms' switch position is afforded by adding a higher voltage battery. One popular model uses a 15 V battery for this purpose, which may be considerably more than the circuit is intended to take. Similarly, when switching ranges, first remove the test probe, to avoid inadvertently imposing a short-circuit as the switch is rotated. Always disconnect the receiver or equipment supply when making ohms tests.

8

SERVICING EQUIPMENT

When asked for a 'priority list' in purchasing servicing equipment for television we must assume that 'television' includes colour automatically. The writer's list would then be:

Absolutely essential

(1) Signal generator
(2) Universal meter (at least 20 000 Ω/V)
(3) Pattern generator
(4) Oscilloscope
(5) E.H.T. voltmeter or adaptor for item (2)

Very desirable

(6) Wobbulator
(7) Component test bridge
(8) Insulation tester
(9) Signal-strength (field-strength) meter
(10) Transistor tester
(11) Valve tester
(12) High-impedance valve or transistor voltmeter
(13) Crystal calibrator.
   This list is not exhaustive but is only a guide.
   When investing in test gear, it must always be borne in mind that the criterion is that the equipment should earn its keep. On the other hand, equipment unintelligently used will never pay. To be satisfactory, the gear must be reliable and always give the same reading in similar circumstances. It must be reasonably easy to handle, so that it does not mislead. Nevertheless, unless test gear is handled and used wisely it can be a time waster. Hours can be spent on the niceties of alignment,

which, although they may appear to have considerable influence on the measured responses, hardly alter picture quality one iota. A good engineer is made efficient and productive by good test gear, but good test gear will not make a bad engineer into a good one. But, what is more important, a trainee engineer, trained to use good equipment properly, has a better chance of becoming an efficient engineer than if he has little gear and has to acquire more instinctive skill before he can show results.

Signal generators

At one time these could be divided into two classes, more aptly named test oscillators and standard signal generators. The former consists of a tunable oscillator, which can be modulated, together with means of attenuating the r.f. output. The second class has all these features, but also has a means of setting the output level to a known and repeatable value on all frequencies. The test-oscillator output varies, perhaps widely, with frequency, rendering comparative tests some-times quite misleading. Years ago the difference in performance was very great, but clever design has produced most efficient test oscillators with remarkably constant output which enables quite good relative checks to be made. It is necessary to stress that the signal generator *must* tune to the fundamental television frequencies. In the writer's opinion the suggestion of using harmonics is a diabolical one designed to mislead the unskilled and unwary!

It would be preferable if service stations used metered signal generators exclusively, but the modern cheaper variety is so good, in its best examples, that cost often overrides. It is, however, suggested that all service stations should have at least one standard metered generator to serve as a check against deterioration of standards of performance.

One essential feature of a signal source is good frequency stability and resetting accuracy. Unfortunately the requirements of television sets working on single sideband are very high, in fact higher than can be reasonably expected from the most expensive laboratory generator. No tunable oscillator of normal type can be expected to hold over a long period to as good as 0·1 per cent, including resetting accuracy, and few manufacturers guarantee better than 1 per cent. The variation of stability of the television set will absorb most of the tolerance available, therefore the source stability should be much better so that the error from it is negligible. This means that it should be certainly within 20–30 kHz of nominal, which is round 0·03–0·04 per cent.

Clearly if we demand such stability from our signal generators we shall either delude ourselves or be disappointed.

The simplest way of overcoming this is to beat the carrier of the signal generator with the station, so synchronising them, and noting how much the generator deviates and how it drifts during working periods. A little experience soon shows how frequently this must be done, and how closely the required channel can be set up without reference to the station. This is not an ideal method, and sometimes it is not possible to use it. The ideal method is a crystal calibrated wavemeter or oscillator.

Universal meters

The ubiquitous Avometer here comes to mind, although there are other good instruments available. For television work it is often essential to have a high-resistance meter of 20 000 Ω/V, such as the Avo Model 8. However, some manufacturers' service data still calls for a 1 000 Ω/V meter such as the Avo Model 7, so that it is normally necessary to have both types at hand.

In choosing an instrument it is very important to be sure that good overload protection is incorporated. The extra cost is saved in every shop sooner or later.

Pattern generators

A modern pattern generator should provide at least the following functions to enable it to be of general use:

(1) cross hatch pattern for convergence alignment
(2) dot-pattern for static convergence
(3) colour-bar signal
(4) grey-scale signal
(5) blank raster signal, red, green or blue, for purity
(6) tunable over all v.h.f. and u.h.f. bands, 405 and 625-line
(7) built in tone oscillator for checking and tuning sound channel.

It is also very desirable that the pattern generator should give a video output as well as an r.f. output so that it can be used separately for checking the operation of colour decoders and so on. It is also of great assistance if the PAL information can be switched off at the generator.

Although it is virtually impossible that an economic generator should provide broadcast standard waveforms it is very desirable if the wave-forms can at least approach this standard otherwise some results can be very misleading in terms of interlace, hum performance, etc.

Oscilloscopes

There is much difference of opinion over oscilloscopes, some finding them of inestimable value, and others not. Probably the reason for this lies partly in the handling, and in knowing how to interpret the results. If the 'scope amplifiers have a reasonably high gain, and the input of the 'scope is connected to a high-impedance circuit, the resulting trace may be modified or completely upset by spurious pick-up, mains hum, etc. The attachment of the 'scope may also disturb the operation of the circuit under test. Only experience can show how to overcome these difficulties in different cases, but a good high-impedance probe can be an investment that very quickly more than pays for itself. Some points to look for in choosing a 'scope are: good sensitivity variable over a wide range, wide frequency range, a good linear timebase and most important, an ability to trigger easily on line and field waveforms and signals which are constantly changing. For many purposes a good h.f. response is unnecessary, but it is invaluable when checking a pattern generator, or when looking for spurious h.f. oscillations on l.f. circuits, and it is essential for realistic measurements of chrominance waveforms (i.e. the bandwidth must extend to at least 5 MHz).

E.H.T. voltmeters

The measurement of e.h.t. voltages has always been a very difficult problem. Very little current is available for a meter, and leakage due to humidity is a particular snag. Two methods, which can be satisfactory, are to use an electrostatic meter, or a very low current (20–25 μA) meter with high series resistance. Both methods are unfortunately costly. However, adaptors are available for extending the range of a meter such as an Avo to read up to 30 kV.

Whichever system is used, the meter should be accurately calibrated; the breakdown tolerance on a 25 kV shadowmask tube is only 1 kV (i.e. meter must be accurate within 4 per cent).

Wobbulators

A wobbulator (sweep generator) or frequency-modulated oscillator, together with an oscilloscope, is undoubtedly by far the best apparatus for lining up television sets to the correct bandwidth accurately and

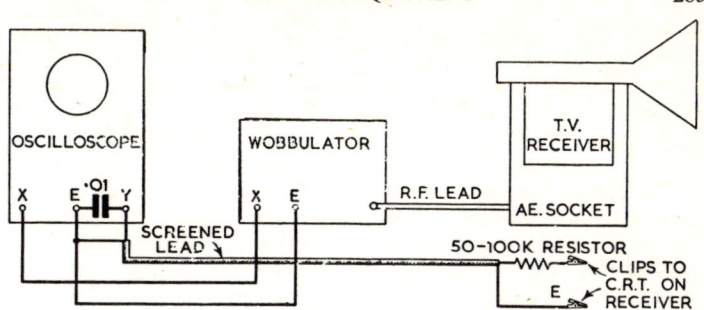

Figure 8.1 Wobbulator connections
N.B. Series earth capacitors to be fitted where necessary.

quickly. As usual, however, there are snags. The apparatus must be right, it must be connected appropriately and must be used correctly. Matters have not been helped in that makers' instruction books do not always emphasise the pitfalls sufficiently.

In the first place, optimum alignment should be made at the working sensitivity of the receiver, or at a lower input signal. As most reasonably priced wobbulators are not fitted with accurate attenuators, this sometimes leads to trouble. Secondly, the oscilloscope must be connected in such a way as not to affect the operation of the receiver. Practically, the output is normally taken from the cathode of the cathode-ray tube. If a clip is fitted to a resistor of 50 000 Ω upwards, which is connected in series with the conductor of a screened cable, with a further clip from the screening for earthing, this can be connected to the 'scope and will not usually upset the receiver at all. If it does, slight readjustment of

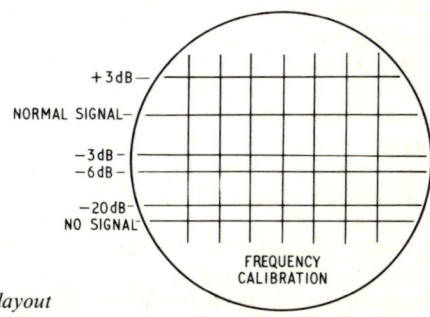

Figure 8.2 Typical graticule layout

leads or resistance should remove the trouble. Next factor of importance is the sweep frequency. This must be low; $16\frac{2}{3}$ Hz ($\frac{1}{3}$ mains frequency) is good, 25 Hz should be the maximum. This is on the limit of a small oscilloscope (and some larger ones), which distorts the waveform if the frequency is too low for it. It is also good to connect a capacitor of about 0·01 μF across the input to the 'scope amplifier.

The method of operation of a wobbulator differs somewhat with different sets. Usually if the set is not far from alignment, injection into the aerial or at i.f., and alignment in order working from the circuits nearest the demodulator through finally to the aerial, is the drill. Sometimes it is necessary to inject into each grid in turn, dealing only with the immediately subsequent circuits in turn, arriving finally at the aerial end. Each type of receiver has its own idiosyncrasies, but experience is the only way to find the optimum method in any particular case. It is a good thing to make a graticule showing the calibration of the instrument in frequency in the horizontal (X) direction and calibrated in dB in the vertical (Y) direction, say at 3 dB up, 3 dB, 6 dB and 20 dB down on the normal deflection (3 dB down is 70 per cent, 6 dB down is 50 per cent, 20 dB down is 20 per cent of normal, but 3 dB up is 140 per cent). Here is a wonderful opportunity for time-wasting, trying to get a perfect curve. It must be remembered that getting the carrier at the 3 dB or 6 dB point is important, but that dips or rises of 2 or 3 dB over the band will hardly affect the picture. As long as the carrier is right and the bandwidth adequate small irregularities are of no consequence.

Another important thing is the frequency setting of the wobbulator. If a signal from the generator is injected into the wobbulator (if provided with a suitable terminal) or through a T-pad with the wobbulator, a kink will be seen on the trace at the frequency corresponding to the generator frequency. The 0·01 μF capacitor across the 'scope will prevent the high-frequency beats from widening the trace, and enable an accurate adjustment to be made. Care must be taken that the signal-generator signal is not too strong, or it may modify the trace from the receiver, and so mislead. A better method is to feed from the T-pad into a crystal rectifier with a load of say 10 000 Ω in series. The 'scope (with 0·01 μF capacitor) is then connected across this 10 000 Ω resistor. The marker can then be used to check the linearity of sweep and calibrate a graticule.

Sweep generators can be purchased to give u.h.f. signals but they are *extremely* expensive. For u.h.f. receiver work one must assume the tuner is aligned correctly and use an i.f. injection signal. Poor results 'off-air' must then be assumed to be due to a tuner malfunction.

Sweep generators can also be purchased with built-in oscilloscopes. These are uneconomic if a good quality oscilloscope is already available. (Note that an oscilloscope for use with a wobbulator or sweep generator must have a separate X input).

Component test bridges

Many faults arise due to components, mainly capacitors and resistors, changing in value. These are best checked on a bridge. The requirements are that the full probable range of values of capacitance and resistance can be measured reasonably accurately, say to within 5 per cent at worst. Even this is a fairly stringent requirement. It is sometimes an advantage to have the facility to measure inductance as well, but for a number of reasons this facility is not as useful as it would appear. Note that basic checks of resistance and capacitance can be made on a multimeter.

Insulation testers

A large proportion of faulty components in television receivers have faulty insulation, and the writer believes that much inferior performance can be attributed to leaky capacitors in particular. In the writer's laboratory all such components are tested on 500 V d.c., with quite startling results. For such tests, an instrument giving a maximum reading of 200 MΩ is the minimum requirement, but high accuracy is not needed. The Wee Megger needs no introduction, and particularly when extreme portability is required, such types with hand-driven d.c. generators have no rivals. It is possible to make a very simple instrument using eight $67\frac{1}{2}$ V deaf-aid-type batteries, a few resistors and a 50 μA meter. The battery drain never exceeds about 50 μA, and the life is extremely long; some have been in use for seven years before running down.

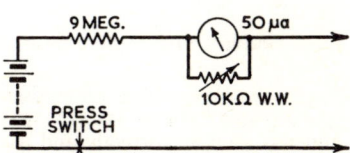

Figure 8.3   Insulation tester

Signal-strength meters

Such a device as a signal-strength (or field-strength) meter is valuable in two main connections. Firstly, in assessing strength of signal at new locations, and secondly, in testing existing aerial installations. It is important to make such measurements not on picture content, but on peak white or synchronising levels.

In recent years meters have become available at most competitive prices; these are usable on v.h.f. and u.h.f., are available in battery versions and often have both meter and sound output monitoring.

Transistor tester

D.C. resistance checking with a multimeter can often indicate whether a transistor is operating or not, but one cannot get any idea as to whether the device is up to its specified gain, leakage, and so on. A number of commercial transistor testers are available for this purpose. It might also be an idea in a less financially endowed workshop to build such a tester; this is reasonably simple and good designs appear from time to time in the popular technical magazines.

Valve testers

A valve tester can be a great help, but it must be borne in mind that it is no reflection on the instrument that it will undoubtedly fail to reveal some valve faults. If it were not for this, a valve tester would rate higher in the priority list. The writer personally prefers a valve characteristic meter, which enables values to be measured on a meter.

In future years the number of valves in use will fall to negligible proportions. Unless a workshop is specialising in 'old-receiver' repairs the use of a valve tester will fall. However this also means that second-hand valve testers have become available at only a fraction of their original cost.

Valve voltmeters (or high impedance transistor voltmeters)

High-impedance valve or transistor voltmeters do approximately all that a Universal Test Meter will do, and much besides. Unfortunately they can be fairly delicate and a shade tricky to handle. If such difficulties are appreciated, and care used in getting results, they are most useful adjuncts to any test shop. The uses are legion, but good engineering skill or supervision is needed. The lack of these and a failure to appreciate the limitations of technique have given this type of instrument a bad name not altogether deserved.

These meters are essential for voltage measurements on high-impedance circuitry and where even a 20 000 Ω/V meter would cause loading breakdown (for example, around the silicon controlled switch (s.c.s.) used in some field oscillators).

Conclusion

Instruments are not the only items of importance in a repair shop. The obvious ones of good bench space and so on need no stressing. But a point overlooked too often is that television sets can be lethal. Every test position must be equipped in accordance with the Factory Acts and regulations.

An essential is a properly screened double-wound transformer of 1 to 1 ratio. The screen should be well earthed. An earth bar may be fitted at the back of the bench, but it must only be connected to earth through a series capacitor of high test voltage and not more than 0·02 μF capacitance. Where earth terminals of instruments would be connected effectively direct to the mains via a live chassis, isolating capacitors of similar type are essential.

The whole load of the bench or test position, for one engineer only, should come from one transformer secondary, connected to a multiplicity of sockets of all likely types and sizes. Two engineers should never share a transformer—they could get two chassis not very far apart connected to opposite sides of the a.c. supply. This problem must be dealt with according to circumstances.

While the use of a double-wound isolating transformer having low leakage inductance has much to recommend it on the grounds of safety, it is unfortunate that some television-receiver fault symptoms and adjustments are affected by the use of such a transformer. For this reason it is often necessary to operate the receiver directly from the mains supply, and in such circumstances the best safeguard is to ensure —by checking with a neon bulb or similar tester of known reliability —that the chassis is always connected to the neutral and not the live mains lead, taking care to re-check each time that the mains lead is re-connected.

The introduction of kits using printed-circuit panels offers an economical method of building up a range of servicing instruments. The home construction of good test instruments has sometimes been regarded as a most tricky operation owing to the likelihood of even minor differences of lay-out and construction leading to considerable variations in results and in calibration. In these kits this difficulty is overcome by the use of printed-circuit panels giving a degree of consistency between instruments difficult to achieve by other means. These kits, including oscilloscopes, valve and multi-range testmeters, signal generators, etc., are supplied with step-by-step assembly instructions.

Kits of basic tools are other obvious essentials in a workshop and it is to be hoped that every servicing technician would have his own kit of essential tools with a workshop stock of heavier tools, power tools, cabinet polishing materials and components. To avoid difficulties it is also hoped that each member of the staff would have available for his own bench position the 'Absolutely Essential' servicing equipment already detailed.

Often disregarded as not being a direct servicing function, but of as great concern, is the availability and storage of manufacturers' servicing material and manuals. The ability of being able to obtain immediately the correct manual can save a very considerable amount of servicing time. In the same way one member of the staff should always be made responsible for updating manuals, etc. as the manufacturer sends out modifications.

9

TELEVISION FAULT FINDING AND ALIGNMENT FAULT FINDING
by 'Telegenic'

Fault finding in television receivers is perfectly straightforward once the circuit functioning is understood. Without that understanding, the process becomes laborious or intuitive. Some workshops have their 'genius' who looks at a receiver when it comes in for repair and coolly states 'that's C_{16} open-circuit'; he may have seen the particular symptoms on that particular type of receiver so often that the chances are that his supposition is correct. Unfortunately the same person may find very considerable difficulty in repairing a receiver with which he is less familiar or which is not displaying the fault symptoms with which he is familiar.

The whole essence of the efficient servicing technician is completely *logical progress* based on the fault symptoms of the receiver and an understanding of the circuit. This does not in any way suggest that fault finding must become a laborious chore, because logical progress in itself suggests the quickest and most reliable way of reaching the destination required—an operating receiver.

Particular advice on fault location in valve timebases has already been given in Chapter 6. This is included as a separate item because of the known difficulty that such timebase faults often give in receiver servicing. Other areas of the receiver are much simpler and it is not felt that such concentrated treatment is required. It would involve, for instance, repeating circuits and circuit details that have already appeared in Chapter 2.

Instead some general guidance on the location of a fault will be given followed by general advice on colour-receiver fault finding and receiver alignment.

It is emphasised that for all receiver fault finding an understanding of the circuit in question pays handsome dividends.

General fault finding

The objective in any servicing is to locate the faulty component, socket, connection or whatever that is causing the receiver to operate in an unsatisfactory way. In all fault finding we first locate the general area of a fault and only then begin to look for an individual component or joint that may be faulty. It is common practice, and perfectly logical, to locate the general area by working towards the source from a known faulty symptom; distortion or loss of vision, for example, would cause us to start at the picture tube and work back through the receiver until we find the area where the fault is located.

We might in this case, find that there was no video on the cathode of the picture tube; so we then look at the input to the video amplifier and find that there is drive present. The conclusion drawn would be that the video amplifier stage was faulty and we could then look at the individual components of the stage to isolate the fault further: substitute valve, check valve voltages, and so on.

In the same situation the procedure would differ if it was found that there was no video from the detector. In this case we would feel that the fault might be in the i.f. strip or earlier. If we have the right test instruments we could then repeat the process—output of last i.f. amplifier, output of first, output of tuner and so on—but the majority of workshops do not have adequate facilities for measuring or examining an r.f. signal. So instead we reverse the process by saying that as the the video amplifier stage appears to be in order we will now inject the correct r.f. signal at points earlier in the chain.

Here, for example, we would start by injecting a signal at the intermediate frequency (vision) at the final i.f. transformer. We should see a detected pattern on the screen. If we do the video detector is in order, and we then repeat the process with a signal at the input of the last i.f. amplifier and move back towards the aerial, stage by stage, *until the signal is lost*. We then know that the fault is in the stage after the injection point.

So, in general we always move from the output point back towards the aerial, stage by stage; if the signal we are looking for is video, audio, line or field signal then we use a suitable detection instrument (meter or oscilloscope); if the lost signal appears to be in one of the r.f. stages (including i.f.) then we inject signals from a signal generator or pattern generator and work back towards the aerial again. The only secret of speedy servicing is the ability to bypass some of these stages of checking because the symptoms clearly indicate such a move—a logical step.

If, for example, there was no sound or vision but a raster was still present on the screen. We know immediately that the h.t. is all right—at least in the timebase part of the receiver—that the heater chain is at least continuous, and we can draw other conclusions. If the receiver was a single-standard u.h.f. one then we know that the fault is before or at the video detector—if it was after, the vision or the sound would be affected but not both. So unless we were very unlucky and we had two simultaneous faults we would neglect unnecessary checks of the video output channel and the sound i.f.

So, provided the circuitry of the receiver you are repairing is known to you there should be little problem. But, there are always snags that can arise. The worst of these is not being able to obtain the correct circuit diagrams for the receiver, when, unless the arrangements are fairly standard, you can waste a great deal of time.

There would therefore be little point in listing all the possible faults that could give rise to various errors on picture or sound. They are only confusing and lead to a general malpractice in the art of servicing. Again, although it is very nice to know that 'such and such components' are regularly failing on one particular model of receiver it does not usually lead to proficient servicing. This is emphasised because the table given below does not fall into this pattern. It is intended to give some assistance *after* a faulty area in the receiver has been located. In general we should use an oscilloscope for tracing video and pulse signals and an oscilloscope or a multimeter (switched to a suitable a.c. range) for tracing audio.

Table 9.1   Fault location chart for monochrome receivers (Chart is for single-standard 405-line or 625-line receivers and dual-standard receivers; valves and transistors are implied but some points must obviously be disregarded if they are not fitted on the receiver being repaired).

Blank screen, no sound

(*1*) *Heaters dead.*   Check mains fuse; heater dropper resistor; mains tapping panel or connections; open-circuit valve or c.r.t. heater; open-circuit power plug to tuner or other subchassis; heater chain thermistor; heater chain rectifier.

Table 9.1 *Continued*

(2) No h.t.   Check h.t. fuse; h.t. mains dropper resistor; h.t. rectifier. Check for an h.t. short-circuit if the h.t. fuse is blown.

Blank screen, sound normal

(1) No e.h.t.   Check e.h.t. rectifier or e.h.t. multiplier; efficiency (boost) diode; line output valve or transistor; line oscillator valve or transistor or driver transistor; line blocking or sinewave oscillator transformer; flywheel sync diodes; drive to line timebase (negative voltage at grid of output valve); boost reservoir capacitor; other timebase components; line output transformer.

(2) C.R.T. biasing.   Check first anode voltage; c.r.t. grid and cathode potentials; grid and cathode biasing and decoupling components including v.d.r. in brilliance circuit if fitted; video amplifier if direct coupled.

(3) C.R.T.   Check for open-circuit heater; low emission; ion trap missing or mispositioned; poisoned cathode (ionised).

Raster ok, no or poor picture, 405 sound ok, 625 poor or no sound

(1) Video amplifier.   Check valve or transistor(s); video detector; video amplifier components.

(2) Vision i.f. strip.   Check valves or transistors; power supply to strip including dropper resistor to transistorised strip; components in strip.

Raster o.k., no or poor picture, no or poor sound on both systems

(1) Common i.f. stage.   Check valve or transistor; interconnecting coaxial lead from tuner to i.f. strip; other components in common stage.

(2) Tuner.   Check r.f. amplifier and oscillator/mixer valves or transistors; switch contacts; other components.

Table 9.1   *Continued*

(*3*) *Aerial.*   Check coaxial plug to receiver; aerial connections; aerial location; receiver input components.

(*4*) *Low h.t. to tuner.*   Check h.t. rectifier; dropper resistor in feed to tuner(s).

Picture o.k., no or poor sound

(*1*) *Audio amplifier and output stage.*   Check for open-circuit loudspeaker leads; valves or transistors; other components.

(*2*) *Sound i.f. and demodulator.*   Check demodulator diode(s); valves or transistors in strip; other components in strip.
    If noise is heard on turning up the volume control the circuits following the control are in order.

Intermittent sound or vision

(*1*) *I.F. circuits, video or audio output circuits.*   Check joints on chassis or printed board; printed circuit board for hairline cracks; cracked or overheating resistors; chassis contacts of screening cans; coaxial cable connections between tuner(s) and i.f. strip; dirty valvebases or pins.

(*2*) *Tuner.*   Check for badly aligned turret contacts or poor tension; locating cam loose or broken; mechanical rigidity; poor aerial socket contact.

(*3*) *Aerial.*   Check connections; mechanical stability; shorting along feeder length caused by cable grip or feeder rubbing on sharp edge.

Uncontrollable brightness

(*1*) *C.R.T. biasing.*   Check grid and cathode bias components; video amplifier valve or transistor(s); d.c. restoration diode.

(*2*) *C.R.T.*   Check for open-circuit electrode or inter-electrode short.

Table 9.1   *Continued*

Varying width and height

(*1*) *E.H.T. generation*.   Check e.h.t. rectifier; efficiency diode; boost reservoir capacitor; line output transformer.

(*2*) *C.R.T.*   Check for varying tube capacitances by replacement.

Lack of height

(*1*) *Field output stage*.   Check valve or transistor(s); cathode (emitter) bypass electrolytic and bias resistor; adequate h.t.; linearity circuit; leaky coupler; output transformer and damping components across primary.

(*2*) *Field oscillator*.   Check valve or transistor; check boost supply to valve (high-value feed resistor, faulty height control, boost reservoir capacitor, also possibility of interelectrode leak in tube); charging capacitor from anode of oscillator to chassis or cathode of output valve.

(*3*) *Field deflection coils*.   Check for possible shorted turns; coil damping resistors; insulation to chassis; thermistor in coil circuit.

Top or bottom compression

(*1*) *Field output stage*.   Check output valve or transistor(s); cathode bypass electrolytic and value of bias resistor; field charging capacitor; leaky coupler; components in linearity circuit; damping across output transformer and output transformer for shorted turns (these last two affect the top of the picture).

(*2*) *Field deflection coils*.   Check for possible shorted turns.

Lack of width

(*1*) *Line output stage*.   Check line output valve or transistor(s); boost diode; preset width control(s); e.h.t. stabilisation circuit; h.t. voltage;

Table 9.1 *Continued*

line output valve screen resistor; adequate line drive; S-correction capacitor(s); line output transformer.

(*2*) *Line deflection coils, width or linearity coil.* Check for possible shorted turns; insulation to chassis; core of width or linearity coil secure; width or linearity coil damping resistors.

Different 405, 625 widths

Line output stage. Check S-correction capacitors; width/stability control setting; stabilisation circuit.

Poor line and/or field hold

(*1*) *Sync separator.* Check valve or transistor; anode or screen resistors; grid leak or base bias resistors; heater rectifier in some Bush-Murphy models.

(*2*) *Video amplifier.* Check valve; screen decoupling; cathode bypass capacitor.

(*3*) *Sync coupling.* Check coupling components between sync separator and timebases including interlace diode in case of poor field hold.

(*4*) *Flywheel sync circuit.* Check discriminator diodes; flyback pulse coupling components; d.c. amplifier if incorporated.

(*5*) *Line oscillator.* Check valve or transistor(s); operation of hold control and its series resistors; oscillator stage components.

(*6*) *Field oscillator.* Check valve or transistor; hold control and associated components; leaky coupling capacitors in stage; interlace diode; heater chain rectifier in some models where the field output stage bias is derived from the heater chain; boost reservoir capacitor; interlace control if fitted.

Table 9.1   Continued

Horizontal line moving vertically

(*1*) *Hum.*   Check h.t. smoothing components; stage decoupling capacitors.

(*2*) *Valves.*   Check for heater-cathode breakdown.

Picture not central

(*1*) *Shift magnets or deflection coils.*   Check setting and position.

(*2*) *Field deflection coils.*   Check for d.c. through coils due to output transformer insulation breakdown.

(*3*) *Line deflection coils.*   Check for d.c. through coils due to output transformer insulation or capacitor breakdown.

(*4*) *Line oscillator.*   Check line hold control setting on receivers with large pull-in range.

Line ringing

(*1*) *Third or spurious harmonic ringing.*   Check width and linearity coils and damping resistors; line deflection coil inductance; line output stage tuning capacitors; line output transformer.

(*2*) *Spurious oscillations.*   Check chokes at top caps of line output valve and efficiency diode; line output valve screen decoupling.

Picture ringing

(*1*) *Excessive video h.f. response.*   Check h.f. peaking components in video amplifier and detector circuits and detector compensating network.

Table 9.1   *Continued*

(*2*) *Tuner*.   Check oscillator setting and level.

(*3*) *Aerial*.   Check for feeder mismatching at array and receiver; receiver aerial input circuit; ghosting.

(*4*) *A.G.C. circuit*.   Check a.g.c. circuit decoupling components.

(*5*) *Vision i.f. strip*.   Check alignment.

Poor definition

(*1*) *Poor video h.f. response*.   Check h.f. peaking components in video amplifier and detector circuits and detector compensating network.

(*2*) *Vision i.f. strip*.   Check alignment.

(*3*) *Lack of signal*.   Check aerial and siting.

(*4*) *Video signal level low*.   Check tuner and i.f. valves or transistors.

(*5*) *Aerial*.   Check for feeder mismatches on short cable run.

Poor focusing

(*1*) *C.R.T.*   Check first anode and focus potential; ion trap (if fitted) position; focus control circuit.

(*2*) *H.T.*   Check h.t. rectifier; boost reservoir capacitor.

Arcing and corona discharge

Check e.h.t. lead; e.h.t. voltage tripler; e.h.t. rectifier heater; internal arcing in efficiency diode; line output transformer.

Table 9.1 *Continued*

Picture distortions

(*1*) *Line pulling*.   Check for weak signal; poor aerial; low gain.

(*2*) *R.F. patterning*.   Check for foreign or co-channel interference; cross-modulation caused by excessive signal (check sensitivity preset); heater chain decoupling; radiating video amplifier; poor i.f. screening can chassis connections; oscillator setting; 4·43 MHz rejector setting (625).

(*3*) *Pincushion distortion*.   Check c.r.t. correction magnets; line linearity control setting.

(*4*) *Line nonlinearity*.   Check line linearity sleeve setting on c.r.t. neck; width control setting; efficiency diode (left-hand edge nonlinearity); line output valve or transistor (right-hand edge nonlinearity); line output valve screen voltage; S-correction capacitors (centre nonlinearity); cathode or emitter circuit of line output stage (centre nonlinearity).

(*5*) *Ghosting*.   Check aerial matching; aerial position; multipath reception.

Negative picture

(*1*) *Vision interference limiter or black spotter*.   Check setting of preset control; diode; bias components.

(*2*) *Low e.h.t.*   Check e.h.t. rectifier; line output valve; boost reservoir capacitor; h.t.; line output transformer.

(*3*) *C.R.T.*   Check for low emission; low heater current.

(*4*) *Signal overloading*.   Check preset sensitivity control setting; a.g.c. circuit; try inserting an aerial attenuator.

(*5*) *Ion trap*.   Check position.

Table 9.1 *Continued*

Sound-on-vision

(1) I.F. and video amplifiers.   Check 3·5 MHz trap setting (405);
6 MHz trap setting (625); sensitivity preset control setting; overloading
—a.g.c. circuit; sound trap.

(2) Tuner.   Check oscillator tuning.

(3) Microphony (increases with volume).   Check for microphonous
valves (particularly in tuner); microphonous c.r.t.

(4) H.T. feedback.   Check smoothing components; decoupling
capacitors.

Vision-on-sound

(1) Sound i.f.   Check first i.f. transformer (405); 6 MHz take-off coil
tuning (625).

(2) H.T. feedback.   Check smoothing components; decoupling
capacitors.

(3) Radiating video amplifier.   Check for 'soft' valve; heater decoupling.

(4) Microphony.   Check for microphonous valves.

(5) Field buzz.   Check field oscillator stage decoupling; f.m. detector
balance (625).

(6) Hum.   Check h.t. smoothing components; valve heater-cathode
leakage.

Sound distortion

(1) Hum.   Check h.t. smoothing components; decoupling capacitors;
valves for heater-cathode leakage.

Table 9.1 *Continued*

(*2*) *Field buzz*.   See vision-on-sound above.

(*3*) *6 MHz buzz* (*625*).   Check a.m. rejector control in ratio detector circuit; detector diodes; 6 MHz take-off coil tuning; alignment of 6 MHz tuned circuits.

General safety precautions

(1) Nearly all receivers use a.c./d.c. techniques: thus the receiver chassis may be *live*. For safety always check the chassis potential and reverse the mains connection if necessary.
(2) Never assume that the receiver mains switch is functioning correctly.
(3) Ensure discharge of the c.r.t. final anode cavity before removing tube.
(4) Do not dispose of old c.r.t.s in public places or in dustbins. Do not attempt to destroy them unless the space they are within is well sealed.
(5) Never overtighten c.r.t. clamps.
(6) Never carry a c.r.t. by its neck.
(7) Always replace implosion screens where they were originally fitted.
(8) Never attempt to use a multimeter at the anode of the line output valve or transistor or efficiency diode cathode—nor on the line output transformer output windings.
(9) Never check for e.h.t. sparks in the line output stage if transistors are employed.

COLOUR-RECEIVER FAULT FINDING

The logic of fault finding already explained in this chapter is carried on to the business of colour-receiver fault finding. Perhaps it is as well to emphasise at this point that servicing a colour-receiver without the manufacturer's manual and without the minimum amount of test gear (Chapter 8) is foolhardy. Also the temptation, that is within us all, to have a little fiddle with this and that control to see the effects on our faulty receiver must be thoroughly resisted; all that you will do is extend the time taking for the servicing by adding another period for setting up the decoder!

The golden rule for fault finding in colour is that first the receiver *must operate perfectly when in monochrome*. Only when this is achieved can fault finding within the chroma sections of the receiver begin. The normal procedures are followed to achieve this perfect result on monochrome, with very few exceptions.

The exceptions occur in an RGB receiver when correct black or white cannot be obtained in the display and on any colour receiver when convergence or purity are faulty.

In the first case, there is the possiblity that one of the colour output stages is faulty and the grey-scale tracking procedure should be followed to see whether this is possible to achieve. If it is not and a fixed hue is shown rather than white then the appropriate amplifiers should be checked to see whether their outputs are satisfactory. (Note, that as given in Chapter 5 an incorrect colour display may be caused by either too much or too little of particular colours). The chroma output stages can be treated as normal video amplifiers with consideration taken of the black level control circuit used. This kind of colour inbalance fault cannot occur in a colour-difference receiver because the colour-difference outputs are zero when the display is supposed to be monochrome.

In the second of our possibilities, the convergence on the monochrome picture is faulty. This is by far the most common demand on the servicing technician and he should obviously be in a position to very quickly run through the purity/convergence procedure should it be necessary. Practice is all that is required. Should the receiver fail to converge it may be possible that the receiver is magnetised and external degaussing should be tried. If this does not alter the position then it may be assumed that there is a fault in the convergence circuitry. This is, in fact, usually obvious during the procedure of convergence because most faults show up as, and indeed are due to faulty controls: common are open-circuit potentiometers due to overheating, fractured cores in coils, and fractures in the printed-circuit board also due to overheating. These should be checked as should be any plugs and sockets at each end of the cable harness to the convergence board, the connections to the deflection and convergence coil assemblies and the connections to the transductor (if fitted).

If this fails to reveal the problem, then fault finding procedures must be brought to bear to follow the convergence waveforms from the timebases through to their coils.

Table 9.2 CHECK TABLE WHEN STARTING FROM 'NO COLOUR' SYMPTOMS: USE COLOUR BAR INPUT FOR THESE TESTS

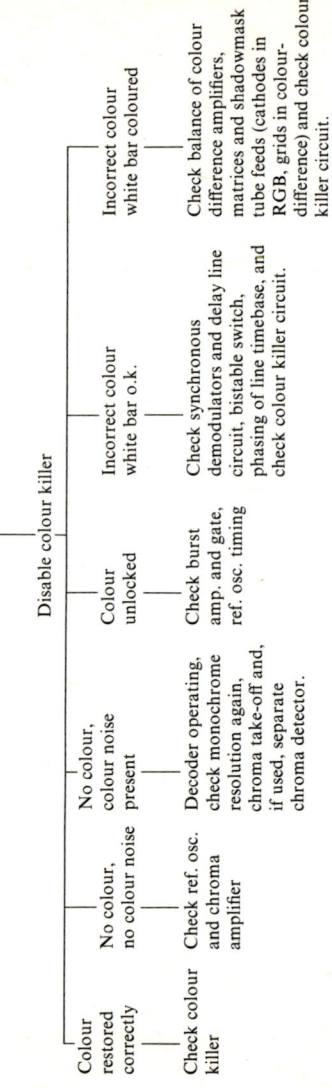

Monochrome pictures o.k. (see text)
No colour

Disable colour killer

Colour restored correctly	No colour, no colour noise	No colour, colour noise present	Colour unlocked	Incorrect colour white bar o.k.	Incorrect colour white bar coloured
Check colour killer	Check ref. osc. and chroma amplifier	Decoder operating, check monochrome resolution again, chroma take-off and, if used, separate chroma detector.	Check burst amp. and gate, ref. osc. timing	Check synchronous demodulators and delay line circuit, bistable switch, phasing of line timebase, and check colour killer circuit.	Check balance of colour difference amplifiers, matrices and shadowmask tube feeds (cathodes in RGB, grids in colour-difference) and check colour killer circuit.

Assuming that the receiver you are repairing has none of these symptoms, or that it had but you have fixed them and that we now have a good monochrome picture with low noise, sufficient picture contrast available and good resolution (remembering that on colour-difference drive receivers there will be a notch filter inserted for rejection of chroma subcarrier in the luminance channel), we are ready to delve into the chroma part of the receiver.

First we must assess the position. Is colour present in any form with the saturation control turned up? or is it completely absent? No colour will usually indicate a fault somewhere before the reference oscillator whilst poor, weak, uncontrollable or patterned colour could be isolated only in general terms as being in the decoder. We can analyse these cases further.

In the complete absence of colour, disable the colour killer in the receiver (this may be done by either a fitted switch marked normal/ disable, a tag to which a flying lead must be attached or by a voltage bias method indicated by the manufacturer in his handbook aimed at putting about $+3$ V on the base of the controlled chroma amplifier). The information given in Table 9.2 should then be of use in locating the correct area of trouble. An input of standard colour bars should be used for this testing and of course the servicing technician should make himself familiar with the correct colours that he should expect.

For all other cases of weak or uncontrollable colour the fault should be located within the decoder using an oscilloscope (fitted with a low capacitance probe) comparing the signals obtained with the manufacturer's waveforms.

Once a faulty stage is located, normal fault finding within the stage takes place. It will be found that with experience some assumptions can be made about the functioning or malfunctioning of the decoder. These are, to a certain extent, intuitive so it is better not to lay them down as direct procedure.

RECEIVER REALIGNMENT

by G. R. Wilding

The most important point before commencing receiver realignment is to establish beyond all doubt that the symptoms giving rise to the assumption that realignment is necessary are truly founded, since incorrect control settings, faulty valves transistors or other components can all cause symptoms very closely resembling those due to misalignment, and a careful check to eliminate all these factors must precede any retrimming.

For instance, an over-advanced sensitivity control can produce sound-on-vision or vision-on-sound that cannot be tuned out in later stages, and absence or near absence of fixed or a.g.c. bias can similarly introduce these symptoms due to the i.f. stages becoming overloaded or even slightly mistuned by the major bias change.

Dry-jointed or open-circuit damping resistors and fixed picofarad trimmers will all materially alter the response of a tuned circuit, or introduce 'ringing', as will defective decoupling capacitors, even if they do not actually introduce instability, while defective i.f. valves drawing grid current can unduly widen the grid circuit response. Instances of vision-on-sound can often be traced to a sound i.f. pentode drawing grid current and thus flattening the response to include the vision frequencies and similar problems can be encountered in transistor receivers.

Reductions in video circuit decoupling capacitors can markedly deteriorate high-frequency response and prevent good resolution of the higher test card gratings, while defective 'peaking coils' will similarly attenuate the top end of the video scale.

To determine picture quality, reception must be appraised on test card only, using a good aerial and paying particular notice to grating resolution, the fine tuner peaking of sound volume with optimum picture definition, absence of 'ringing', sharp delineation of all outlines and complete absence of sound-on-vision or vision-on-sound.

If only one coil, capacitor or other small component in the receiver section has been replaced, or if, as sometimes occurs, the owner has maladjusted only one or two coil cores, it is usually possible to restore perfect alignment by careful trimming on the test card alone, provided that the sound and vision balance is constantly checked and that no other slug or trimmer is touched.

Careful adjustment, and on the test card only, is imperative.

However, where it is apparent that a complete re-alignment is essential—invariably produced by a previous attempt at improving reception since television alignment drift is so very slight today—the maker's service manual and alignment instructions should be obtained and followed implicitly.

Manufacturers usually indicate two separate systems of receiver alignment: (*a*) by tuning the response of individual circuits to particular frequencies using the output from a variable-frequency signal generator

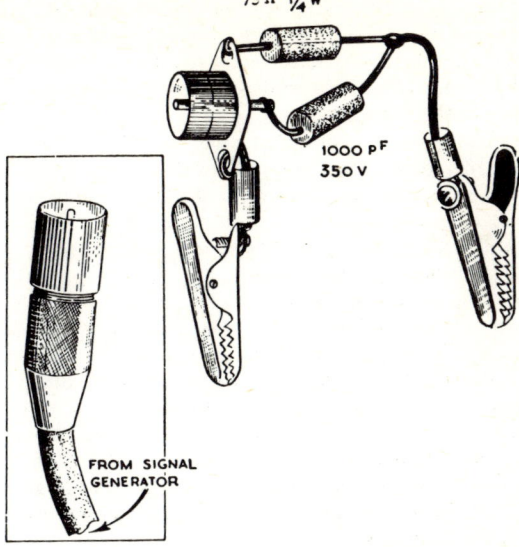

Figure 9.1 (a) Termination device

The signal-generator termination device shown above is intended for use with a signal generator having an output inpedance of the order of 75 Ω, and should be connected in circuit when injecting a signal directly to the grids of the valves: it will not normally be required when injecting signals to the aerial-input socket of the receiver. The device shown below is for damping the primary winding of intermediate-frequency transformers whilst the secondary winding is being adjusted, and vice versa. It is important that the crocodile clips make good electrical connection with the receiver circuitry. For convenience, both units should be made up and kept available for use when required.

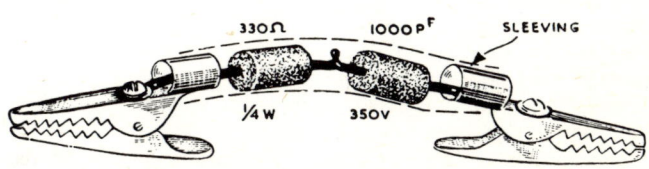

Figure 9.1 (b) Damping device

and obtaining signal indication on an output meter at the relevant video or sound test points; or (*b*) by visually matching the response of the tuned circuits as displayed on an oscilloscope, with injection of the output from a frequency modulated oscillator or wobbulator, to those shown in the service manual.

Method (*a*) is most commonly used in service workshops, and apart from the signal generator and output meter (which can conveniently be a normal multi-range testmeter) a small 3 V or 4·5 V battery will be needed for biasing purposes, plus the recommended resistors and capacitors to terminate test equipment or for circuit loading or stabilisation (*Figure 9.1*).

The complete alignment information for any particular model is both lengthy and involved and for perfect results demands painstaking care.

Very careful attention should be paid during realignment to the resetting accuracy of the frequency generator used, and because the absolute accuracy of it cannot be guaranteed *all of a receiver section* must be realigned even if only one part is out of specification.

Realignment on dual-standard receivers is without doubt a nightmare and *very* great consideration should be given to its necessity; the stability of modern receivers is so good that realignment should be virtually never necessary.

If it is felt, after exhaustive tests, that realignment of a colour receiver is necessary, this should *never* be undertaken with the spot frequency method. Neither should it really be attempted using a wobbulator unless facilities also exist in the workshop for measuring the group delay of the circuits being adjusted. In the great majority of these cases, realignment is not necessary and even where it is a replacement module from the manufacturer is a great deal cheaper in workshop time.

10

AERIALS AND INTERFERENCE

The requirements of an aerial system for television reception are considerably more exacting than those for sound broadcasting purposes. This is because of:

(1) the lower power, and correspondingly lower field strength of television stations;
(2) the greater bandwidth required, and consequent lower gain per stage in the receiver;
(3) the higher circuit and insulation losses, and greater valve and transistor noise on v.h.f. and u.h.f.;
(4) the greater susceptibility of v.h.f. signals to electrical and ignition interference;
(5) the necessity to avoid receiving transmissions by multiple paths in order to reduce 'ghost' images;
(6) the greater susceptibility of the eye, as compared to the ear, to interference and signal variations.

On the other hand, it should be recognised that the use of a directional aerial system, tuned for optimum pick-up on a limited range of frequencies, and coupled to the receiver by a matched transmission line, represents basically a much more efficient type of arrangement than is customarily employed for sound broadcast reception.

The choice of an aerial and transmission line for any particular installation will depend upon the distance from the transmitter; its height and freedom from screening; the level of local interference; the length of feeder cable necessary; the sensitivity and input impedance of the receiver; and any restrictions imposed by the landlord or local authorities. It should be emphasised, however, that when planning an installation it is better to provide too much rather than too little signal input; for while it can be a relatively simple matter to incorporate an attenuator pad in the feeder to reduce an excessive signal, the raising of the signal level by even a few decibels may require the complete re-planning of the installation. Unlike valve or transistor amplification, additional gain in the aerial does not introduce 'noise'.

Aerial terminology

Half-wave Dipole. The fundamental form of a resonant aerial is a single conductor with an electrical length equal to half the wavelength on which optimum reception is required.

Aerial Impedance. The natural impedance varies with the current distribution along the aerial. For a half-wave dipole the impedance at the ends is several thousand ohms, and approximately 72 Ω at the centre. Additional director or reflector elements will tend to reduce the impedance, and on a four-element array the centre impedance may be of the order of 25 Ω.

Field Strength Pattern. The variation of reception around an aerial is normally shown graphically by means of 'polar diagrams', which are circular charts with the angle (0–360 deg.) indicating the direction for which the signal strength is plotted, and the length of the radial arm indicating its magnitude.

Aerial Gain. This term expresses the increase in signal strength for one type over another, a standard half-wave dipole usually being adopted for reference purposes. The gain is measured in the direction of optimum reception, and is usually expressed in decibels. Thus an aerial with a 6 dB gain would provide double the voltage, or four times the power, across the input circuit of the receiver.

Front-to-back Ratio. This is the term used to denote the ratio between the pick-up, when the aerial is orientated for optimum response, to that for the position of minimum response. It is usually expressed in decibels.

Decibel (dB). The one-tenth part of a bel, this latter being the common logarithm of the ratio between a second power, or intensity, and a first one. The decibel, therefore, is a unit for expressing the gain or loss in an electrical or acoustic circuit when the input is known, or of defining any power in relation to a predetermined basic level of that power.

The Dipole

This, the simplest form of tuned aerial, consists in practice of a metal rod, divided at its centre by an air gap of about 20 mm for connection of the transmission line, and of overall length approximately half that of the wavelength on which optimum reception is required. Owing to end effect with the ground, the length is not an exact half-wave. The formula below will enable the length to be determined approximately:

$$\text{Length (cm)} = \frac{13\,716}{\text{Frequency (MHz)}}$$

This is based on the use of 9·5 mm tubing. In the case of 12·5 mm tubing the formula is $13\,385/f$, and in the case of 6·5 mm tubing $13\,850/f$. The figure varies with the ratio of width/length of the element used: the thicker the element, the less its length. The calculation of exact lengths is a matter of some complexity, and in practice a certain amount of trial and adjustment is often used. In general the larger diameter tubing should be used because this increases the bandwidth of the aerial. Mechanical strength and the aerial support system may dictate a smaller diameter. The impedance at the centre is approximately 72 Ω, but this may be affected by the presence of nearby objects or additional elements.

The signal being received produces a standing wave in the dipole, which results in there being zero current but maximum voltage at the extreme ends of the rod, and maximum current but low voltage at the centre. The signals are fed to the receiver via a feeder cable which in order to minimise loss should be fairly accurately matched in impedance to that existing at the point of connection to the aerial system, and also to the input circuit of the receiver. Where the co-axial type of feeder is employed, the centre conductor should be connected to the upper half of the rod and the outer conductor to the lower section.

The simple half-wave aerial is intended for use in areas where reception conditions are good; where interference is to be met with, its non-directional properties are against its use.

Dipole plus reflector: the H type

By mounting another slightly longer metal rod at one-quarter or one-eighth of the signal wavelength behind the dipole, the aerial can be

Table 10.1   TYPICAL DIMENSIONS OF V.H.F. TELEVISION AERIALS

Channel	Mean Freq. (MHz)	Dipole (m)	Director (m)	Reflector (m)	Spacing $\frac{1}{4}\lambda$* (m)
1	43·5	3·226	3·073	3·378	1·689
2	50	2·819	2·680	2·959	1·448
3	55	2·565	2·438	2·692	1·346
4	60	2·349	2·235	2·463	1·232
5	65	2·158	2·057	2·260	1·130
6	178	0·790	0·762	0·838	0·419
7	183	0·775	0·737	0·813	0·406
8	188	0·750	0·711	0·787	0·394
9	193	0·724	0·686	0·762	0·381
10	198	0·699	0·673	0·737	0·368
11	203	0·686	0·660	0·724	0·362
12	208	0·673	0·648	0·711	0·356
13	213	0·660	0·635	0·686	0·343

*Halve these dimensions for $\frac{1}{8}\lambda$ spacing.

Note. The exact dimensions are affected by the ratio of element diameter to overall length and will therefore vary according to the type of tubing used.

made directional. Typical dipole and reflector lengths are given in Table 10.1.

The spacing of the two rods is not critical; spacing of $0 \cdot 1\ \lambda$–$0 \cdot 25\ \lambda$ is usual. The spacing distance, however, will affect the dipole impedance and its polar diagram. The rod has no electrical connection with either the dipole or any other part of the aerial system.

The reflector increases the pick-up efficiency of the dipole in the forward direction (as a rule it is about 3 dB better than the standard dipole) and reduces it in the rear, the front-to-back ratio being of the order of 9 dB. Where interference radiation is strong and lies behind and not across the transmitted signal's path, effective screening can be achieved by putting the reflector between the interference source and the dipole. When this has been done, the aerial may no longer be 'pointing' towards the transmitter, but this makes little difference, provided that the direction of the aerial is not more than about 30 deg. from the correct position.

From a signal-strength viewpoint, and also as regards interference-free signals, the H type aerial should prove satisfactory in all but the most distant and difficult Band I locations.

Yagi aerial array

Adding a further reflector element adds very little to aerial performance. If an element slightly less than a half-wavelength is added in front of the dipole, however, the forward gain and directivity of the aerial are improved. Such an element is called a director, and the combination of reflector, dipole and director is termed a Yagi array. The addition of further director elements further improves performance with respect to gain and directivity, although bandwidth is reduced. Various techniques, such as making the directors progressively shorter or using a greater number than the optimum (found by experiment) and varying their spacing, may be adopted where the requirement is wide bandwidth. Like the reflector, director elements are not electrically

Table 10.2   PROPERTIES OF VARIOUS TYPES OF BAND I TELEVISION AERIALS

TYPE OF AERIAL	RELATIVE GAIN TO A DIPOLE	POLAR DIAGRAM (RELATIVE)	PROBABLE MAXIMUM RANGE (km)	OCCASIONAL RANGE (km)
DIPOLE	0 dB (I.E. THERE IS NO GAIN SIGNAL PICK UP IS x 1)		29	80
V DIPOLE	−6·0 dB		22	56
DIPOLE WITH REFLECTOR	ABOUT 3dB		56	110
ARRAY	ABOUT 6dB		56+	110+

Note: as a simple, but approximate, rule 'doubling the plumbing doubles the power', so doubling the number of aerial elements in an aerial gives about 3dB increase in gain.

connected to the dipole, from which the signal is taken. Approximate reflector length, based on the use of 9·5 mm tubing, is given by

$$\text{Reflector length (m)} = \frac{14\ 638}{f(\text{MHz})}$$

Approximate director length, again with 9·5 mm tubing, is given by

$$\text{Director length (m)} = \frac{13\ 335}{f(\text{MHz})}$$

and the director lengths can be made progressively shorter by about 5 per cent to increase bandwidth.

Folded dipole

In this arrangement two half-wave dipoles are connected at the ends and run parallel to one another, some 10 cm or so apart, one dipole being broken at the centre for connection to the transmission line. The polar diagram and gain are substantially the same as for a simple dipole, but the system has a broader bandwidth, while the impedance is approximately four times greater. These characteristics are of value for multi-element arrays in order to simplify matching to the feeder—the additional elements reducing the impedance back towards 72 Ω.

Room aerials

One result of the greater sensitivity of modern receivers is that telescopic and other simple room aerials are capable of providing reasonable results over quite large areas, although—particularly on Bands III–V—pockets of poor signal strength are likely to be found even close to the transmitters. Many of these aerials are basically Band III dipoles which also provide a roughly equivalent signal from Band I stations whose signal strengths tend to be higher.

The best location for these aerials can be found only by trial and error; moving the aerial to find a place where there is minimum fading caused by people moving in the room (or beyond a party wall) and freedom from 'ghost' images.

Loft aerials

Special loft aerials are sometimes used, having directional properties. An example is the inverted V type, which possesses sharp minima at right angles to its plane, an advantage for the removal of 'ghosts' or the elimination of local interference. It comprises two quarter-wave rods set at 45 deg. to the vertical, each half of the V being connected to the feeder cable, thus making it independent of the angle of polarisation.

High-gain 'bi-square' loft aerials are also available for long-distance reception on Band I for use in locations where outdoor aerials would be difficult to erect.

Slot aerials

This type of aerial may be used as an alternative to the simple dipole, or dipole and reflector, to which it has roughly similar performance. For installation in the loft of a house, it has the advantage that, for vertically polarised transmissions, it is long rather than high. It consists of a vertical sheet of conducting material, such as wire netting, with a slot running horizontally in the centre and the feeder connected to the mid-points of the long sides of the slot. Typical dimensions for Channel 1 are: netting, 4 m long × 1·5 m high (these dimensions are not critical); slot 3 m long × 0·3 m high.

Skeleton slot aerials

A form of the slot aerial that is in common use for Band III reception is the 'skeleton slot'. As its name implies, this type of aerial was developed from the normal slot by gradually reducing the metal surround. It has been found that results substantially the same as those of the normal slot may be obtained even when the surround is reduced to a mere rim of metal, which in practice may take the form of metal tubing (about 12·5 mm diameter) enclosing the 'slot'. This provides an aerial which is comparatively simple mechanically, and which does not offer the wind resistance of conventional slot assemblies. Directional arrays may be formed by adding directors or reflectors, though these will be mounted at 90 degrees to the major axis of the slot. The feeder is normally matched to the mid-points of the slot (where the impedance may be as high as 600 Ω) by means of a quarter-wave stub or a linear transformer section.

Modified skeleton slot aerials also give good performance on Bands IV/V and have been used more on these Bands than on Bands I and III.

AERIALS FOR BAND III

On the higher frequencies of Band III the voltage induced in a dipole element is appreciably less than that induced in a Band I dipole in an area of similar signal strength, while the losses in feeder cables will be approximately doubled; furthermore, the sensitivity of a receiver will not be so good on Band III as on Band I. For all these reasons, greater care has to be taken if good signals are to be presented to the receiver on Band III. On the credit side, however, the shorter elements make multi-element arrays relatively simple to construct and, in fact, arrays of ten or so elements mounted on a single cross-arm are available. These are highly directional, and great care should be taken to orientate them correctly in order to obtain the optimum results.

When planning aerial installation for both bands, the following questions need to be answered before any decision can be made as to the best type for a given location:

(1) Are the transmitters co-sited or located in different directions from the receiver?

(2) Is the location within the primary service area (i.e. high signal strengths) of one or both stations?

(3) Is the feeder run comparatively short (not more than about 15 m) or will a long cable run be required? Loss of signal in short runs is usually unimportant, but this will not be the case with long runs on Band III.

Undoubtedly, one of the major sources of difficulty with Band III aerials is the greater number of 'ghost' images that occur on these transmissions. Hills, gasholders, spires, steel structures and many other reflecting surfaces may give rise to strong signals, arriving slightly out of phase with the direct signals, and thus producing ghost images slightly displaced from the main ones. These can often be eliminated by very careful orientation of the aerial, sometimes suffering some reduction in strength of the main signal.

On Band III, poor aerial siting or installation can impair picture quality even within a comparatively short distance from the transmitter.

Manufacturers have drawn attention to the advantages of an aerial sited above the chimney stack and as far as possible from obstructions; or failing this erected so that the chimney is to the side of, or behind, the aerial. Loft aerials for Band III are rendered useless by the proximity of a metal water tank. The following are among the suggestions put forward by a prominent manufacturer:

Install an outdoor aerial.

Have it on the highest point possible (e.g. a chimney).

Use a good quality low-loss feeder cable.

Use the sensitivity (local-distant' pre-set) control on the receiver to obtain a good picture on the weaker signal.

If necessary use an attenuator on the stronger signal to equate the two signals.

Remember that multi-element aerials are very directional, and even a few degrees off beam will make a big difference.

When orientating an aerial, avoid the effect of the receiver's a.g.c. circuits masking increases in aerial signal by using an attenuator at the input to the set.

Filter units for separating or combining Band I and Band III signals —known variously as 'cross-over filters', 'diplexers', 'splitters', etc.— have a number of uses. For example, such a unit is necessary where separate Band I and Band III aerials are used with a receiver having a single input socket.

Wide-band v.h.f. aerials

Because of the interference problems on Band I and the extension of services mentioned in Chapter 1, in some areas B.B.C.-1 programmes are transmitted in Band III. This means that in these areas two television programmes are available in Band III, and manufacturers have introduced wide-band v.h.f. aerials to provide balanced reception of the two programmes (ordinary v.h.f. aerials receive only one channel satisfactorily, discriminating against others). The alternative approach is to use two Band III aerials, preferably with a diplexer and single downlead. Choice of arrangement depends on the relative signal strengths and the positions of the transmitters relative to the point of reception: generally where the transmitters are co-sited a wide-band aerial will be better. The areas involved are South Wales, North Wales, S.W. Scotland, West Lancashire, East Lincolnshire and Bedford.

AERIALS FOR BANDS IV/V

The u.h.f. system is planned on the assumption that the minimum acceptable field strength on Bands IV and V is 70 dB above $1\mu V/m$. With a high gain aerial system and average feeder losses, this should provide a receiver input of about 2–3 mV. The range at u.h.f. obtainable from high power transmitters is only about two-thirds of that of an equivalent v.h.f. transmitter, representing an area coverage per station at u.h.f. of under one-half of that at v.h.f. Considerable variations in signal strength even quite close to the transmitter occur at u.h.f. In particular, hilly districts and positions in towns among high buildings give rise to pockets of unsatisfactory reception, so that fill-in relay stations and wire distribution systems are often necessary.

The u.h.f. network is planned on the assumption of the use of an aerial having eight elements. In fringe areas 10–12 elements will usually be needed, and stacked arrays have been produced with up to about 40–50 elements (stacked in groups of two or more Yagi arrays). Close to the transmitter, a 5-element aerial may prove satisfactory, and even set-top types may be suitable where the signal strength is very high. A u.h.f. aerial should never be combined with any other kind of aerial.

A half-wave dipole for the lowest Band IV channel (21) is about 33 cm, reducing to only 18 cm for the top of Band V (channel 68). The dimensions are approximate and aerial bandwidth at these frequencies

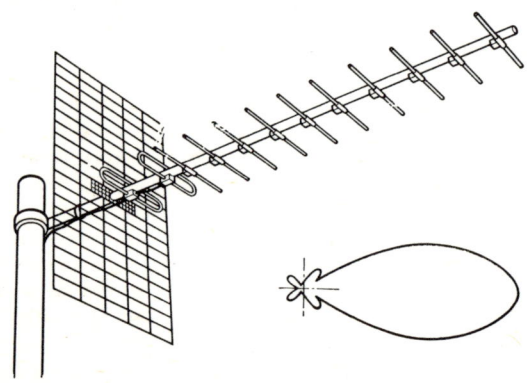

Figure 10.1   U.H.F. aerial with mesh reflector element, showing directional polar response

will be wider than for v.h.f.—the higher the frequency the wider the aerial bandwidth. This is what is required for u.h.f. reception, since it is intended eventually to provide four u.h.f. transmissions in each area (see Chapter 1). The bandwidth of a u.h.f. aerial can also be increased by the design and spacing of the parasitic elements.

As in the case of Band III, the folded dipole is much used for Bands IV/V in order to keep the impedance of the feeder connection point to a suitable value for matching to the coax with the large number of director elements used. A number of variants of the folded dipole have been evolved for use at Band IV/V frequencies, and the skeleton slot also provides good results. Some aerial designs employ baluns for matching to the coax. and for impedance matching. The conventional single reflector element is not suitable at Band IV/V frequencies; to provide gain over the wide bandwidth and maintain front-to-back ratio, the reflector element is often a mesh or corner arrangement.

U.H.F. aerials should be mounted as high as possible: raising an aerial may increase the gain by as much as 2 dB/m. Near a transmitter, however, aerial signal may be increased by lowering the aerial as ground reflection can cause signal cancellation at certain heights. Screening may also produce this effect. Because of the highly directional characteristic of multi-element arrays it is necessary to beam them carefully in the direction of the transmitter. The aerial may need to be twisted on its axis to obtain maximum signal as the polarisation of the signal can be altered by screening effects, etc. Tilting the aerial upwards can increase the effective aerial height. If a multi-element aerial is mounted in a position exposed to high winds or close to tall trees, picture flutter may be caused through aerial movement.

The advice given on aerials for Band III also applies generally to the selection and installation of aerials for Bands IV/V. Sharp bends in the feeder must be avoided and the best low-loss type used. Never connect extra lengths of coax. to extend a feeder. Always ensure that coax. sockets are correctly fitted, with the inner connector properly soldered in.

Particular u.h.f. aerial problems

The problems of maintaining bandwidth in an aerial are more acute at u.h.f. than on any other band. In a number of reception areas the four channel allocations from a particular transmitting station cover more than the normal 88 MHz (11 channels). To allow for the manufacture of only a reasonable number of variations of aerial, aerial groupings have been devised. These groups are colour coded (see Chapter 1) according to their particular identification (Groups A–E). This allows the user to recognise quickly the correct aerial for his particular area of reception. (These groups are also indicated beside the station name in Table 1.7). In terms of the channels covered, the last two Groups, D and E cover rather more than the normal range with Group E the highest where the requirement is for 240 MHz—i.e. about 30 per cent of the carrier frequencies involved.

Bandwidths of this order can only be attained using carefully controlled design and manufacturing techniques. Of principal concern are maintaining the polar diagram over the bandwidth, maintaining the aerial impedance over the bandwidth, and maintaining the bandwidth itself. The results of not concentrating on these things would be different signal levels on different channels and perhaps different aerial gains across one channel—causing chrominance-luminance gain inequality.

Each of the elements in a Yagi aerial acts as a tuned circuit. The dipole (driven or radiator element) is tuned to resonance at the centre frequency of the band being covered; above that frequency—like any tuned circuit—the dipole becomes more inductive and less capacitive, below resonance the reverse occurs. At the feeder terminals, the impedance therefore changes from being purely resistive at resonance to being inductive above resonance and capacitive below resonance—with consequent mismatching and loss of signal (i.e. loss of gain). To reduce this effect to a minimum, the Q of the tuned circuit dipole must be reduced, and this can be done by shaping the halves of the dipole differently to give a different distribution of inductance and capacitance —thus the *bow-tie* and other varieties of shape of the dipole in a u.h.f. Yagi.

But the effects of the non-driven elements—the directors—are no less severe. They operate in getting additional aerial gain by concentrating the beam pattern; but this is achieved by increasing the signal coupling along the axis of the aerial, each director (parasitic element) acting as a tuned circuit coupling each back to the dipole. The overall effect is the increase in gain required but there is also a large increase in the resultant Q at the dipole. This can only be reduced by 'stagger tuning' each of the directors with respect to one another; but just as this is done in the i.f. amplifier of a receiver, for example, there is always some cost to be paid in the overall gain.

This is not to say that the price of gaining a flat bandwidth is at the cost of the polar response. It need not be so because our directors are still in line still giving gain. It is the way that gain is used that is being varied. The gain-bandwidth product is always a constant in any electronics situation, so that as we demand extra bandwidth the gain across the band reduces in proportion.

The 'stagger tuning' of the directors can be achieved in a number of ways. Most popular has always been a progressive shortening of their length away from the dipole (see earlier under Yagi Aerial Array), and this artifice is still used but often in conjunction with another technique. One of the simplest is to vary the tuned frequency of each element by connecting it to the aerial boom at a point other than its centre. The slight imbalance on one side or the other changes the resonant frequency but is more commonly referred to as *phasing* the elements because the impedance at the boom point is artificially thrown out of phase.

Other techniques involve 'swamping' the director impedance effects by using a high impedance dipole such as a skeleton slot (see earlier in this chapter) followed by a balun transformer for matching the high-impedance dipole to the low-impedance feeder, and shaping the directors as 'four-pronged' elements or squared-off S shapes.

Another approach, and one that has been most successful is the use of the log-periodic aerial.

Log-periodic aerial

This is really a form of completely staggered aerial. The principle is that we have a centre boom that is not a single tube or solid but constructed as a parallel transmission line. The feeder is connected across this line.

At suitable points on the line (determined by the feed impedance and the gain required) a dipole is constructed—one half of the dipole on one side of the line, the other half on the other side of the line. The dipoles are alternated so that the half-sections on one side of the line go alternately up, then down (*Figure 10.2*).

The aerial operates by virtue of the fact that as all the dipoles are connected at the same time and as all are different lengths, only one will be at resonance at any particular frequency. At that frequency all the other dipoles act as parasitic matching and director elements.

By giving a logarithmic spacing to the elements and a logarithmic change of element length (hence the name of the aerial), it can be shown that the changeover from one element to the next is quite smooth as frequency changes over the band—this is the purpose of alternating the

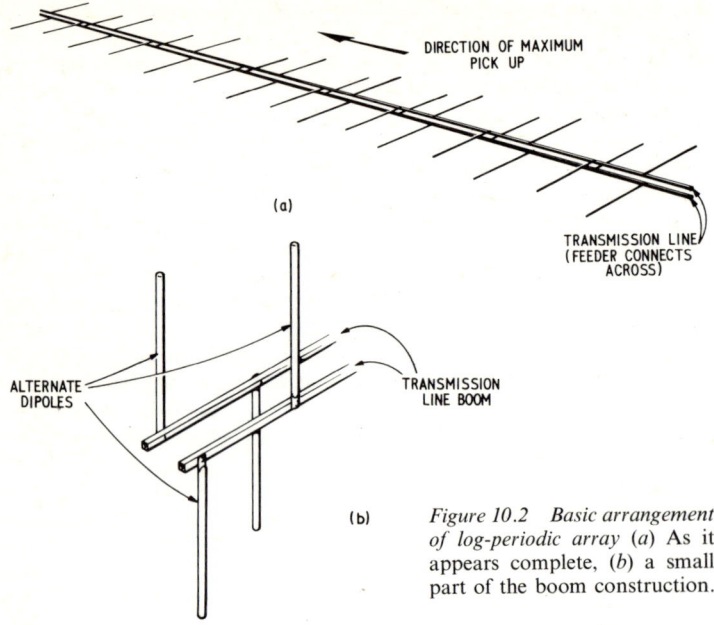

Figure 10.2   Basic arrangement of log-periodic array (a) As it appears complete, (b) a small part of the boom construction.

dipoles on either side of the boom; if this was not done the phasing of the elements would be incorrect.

This design gives very good constancy of drive impedance—as good as the manufacturer's standards allow—whilst the polar response is still basically that of a Yagi aerial.

The overall result is an aerial array which has the good directivity and front-to-back ratio of the Yagi whilst having a very large possible bandwidth—at least sufficient to cover Group E. There is of course some cost in gain but this can be made up for by the number of elements used being increased at any particular location.

FEEDERS

At radio frequencies a length of feeder cable may be considered as a series of resonant tuned circuits. By variations in ratio of inner to

Table 10.3   TYPICAL ATTENUATION LOSSES IN FEEDERS

Type of Cable	Impedance (Ω)	Attenuation loss (dB/100 m)	
		50 MHz	200 MHz
Solid polythene coaxial	75	8·8	19·4
Cellular polythene coaxial	75	7·0	14·9
Semi-air-spaced coaxial	75	4·9	10·7

outer conductor and conductor spacing, cables having wide limits of characteristic impedance are possible.

While the terminating impedance of a television dipole aerial is about 72 Ω, that of a multi-element array for use in fringe areas, etc., should also be about 70 Ω because of the use of the higher impedance folded dipole. Without a folded dipole a 10-element Yagi might have an impedance of only 20 Ω.

The screening provided by the coaxial form of cable construction has advantages in reducing the influence of nearby conducting surfaces on the feeder and reducing radiation from it. Consequently coaxial cable, with a characteristic impedance of about 75 Ω, has been adopted as standard in the U.K. and the input stages of receivers are designed to have an impedance of about 75 Ω, unbalanced, and coaxial aerial input sockets are provided. 75 Ω coax. thus matches to the receiver and to the aerial, but at the aerial end the fact that coaxial cable is unbalanced while a dipole is balanced may need to be taken into account, e.g. by the use of a balun (balance-to-unbalance transformer). In particular these are in use on some commercial u.h.f. aerials where a balun is used to match the high impedance of a skeleton slot radiation element.

Sections of transmission line may be used as circuit elements for various applications, e.g. traps and balance converters.

ATTENUATORS

The two types commonly used are the 'pi' and the 'T'. The 'pi' is more suitable for test purposes, as it uses higher resistance values. *Figure 10.3(a)* shows the circuit, and resistance values for different attenuation requirements are given in Table 10.4. Note, however, that where

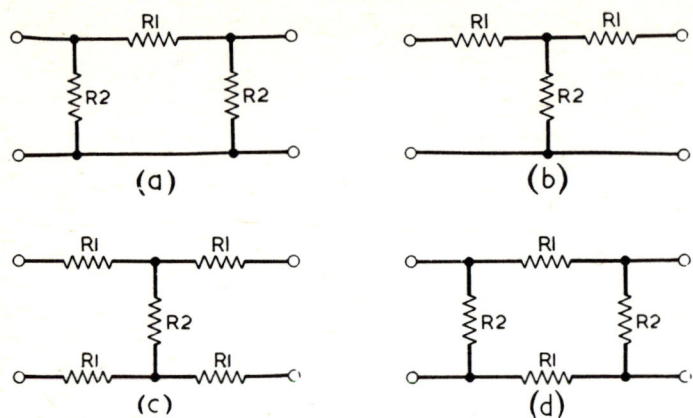

Figure 10.3   Attenuators. (a) π type; (b) T type; (c) T type for balanced twin feeders; (d) π type for balanced twin feeders

attenuation greater than 10 dB is required it is preferable to use cascaded networks rather than a single attenuator circuit.

The T type circuit is shown in *Figure 10.3(b)*. Above 20 dB attenuation this type is not suitable, as the resistance value of R_2 becomes too small. Resistance values are given in Table 10.5

The circuits given are for use with coaxial feeders. Where balanced twin feeders are used the value of R_1 should be halved and the circuits are as shown in *Figure 10.3(c)* and (*d*).

Table 10.4

Required Approximate Attenuation, dB	R_1, Ω	R_2, Ω
10	150	100
20	470	100
30	1 500	82
40	3 900	82
50	10 000	82
60	39 000	82

Resistors should be of ¼ W normal rating, with tolerance of ± 10 per cent, non-inductive.

Table 10.5

Required Approximate Attenuation, dB	R_1, Ω	R_2, Ω
10	39	56
20	68	16

It should be noted that attenuators are difficult to construct at u.h.f. because of the 'skin effect'—increasing resistance with frequency.

AERIAL INSTALLATION

The recognised method of attaching an aerial to a chimney stack is by the use of a right-angle bracket engaging on a corner of the stack and tensioned against it by a wire lashing located around the stack. The object is twofold. Firstly, it avoids hammering holes in the brickwork and thus weakening the stack, particularly in the case of old property. Secondly, it avoids the legal implications of landlord's fixtures, and the aerial may therefore be subsequently removed by an outgoing tenant if he is the owner. Brackets are usually cast in high-tensile aluminium alloys to avoid the need for protective finishes.

Suitable lashing wire is galvanised 7/1·6 mm high-tensile steel wire, the tensioning being taken up by J bolts, which are located on the corner bracket and are hooked into ferrules spliced on to the lashing wire. To avoid chafing, the lashing wire passes over small angle brackets located at the three remaining corners of the stack, held in tension by the lashing wire. Single-flue chimneys are generally considered unsatisfactory for Band I aerials; two-flue chimneys are suitable for most normal installations, but four-flue chimneys should be used for heavy multi-element arrays unless the mast is 'stayed'. Wherever possible the aerial elements should be kept clear of the chimney outlet to avoid damage when the chimney is swept and deterioration due to smoke contamination.

Feeder cables tend to deteriorate rapidly where long stretches, either vertical or horizontal, are left without proper anchoring and are thus subject to continuous strain from their own weight or to grazing by wind action. Where cables come down over tiled roofs suitable clips, such as lead-headed wall nails, should be used to anchor the cable by careful insertion under the tiles. Preferably the cable should be fastened at intervals not greater than 1 m for vertical runs and 0·3 m for horizontal runs. Where soft lead or other metal nails or electricians' cleats are used to secure the feeder to roofs and outer walls, a small piece of fibre or tape should be wrapped around the cable to prevent the fixing metal from puncturing the outer sheathing. The fixing cleat should not be hammered home too hard, otherwise the cable will be subject to excessive pinching and the sheathing may be punctured, allowing moisture to enter, with consequent loss of signals. The feeder should be looped slightly away from any woodwork so as not to interfere with painting. Where the feeder is routed externally to the mast it should be taped to the mast using waterproof tape. It should be taken behind guttering and on the gutter board or preferably stand-off brackets should be used.

At the point where the feeder enters the building a water-drip loop should be formed in the feeder and entry holes should be drilled at an angle of 45 degrees downwards from the inside.

No matter how good the aerial installation, considerable deterioration takes place under continuous exposure to a smoky atmosphere and variable weather. Therefore to maintain its efficiency, regular inspection should take place preferably at intervals not exceeding two years.

Multi-receiver installations

Where there is only one convenient mounting point for two houses or flats, it is often more satisfactory to feed two receivers from a single aerial rather than to place two aerials in close proximity. This can be done without difficulty at v.h.f., provided that the signal strength in the area, for the type of aerial system employed, is sufficient to permit a loss of 6 dB from the signal which would be fed to a single receiver. Star networks which enable two receivers to be fed from a single coaxial or twin-feeder line are shown in *Figure 10.4*.

Where it is desired to operate a large number of receivers from one aerial without loss of signal, as may be the case in a service workshop, a block of flats or an hotel, or where it is required to distribute a u.h.f. signal, it is essential to ensure that there is no interaction between receivers, and that all outlets receive a satisfactory signal. This will normally entail the use of a distribution amplifier in conjunction with an efficient aerial, suitable cable runs and correct socket outlets. The

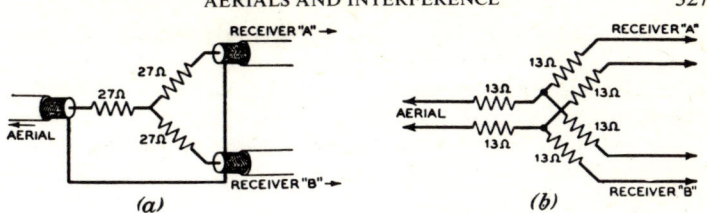

Figure 10.4   Star networks. (a) Coaxial feeder, (b) twin feeder

distribution amplifier should be fitted as near to the aerial as possible, and suitably accommodated in a dry, weatherproof room or enclosure, properly ventilated and reasonably free from dust. Flue gases or smoke should not be allowed to enter the amplifier enclosure, as otherwise rapid corrosion of metalwork may be experienced. In all distribution work the lowest-loss cable available should be used.

Fringe-area equipment

Since valve noise becomes increasingly important as the input to the first stage falls, losses in the feeder line are often the decisive factor in determining whether or not a satisfactory picture can be obtained in a fringe area.

Since special feeder cables are considerably more expensive than the standard types, it will be necessary, in many border-line cases, to weigh carefully the relative costs of the various types of aerial arrays, high masts, preamplifiers, cables and the like. For example, it may be found more economical to use a more complicated array with standard cable than a simple aerial with special cable, or vice versa.

In particular, the use of a mast-head amplifier should be limited to those situations where it is impossible to improve the aerial any further.

'Ghost images'

Reflection of signals from natural and man-made objects may cause a double or multiple image to appear upon the screen of a receiver; and in certain districts such conditions may prove most troublesome and difficult to overcome. Hill faces are probably the most frequent cause of 'echo' signals, but almost any reflective surface, such as trees, buildings, gasometers, or factory chimneys, may give rise to 'ghosts'. By measuring the displacement of the spurious image, a rough estimate may be made of the distance of the offending objects.

The approximate image-displacement values and distances of the reflecting structure from the aerial for a 49 cm (19 in) cathode-ray tube are given in Table 10.6.

Intermediate angles between the rear and side-on positions will give intermediate values between those shown in the middle and right-hand columns. Objects slightly in front of the side-on view would be at a distance greater than that given by the middle column.

The cure of double images is still largely a matter of trial and error in the positioning and orientation of the aerial: the principle being the

Table 10.6

Displacement of Image	Object to Right or Left	Object Immediately Behind
0·2 mm	4·8 m	2·4 m
1 mm	24 m	12 m
2 mm	48 m	24 m
5 mm	120 m	60 m
10 mm	600 m	300 m
50 mm	3 km	1·5 km
100 mm	6 km	3 km

adjustment of the system to give minimum pick-up of the 'ghost' reflection. It is, for example, useless in areas prone to 'ghosts' merely to point the aerial towards the transmitter. Instead, the aerial should be carefully adjusted when a programme, preferably test card, is being received. In some cases it may prove easier to reject the direct signal and concentrate on receiving the reflected signal, for example by turning the aerial on its side so as to receive horizontally rather than vertically polarised waves. With indoor aerials, a change of position of only a metre or two may make a considerable difference to results.

Surprisingly, ghosting has been less of a problem on u.h.f. than on v.h.f. This is because the attenuation on a delayed-path u.h.f. signal is usually greater due to the higher frequencies, and because signals which are reflected in the opposite polarisation are better rejected by the higher accuracy of a correctly polarised u.h.f. aerial.

More problems are encountered on u.h.f. by very short-term ghosting giving 'soft' vertical edges to pictures and these (as Table 10.6 would indicate) are due to very close objects causing reflections. Often these can be eliminated by care over the positioning of the aerial, many short-term reflections being due to water tanks, adjacent roofs, nearby chimneys and so on. If no position or tilting affects the ghosting the feeder and aerial should be checked for dry joints and damaged cable.

INTERFERENCE

Interfering signals may arrive at the receiver either by direct radiation from the source or by conducted radiation along power mains and overhead wiring, or by a combination of the two.

The Engineering Branch of the Post Office is prepared to assist licence holders in the tracing of man-made interference and to offer advice on its suppression. Applications for assistance, which is provided free of charge, should be made by licence holders on Form T.466G 'Electrical Interference Questionnaire', obtainable from any Head Post Office.

It is, however, of considerable importance that the installation and servicing engineer should be able to distinguish between local interference and receiver faults, to recognise the various types of interference from the symptoms they produce upon the picture and to know what can be achieved by the installation of suppression devices to minimise the effects of such interference.

In practice, interfering signals fall into two main categories, which require entirely different treatment: the impulse type of interference producing spots or lines of peak white (on 405 lines) across the screen, and continuous wave signals on frequencies falling within the acceptance band of the receiver and producing heterodyne interference in the form of a 'herring-bone' pattern of alternate dark and light bands running diagonally across the screen.

Interference is much less of a problem at u.h.f.; and with the negative picture modulation used on the 625-line system impulsive interference causes black spots (since the polarity of the signal is changed at the video detector) which are much less noticeable than peak white ones. However, even though u.h.f. receivers are always fitted with flywheel sync., severe interference can cause mistriggering of the timebases.

Impulse interference

Electrical apparatus which utilises commutation (e.g. d.c. motors and generators), vibrating contact points (e.g. electric shavers), spark discharge (e.g. automobile ignition) or any mechanism whereby an electric spark, however minute, is produced will radiate r.f. waves, covering a wide frequency spectrum, unless preventive measures are taken. Such signals will cause crackling on sound and a series of white spots or bright streaks of light on the screen. In this category must also be included switching circuits, such as thermostats and dirty light switches, where slight arcing may take place. As with all forms of interference, the effect will largely depend upon the ratio of the levels of the interfering signal to the picture signal, and will thus be more severe in 'fringe' areas.

By far the most satisfactory cure is the suppression of such interference at the source: ignition interference, for example, can be greatly reduced by the fitting of a resistor of about 15 000 Ω in the lead from the distributor to the induction coil mounted as closely as possible to the

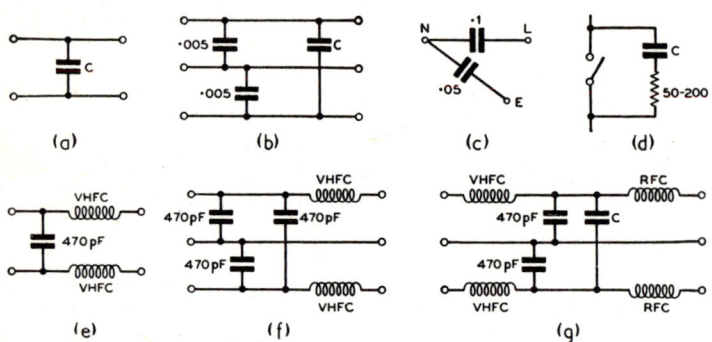

Figure 10.5   Basic interference suppression devices
(*a*) For two-core cable appliances. (*b*) For three-core cable appliances. (*c*) For three-pin socket. (*d*) For thermostats. Types (*e*) and (*f*) are for the suppression of television interference from two- and three-core appliances. (*g*) All-wave filter for broadcast and television interference suppression. The value of C may vary between 0·01 and 0·5 μF. Values given for type (*c*) are the largest permissible. Capacitors used in mains suppression circuits should have an a.c. voltage rating of at least 300 V. If the a.c. rating is not quoted, the d.c. rating should be at least 750 V.

distributor; while most small electric motors as fitted to domestic appliances will respond satisfactorily to the fitting of a capacitor, or capacitor-choke filter, which will provide an alternative path for the r.f. pulses, and which should be positioned close to the offending apparatus. Most home appliances are fitted with built-in suppressors, but for other apparatus a wide range of proprietary filters are made; typical circuits are shown in *Figure 10.5*

Where the source is unknown or cannot readily be suppressed, the effects can be reduced by: (1) the provision, on the sound and vision receiver units, of interference limiter circuits, designed to cut off the high amplitude peaks of the interfering signals—typical circuits for such limiters are shown in *Figure 10.6(a)–(d)* (in practice the valve

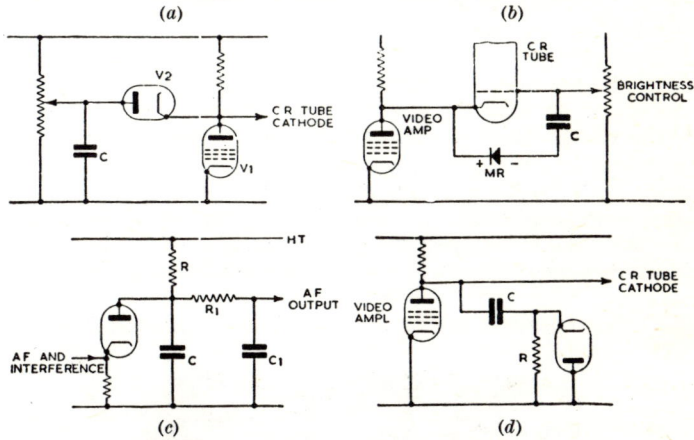

Figure 10.6   Interference suppression circuits
(*a*) Basic vision-interference suppression circuit. (*b*) Automatic form of peak limiter. (*c*) Typical pulse-width limiter circuit. (*d*) Vision-interference limiter operating on a time-constant basis.

diodes are sometimes replaced by crystal diodes)—or alternatively, in 405-line receivers, by inverting the pulses so that they produce black instead of white spots; (2) ensuring that the most efficient aerial system is employed, with the elements as far away as possible from the source of interference or from wires and guttering that could form conduction paths; (3) where the source of the interference is known, as, for example, the cars passing along a main road, a directional aerial system

Figure 10.7   Ignition interference

may be orientated so as to provide minimum pick-up from this direction; (4) the use of interference filters in the mains supply leads to the receiver. Method (4) is unlikely to prove as effective at television frequencies as for normal broadcasting frequencies owing to the greater ease with which the leads before and after the filter can act as aerials and thus allow the interference to bridge the filter; nevertheless, a number of filters especially designed for use at television frequencies are available, and will often bring about a considerable improvement, particularly when used in conjunction with methods (1)–(3).

An effect somewhat akin to ignition interference may be caused by

Figure 10.8   Interference caused by electric motors

corona discharge ('brushing') from points at e.h.t. potential, but it can easily be recognised by its more continuous nature. Cleanliness in the e.h.t. circuits is the most effective cure but high-voltage silicon grease may be used when the problems are severe.

Heterodyne interference

The second and less common group of interfering signals are those in which oscillation is continuous, as opposed to trains of damped oscillation, and these are usually tunable over a comparatively narrow band; they produce heterodyne interference, against which peak limiters and suppression filters of the type so far described are ineffective. The most common causes of such interference are diathermy apparatus, adjacent-channel interference, harmonics of short-wave broadcasting, communication and amateur transmitters, and radiation from the local oscillator of other television or short-wave receivers. Susceptibility to certain forms of this interference, partcularly, for example, to break-through on the intermediate frequency of the television receiver, will depend very largely upon the inherent design of the receiver and the choice of the intermediate frequency.

The interference may arise from radiation (usually harmonic) taking place at frequencies within or close to the television channel (in this case suppression at the source or careful orientation of the receiving aerial are likely to be the only effective cures); to image (second channel) response in the receiver; from 'blanketing' or swamping of the receiver by very strong local signals, or cross modulation, sometimes produced by rectification in the aerial system or local metalwork; from break-through of a signal on a frequency close to the intermediate frequency; or from a combination of these. By fitting suitable traps and filters in the offending transmitter, and by careful screening, harmonic radiation can be much reduced. Rejection of fundamental or intermediate frequency signals can often be improved by screened traps or filters in the aerial circuit of the receiver, close to the receiver, or alternatively as close as possible to the control grid of an r.f. amplifier or mixer valve. Such a wave trap resembles those used in the early days of broadcasting, and is tuned to the unwanted frequency (provided that this does not lie within the required television channel). A coil with 10 turns of 18 s.w.g. wire, spaced wire diameter, and with an internal diameter of 7·9 mm, tuned by a 3–30 pF trimmer, will have a tuning range of about 40–50 MHz and for offending signals on other

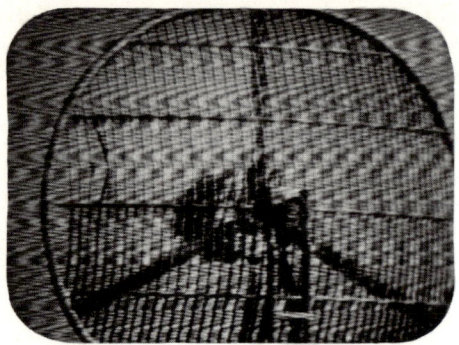

Figure 10.9   Modulated r.f. interference

frequencies the number of turns should be increased or decreased accordingly. For the rejection of signals at intermediate frequencies, high-pass filters with a pass band of about 40–60 MHz may be fitted in the aerial lead of the receiver, and will normally prove effective provided that the screening of the i.f. stages is adequate to prevent direct pick-up. Here again, careful positioning and orientation of aerials may prove of assistance.

Diathermy

This is a particular form of heterodyne interference, the cause in this case being the harmonic output from the relatively unsmoothed valve oscillators used in electro-medical apparatus. In addition to the herring-bone pattern, the sound channel is often affected by harsh crackling or low-pitched hum. The most satisfactory cure for this form of interference is the complete electrical screening of the offending apparatus; though in some cases relief can be obtained by slightly changing the frequency of oscillation of the equipment so that the harmonics no longer fall in the television channel concerned.

Freak propagation

Owing to the fact that television channels are shared on a geographical basis, herring-bone interference patterns may sometimes be caused by

Figure 10.10   Severe diathermy interference

signals being received from a distant station, normally unviewable. The most common cause of such a propagational condition is a period of 'Sporadic E', when the E layer becomes highly ionised and reflects signals up to about 70 MHz over distances between 300 and 1600 km. Pronounced temperature inversions, such as occur during summer evenings, may also cause slight interference. As such conditions seldom last for long, it is usually considered unnecessary to take precautions against this form of interference.

Aircraft flutter

Reflection of signals from aircraft may provide alternate augmentation and attenuation of the signal as the phase of the reflected signal changes in relation to that of the direct signal; this will cause the contrast of the picture to change rapidly from normal to low and then to high, the cycle being repeated in rapid succession for as long as the aircraft is in the neighbourhood.

   The degree to which a particular receiver installation is susceptible to this form of interference will largely depend upon the aerial system: a system which receives only vertically polarised signals (e.g. u.h.f. relay station reception) will generally be less affected than one capable of responding to the horizontal component. The effects may also be diminished by reducing the d.c. coupling to the cathode-ray tube, or by the use of automatic picture control.

Figures 7.7–7.10 are 'Tele-Snaps' by John Cura.

11

CATHODE-RAY TUBES

(Colour tubes are discussed in Chapter 5 but basic data are included in Table 11.2.)

At one stage, almost all cathode-ray tubes used in television receivers employed magnetic deflection and focusing of the electron beam. For many years, however, electrostatic focusing has been used, thus eliminating the external focusing magnets and making the tube easier to set up.

Figure 11.1 shows the arrangement of a tetrode tube with magnetic focusing. Electrons are emitted by the indirectly heated cathode K and

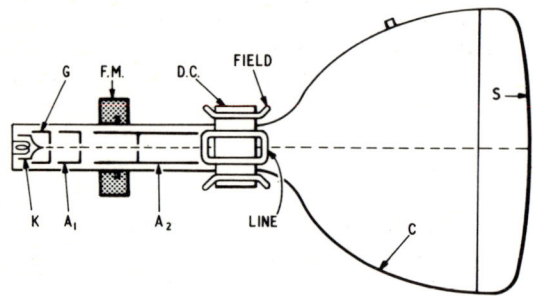

Figure 11.1   Arrangement of a typical tetrode c.r.t.

are attracted by the first anode A_1, which is connected to a point positive with respect to the cathode. A_1 consists of a skirted disc with a small central hole through which most of the electrons pass up the tube to strike the fluorescent screen S, which is composed of 'phosphors' that emit light under electron impact (e.g. zinc sulphide, blue; or zinc cadmium sulphide, yellow). To obtain an approximation to white light, mixtures of phosphors are often used. Surrounding the cathode is the cup-shaped grid G, also with a small central hole. An electric field is produced between grid and cathode by virtue of their potential difference, and this controls the number of electrons drawn through the first anode—and hence the brightness of the spot. A grid voltage 25-100 V negative to cathode will cut off the electron flow altogether. The second anode, A_2, connected internally to the graphite coating C inside the wall, gives further acceleration and governs the final electron velocity: its voltage is 16-20 kV, determining the ultimate possible brightness of the screen fluorescence.

Focusing

The electrons, travelling at high velocity, must be focused to a small spot, e.g. 25 thousandths of an inch in diameter on a 14 in tube. Now a flow of electrons, whether in a wire or not, is a current; and when a current flows across a magnetic field, it experiences a force which tends to move it. In the cathode-ray tube the electron stream thus experiences the deflecting force when passing across a magnetic field. By means of a short coil, or a permanent-magnet arrangement, a field, as shown in *Figure 11.2(a)*, is produced. Electrons passing along the tube axis do not cross this field, and so are unaffected, but those entering at P are deflected into a helical path whose radius is proportional to the strength of the field, and whose pitch is proportional to the *tangential* velocity of the electrons. Since the *axial* velocity is not affected by the magnetic field, all electrons may be brought to a focus on the screen in a spot corresponding to the origin—the cathode. It is thus important for the cathode to be small so that the focused spot is well defined.

In electrostatically focused tubes (see *Figure 11.3*), A_2 and A_4 are connected internally and to the graphite coating C, to which the e.h.t. is applied, and the focus potential applied to A_3.

Deflection and modulation

The spot must be made to traverse the screen rapidly from side to side and more slowly from top to bottom to produce the 'raster'. For line scanning a varying magnetic field is required which varies linearly to deflect the spot left-to-right and then rapidly right-to-left. Simultaneously, there must be a corresponding top-to-bottom and return motion for fields. D.C. in *Figure 11.1* shows the deflecting coils: the deflection is at right angles to the direction of the magnetic field, so that

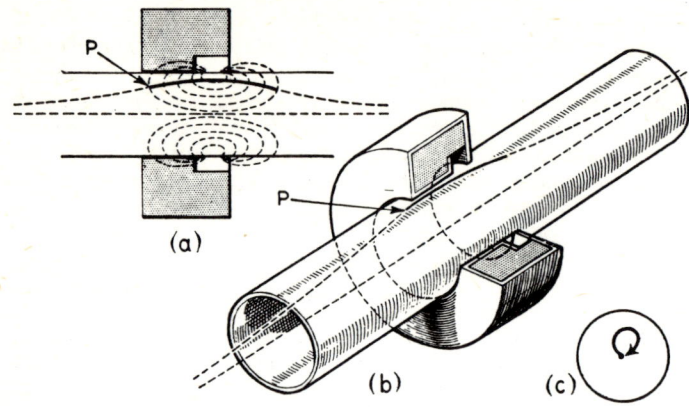

Figure 11.2   Focusing by short coil. (c) Shows an end view of the electron path

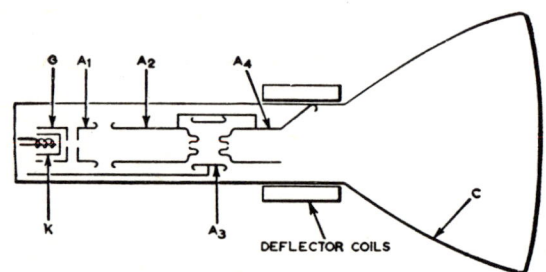

Figure 11.3   Typical tube using electrostatic focusing

the vertically-disposed field is that of the line-deflection coils. In this way the spot will trace out the 'raster', which corresponds to the pattern traced by the camera tube.

To reproduce the 'picture' as seen by the camera, the instantaneous brightness of the spot is modulated by the instantaneous variation of the cathode-grid potential from its mean or quiescent value, which is determined by the 'brilliance' control. The tube may be modulated either by driving the grid positive with respect to the cathode, or by applying a negative-going signal to the cathode, the latter being adopted in practice. The brilliance control may be in the grid or cathode circuit.

Ions

In addition to the electrons, positively and negatively charged ions are also produced in the cathode-ray tube. The former are attracted to the grid and cathode, but the latter travel to the fluorescent screen. They are between 5 000 and 500 000 times heavier than electrons, and though like electrons they are attracted by the first and second anodes, they are much less deflected by magnetic fields. The deflection of electric particles by a magnetic field is directly proportional to the strength and the length of the deflecting field and its distance from the screen, and inversely proportional to the particle velocity. But the latter varies as the square root of the quotient of the final anode voltage and particle mass: hence the heavy negative ions are less deflected, and may destroy the fluorescent property of the screen by bombardment over a small central area. The result is a dark spot referred to as 'ion-burn'. Ion-burn may be avoided by removing the ions from the beam, either by passing the beam through an electrical and then a magnetic field so that the ions are deflected in the first and not in the second (see *Figure 11.4*)

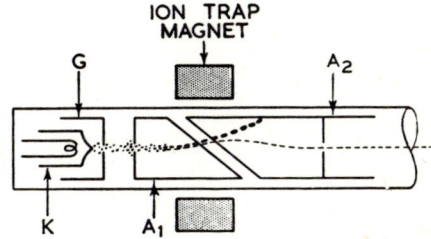

Figure 11.4 'Undeflected-beam' type of ion trap

or by a 'bent-gun' in which the beam is initially projected at about 10 degrees to the tube axis and an 'ion-trap' magnet deflects the electrons back to the tube axis. The ions, however, carry nearly straight on to the second anode (see *Figure 11.5*).

A different treatment of the problem is to prevent the ions in the beam from reaching the fluorescent material by backing it with a very thin metallic film. Aluminium is the usual metal, and considerable success in reducing ion-burn has attended 'aluminising', which was originally introduced mainly to improve brightness and contrast. In some tubes both methods of reducing ion-burn were used.

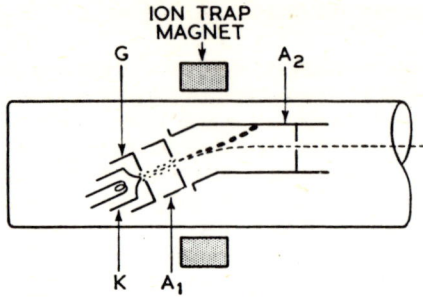

Figure 11.5   'Bent-gun' type of ion trap

Because of the thinner stem used to reduce the power needed in the deflection coils, ion traps are not fitted in 110 degree deflection tubes, these being metal backed.

Aluminising

In a non-backed tube about half the light is visible at the front of the phosphor. The remainder appears at the back, and even with an internal graphite coating some of this light is reflected back to the screen, where it reduces contrast by faintly illuminating the whole screen. With aluminising, the useful output is increased, as the aluminium film reflects the backward output, improving both brightness and contrast.

Tinted screens

Dark- or tinted-glass screens have also been introduced to improve picture contrast when viewing in strong ambient light. Ambient light illuminates the 'black' portions of the screen and reduces contrast. With tinting, the small loss of brightness (as the phosphor is seen through the tinted glass) is compensated by increasing the brilliance control. The 'black-level' seen by the viewer is determined by the ambient light reflected from the screen, and as this has to pass through the tinted glass twice (once to reach the phosphor and once to reach the viewer) the result is a darker 'black' and improved contrast.

Multi-electrode guns and electrostatic focusing

The guns in very early picture tubes were of either triode or tetrode design, but pentode, hexode and heptode types are now used exclusively. These offer the advantage that a smaller beam diameter can be produced in the region of the deflection coils, making for more uniform focusing over the entire area of the screen. In earlier tubes there was often a tendency towards 'deflection defocusing' at the edges of the screen.

With electrostatic focusing a potentiometer-type focus control can be fitted, but, as this method of focusing seldom changes, subsequent readjustment may, instead, be made by altering the connection of the focusing electrode to the tube base, permitting a choice between the fixed potentials represented by 'chassis', the normal h.t. line and the 'boost h.t.' line.

To permit tubes to be produced with shorter necks, a form of pentode gun known as 'tripotential' construction has been developed which allows beam acceleration and focusing to take place simultaneously. Since the focusing of these tubes is more critical than with the longer-neck tubes, a pre-set focusing control is usually provided. The main focusing electrode is the A_2 cylinder (at about 100 V positive to cathode) between the first and final anodes.

Wide-angle deflection tubes

The length of the picture tube is the main factor in determining the depth of the receiver cabinet, and considerable reduction in cabinet size for a given size of picture has been made possible by increasing the deflection angle of the electron beam. In very early tubes this angle was about 50 deg., requiring a long bulbous section of tube, but the angle has been progressively increased, first to about 70 deg., then to 90 deg. and more recently to 110–114 deg. The overall length of a modern 23 in tube is thus less than that of an early 9 in tube.

The main disadvantage of the wide-angle deflection tube is the greater scanning power required, although increases in the efficiency of scanning components and new circuit techniques can provide the extra power without unduly adding to the power dissipation of the timebases.

One means of making a given amount of scan power more effective is to place the deflection coils nearer to the electron beam, and the necks of 110 deg. tubes are narrower than for earlier tubes. For this reason, these tubes have a smaller eight-pin base, known as the B8H. Since these pins are more readily damaged than with B12A bases, care should be taken when removing or replacing the socket.

19 and 23 in wide-angle tubes

The introduction of the 17 and 21 in 110 deg. deflection tubes led to pictures in which a considerable departure from the true flat rectangular shape of the picture had to be tolerated; this was necessary partly to provide a reasonably strong bulb and partly to simplify scanning problems. These tubes inevitably cut the corners of the picture.

Picture tubes were subsequently introduced with the object of providing squarer, flatter pictures. Although the diagonal dimensions of these tubes are increased from 17 to 19 in, and 21 to 23 in, the actual width is increased only a little. Deflection angles are usually between 110 deg. and 114 deg. The 23 in tube has some 20 in^2 of extra screen area compared with wide-angle 21 in tubes and a much greater radius of curvature of the screen to provide flatter pictures.

Aspect ratio of wide-angle tubes

The 110 deg. and 114 deg. wide-angle tubes cannot be used to give 4:3 aspect ratio pictures without undue loss of screen area, and are usually fitted with 5:4 masks. This means that the set should be adjusted for a horizontal overscan of up to 8–9 per cent. To estimate the degree of overscan required, picture height should be set up correctly and then width adjusted to give a true circle in the centre of test card.

Squared-off tubes

Modern developments in glass technology now permit the corners of a picture tube to be 'sharper' than before. This led in 1970 to the introduction of a new range of picture tubes (squared-off tubes) which have the correct picture aspect ratio of 4:3. The increase in corner angle also gives a larger screen diagonal (20 in, 22 in, 24 in, 26 in).

Tube protection

Picture tubes represent a potential danger in that under certain circumstances implosion can occur, the resultant broken glass, moving at considerable velocity, being a very real danger. The traditional method of protecting viewers against this risk is the provision of a glass safety screen at the front of the set. However, to prevent electrostatic dust collecting between the tube screen and the glass guard, to reduce the number of reflections produced by different surfaces, and to make possible different forms of cabinet construction, several alternative methods of providing protection have come into general use in recent years. These include twin-panel tubes, special 'protected' tubes, and the use of plastics guards directly over the tube face. Tubes are today classified as 'protected' or 'unprotected', and it is vitally important not to replace a protected tube with one that requires separate protection.

In twin-panel tubes, a protective glass panel moulded to the contours of the tube face is bonded directly to the tube. In the case of protected tubes, or P-type as they are sometimes called, which include Panorama, Rimband and Rimguard tubes, the protection is built into the tube by using special constructional techniques in which reinforcement is provided at those points of the bulb where implosion could otherwise start. Many twin-panel and protected tubes are provided with mounting lugs so that the tube can be directly mounted in the receiver.

The plastics guards, generally known as Fenbridge Guards or Cornehl Hoods, are made of optical quality flexible p.v.c. some 0·04 in thick, the inner surface being moulded with a 'dew-drop' pattern to prevent adhesion or 'Newton's rings' and the outer surface being semi-polished. These guards are made with different values of light transmission and in different colours in accordance with setmakers' requirements. The two main types are Fenbridge caps, which are fitted to the face of the picture tube by means of a clamp band around the perimeter of the tube face, and Fenbridge polyflex, which is fitted to the cabinet as a flat membrane and follows the tube shape when the tube is inserted into the set.

Small indentations in a Fenbridge guard can be removed by warming with a hot-air blower such as a hair-dryer. Minor scratches can be polished out by using jewellers' rouge or a non-abrasive polish such as Silvo: an abrasive polish must not be used, and the whole area of the screen, not just the damaged area, should be treated. A replacement guard should be fitted in cases of more extensive damage.

The following recommendations on fitting replacement Fenbridge caps are made by Mazda.

Replacing the Picture Tube.   (1) Wait until the new tube is available before removing the faulty one, and wear goggles when handling unprotected tubes. (2) Remove old tube with guard attached, then remove guard from tube. (3) Clean screen of new tube. (4) Clean the inside surface of the Fenbridge guard, removing dust by blowing (a cycle pump is suitable) and other matter with a moistened finger tip— *never use a duster or rag*. (5) Lay cap face downwards on a soft surface on the bench. Lay clamping band on the bench around the guard. Insert the tube screen into the guard, and then pull up the clamp band into position. (6) Tighten the band until it just begins to bite. Tension the guard by pulling on the four corner 'ears' in turn, then on each of the smaller side ears (a hook through the ear eyelet is best). (7) Fully tighten the clamping band and clip the small ears to the band in the manner used by the setmaker concerned. (8) Refit the tube and guards together into the set, and fix the corner mounting lugs to the cabinet. Some setmakers also fit the small ears to the cabinet.

Replacing the Guard.   Remove the picture tube, with damaged guard attached, from the cabinet. Remove guard from tube and clean tube face. Remove replacement guard from anti-shrinking polystyrene former, and warm if necessary to increase flexibility. Proceed as under (5) and (6) above. Use a hot-air blower to shrink out any pockets of non-contact that might remain. Proceed as under (7) and (8) above, clipping ears and refitting tube in set.

As a safety precaution many models using P-type picture tubes are fitted with a high-value resistor (e.g. 2·2 M) between the picture tube clamping straps and chassis.

Flashover protection

With the higher e.h.t. and boost voltages used in modern receivers the risk of flashover between tube electrodes is increased. Many receivers incorporate spark gaps in the form of low-resistance paths to earth on the copper side of a printed board to provide protection against this. An alternative approach is the use of a specially designed tube base moulding, called a Spark-guard, which incorporates within it two critically spaced spark gaps. These gaps are between the first anode and focus electrode pins and a metal plate, which is provided with a separate external connecting tag which is intended to be connected to the external conductive coating of the tube. It is recommended that a 2·2 k (minimum value) isolating resistor is connected in the supply leads to the first anode and focus electrodes. These resistors should be fitted close to the base and be of at least $\frac{1}{2}$ W rating.

A coating of silicon-based insulating lacquer is applied in the shape of a disk around the picture tube final anode connector in some modern tubes to prevent surface leakage across the glass from the connector to the external conductive coating in humid conditions. Such leakage in severe cases can cause interference on sound or vision.

Polaroid screens

A few makers have used Polaroid protective window screens to improve contrast ratio. These filters are part of the implosion guard, which takes the form of a laminate of two sheets sandwiching a thin film of controlled thickness to provide a 45 deg. change of polarisation of light passing through the guard. Whereas light from the screen passes only once through the filter and is little affected, room light must pass through twice and is appreciably attenuated. If one of these screens is removed it must be replaced with the correct side outwards. This can be found by placing a silver coin on the back of the screen: if viewed through the screen from the correct side the coin will have a bluish tint.

Handling cathode-ray tubes

Although the danger of implosion when handling tubes is relatively slight, provided that care is taken not to let the neck strike the chassis or service bench, it *can* occur, and it is advised that precautions to guard against the effects of such an implosion should always be taken. The high vacuum of tubes means that the pressure on the envelope is very high, amounting to more than a ton for relatively small tubes. When a tube is dropped or comes into sharp contact with some other object, the glass may shatter, and be spread in all directions at high velocity. It is therefore highly advisable to wear gloves at all times when handling tubes, and protective goggles.

The tube should always be lifted and carried by placing one hand beneath the face, this hand taking the weight and fulfilling the lifting action. The other hand should be placed on the flare to steady the tube. This avoids placing any strain on the junction between the cone and the neck, which is mechanically the weakest part of the tube. With the now obsolete metal-cone tubes, the supporting hand must be beneath the face and not merely hooked under the lip, as this would impose a strain between the glass face and the metal cone.

The tube must never be placed face downwards on the bare surface of a bench, as this is likely to cause scratches which will mar the picture. It is best to place the tube on a piece of felt or other thick material, or, failing this, a few sheets of paper.

The risk of implosion is greatly increased if the glass surface is scratched or if the rim of a metal cone is knocked. It should also be remembered that the internal and external coatings of all modern tubes form a capacitor, and if the tube is handled when this is in a charged condition, it is possible to receive a shock sufficiently strong to induce the recipient to drop the tube; this risk can be eliminated by always connecting the final anode to chassis before the tube is removed from the cabinet. Also, do not touch the outer coating with damp hands.

Viewers should be protected from implosion by a strong glass or plastics screen (unless the tube is of the protected type): suitable screens include $\frac{1}{4}$ in armour-plate glass (not ordinary plate glass) or $\frac{3}{16}$ in-thick flat perspex sheet.

In time, a film of dust will be formed on the face of the tube by electrostatic attraction. This can be removed by wiping with a soft, slightly moistened cloth; the face must then be dried thoroughly. Anti-static preparations may be used, but the makers' instructions should be carefully followed.

When refitting, never use force when inserting the tube into the deflection yoke, and take care not to tighten tube straps and clamps excessively: remember that while this might not by itself cause the tube to break, a tube subjected to excessive clamp pressure is much more liable to break should it be accidentally struck.

Again, it cannot be overemphasised, *never carry a picture tube by its neck*.

PICTURE-TUBE SALVAGE

The replacement of a picture tube represents a considerable item of expenditure to the average owner, and service engineers are often asked if they can avoid the need to purchase a new tube, even at some cost to picture quality.

Makeshift repairs are often inadvisable, as any temporary improvement may soon be reversed, with the result that a new tube has to be purchased after all, and money spent on extending the life of the original tube will have been wasted. There are, nevertheless, several methods of picture-tube salvage which have become fairly well established in practice, and often give reasonably satisfactory results for a worthwhile period.

With modern tubes the need for this type of work is less; readers may, however, find the following information helpful in dealing with older receivers still in use.

Low-emission tubes

Where the picture has become 'fuzzy' with little contrast or brilliance and with a tendency to turn negative when either of these controls is advanced, this is often a sign of low emission. Two methods of temporary rejuvenation of the cathode emission, provided that the tube is still 'hard', are fairly widely used, although in both systems it should be clearly recognised that there is the risk of the attempted 'cure' causing the complete breakdown of the heater.

The first is to run the tube for a short period with the heater considerably over-run and with all other voltages removed (sometimes a potential of about 100 V is put on the grid), by means of a tapped transformer or auto-transformer. The other method, which has probably been most widely used, is to install a permanent 'boost' transformer or auto-transformer providing from 20 to 50 per cent additional heater voltage. The installation of a boost transformer has frequently extended the useful life of tubes by many months, particularly with the older type of low-voltage heater, although in other cases the life may be prolonged for only a short period; the extra life averages roughly about four to six months.

With the present, relatively low costs of refurbished monochrome picture tubes, this heater booster is barely economic.

Electrode short-circuits and miscellaneous faults

Heater-cathode short-circuits, often of an intermittent nature, are not infrequent, and may result in flashing, uncontrollable brilliance, hum bars or absence of raster. Where a heater-cathode short-circuit has been traced, isolating transformers specially made for this use provide a means of extending the life of the tube. Such transformers must have a very low interwinding capacitance, as otherwise there will be considerable loss of high-frequency video signals (which, with the heater–cathode short-circuit, will appear on the heater line), and hence deterioration of picture quality. The use of an isolating transformer is, of course, practicable only on a.c. mains. With the aid of an isolating transformer, tubes with heater–cathode short-circuits can often be used quite successfully for relatively long periods; in fact, there is little evidence that the life of a defective tube fitted with an isolating transformer differs appreciably from that of a normal tube.

Similar fault symptoms may sometimes be traced to grid-cathode short-circuits. Where the tube has a tetrode gun, it is possible, though often at some cost to picture quality, to strap the grid to the cathode and then to rewire the tube with the first anode acting as control grid.

Intermittent inter-electrode short-circuits often occur only when the tube is at full operating temperature and can sometimes be eliminated by slightly lowering the heater voltage; a simple method with series-connected tubes is to wire a suitable resistor in parallel with the heater of the tube.

It is worth noting that premature tube failures and inter-electrode short-circuits are frequently caused by the over-running of the heater and consequent high cathode temperature. It is recommended, when replacing a faulty tube, to check the heater *voltage* in a parallel-fed receiver and to check the heater *current* in a series-fed receiver. Measurements should be made after checking that the mains-tapping is correctly adjusted. Currents should be within 5 per cent of the rated figure and voltages within 7 per cent.

A picture-tube fault that may occasionally develop in fairly new tubes is a change in the grid cut-off characteristics; this may result in excessive brilliance even with the brilliance control set at minimum. This type of fault can sometimes be overcome by adjusting bias levels; sometimes by simply connecting a high-value resistor (e.g. 10 M) between the first anode and chassis.

Scratched tube faces can often be repolished by the tube manufacturers provided that the scratches are not too deep.

Although not strictly speaking a picture-tube fault, it is worth noting that as picture tubes age any deficiency in e.h.t. voltage tends to produce results akin to that of a 'soft' tube or failing emission, and before deciding that a tube has no further useful life it is advisable to check the e.h.t.

In addition to such work, there are a number of firms who specialise in salvage work for the trade. This includes fitting new cathode assemblies after opening the tube; or alternatively re-activating the cathode followed by a re-vacuuming process by heating the 'getter' with an r.f. heater.

Some firms are also able to offer a refurbished shadowmask tube service.

Picture-tube type numbers

Most picture tubes available in this country are given type numbers which follow the 'European' coding system. The first letter indicates the type of focusing employed, M indicating magnetic and A electrostatic. The second letter indicates the screen properties—the letter W will usually be found, indicating white phosphor. In the latest type numbers this second letter, in accordance with the European standard system, is given at the end of the type number instead of following the initial letter. The first numbers indicate the screen size, measured in centimetres, diagonally for a rectangular screen.

A second set of numbers, separated from the first by means of a hyphen, is a serial number to indicate the particular design. Thus a tube bearing the number A47-13W would be a 19 in electrostatically focused tube with white phosphor screen. The number would formerly have been designated AW47-13.

When the screen's property letter is given as X, this indicates a tri-colour (shadowmask) tube. Thus the A56-120X is a 22 in (56 cm) shadowmask colour tube.

Older Mazda tubes are given numbers commencing CME or CRM. CME numbers indicate electrostatic focusing and magnetic deflection; CRM numbers indicate magnetic focusing and deflection. The first two figures of the number indicate screen size in inches. Thus the CME1902 tube is a 19 in one with electrostatic focusing and magnetic deflection.

The tables that follow give basic data on monochrome and colour picture tubes. The tables give only replacement and current types. For obsolete tubes an equivalent may be found in Table 11.3. For very old tubes there may be no equivalent or method of using any alternative tube.

For complete data on any tube the reader should contact the manufacturer.

Table 11.1 MONOCHROME PICTURE TUBE DATA (THIS LIST CONTAINS CURRENT AND REPLACEMENT TYPES ONLY)

Type	Size (in)	Deflection Angle (deg)	Heater Volts	Heater Amps	Final Anode kV (max)	First Anode kV (max)	Base (see end of Table)	Notes (see end of Table)
BRIMAR								
A47–13W	19	110	6·3	0·3	20	0·7	B8H (1)	A, E, M, R
A47–17W	19	110	6·3	0·3	20	0·7	B8H (1)	A, E, M, R, RG
A47–25W	19	110	6·3	0·3	20	0·7	B8H (1)	A, E, M, R, RG
A59–12W	23	110	6·3	0·3	20	0·7	B8H (1)	A, E, M, R, RG
A59–13W	23	110	6·3	0·3	20	0·7	B8H (1)	A, E, M, R
A59–25W	23	110	6·3	0·3	20	0·7	B8H (1)	A, E, M, R, RG
AW47–90	19	110	6·3	0·3	17	0·5	B8H (1)	A, E, M, R
AW47–91	19	110	6·3	0·3	20	0·7	B8H (1)	A, E, M, R
C17AA	17	110	6·3	0·3	17·6	0·5	B8H (2)	A, E, M, R
C17AF	17	110	4	0·3	17·6	0·75	B8H (2)	A, E, M, R
C17FM	17	70	12·6	0·3	17·5	0·41	B12A (4)	A, IT, M, R
C17PM	17	70	6·3	0·3	18	0·5	B12A (6)	A, E, IT, M, R
C17SM	17	90	6·3	0·3	18	0·5	B12A (6)	A, E, M, R
C19AH	19	114	4	0·3	16·5	0·7	B8H (2)	A, E, M, R
C19AK	19	110	6·3	0·3	16	0·5	B8H (1)	A, E, M, R
C21AA	21	110	6·3	0·3	16	0·5	B8H (1)	A, E, M, R
C21AF	21	110	4	0·3	17·6	0·75	B8H (2)	A, E, M, R
C21TM	21	90	12·6	0·3	20	0·5	B12A (4)	A, IT, M, R
C23AK	23	110	6·3	0·3	16	0·5	B8H (1)	A, E, M, R
EMISCOPE								
5/3T	17	70	8·5	0·3	17	0·6	B7B	A, E, M, R
7205T	14	90	12·6	0·3	14	0·4	B12A (2)	A, E, IT, M, R
7404A	17	70	12·6	0·3	16	0·4	B12A (1)	A, IT, R
7405A	17	110	12·6	0·3	16	0·4	B8H (1)	A, E, M, R
7406A	17	110	12·6	0·3	16	0·5	B8H (2)	A, E, M, R
7502A	21	90	12·6	0·3	20	0·4	B12A (1)	A, IT, M, R
7503A	21	110	12·6	0·3	16	0·4	B8H (1)	A, E, M, R
7504A	21	110	12·6	0·3	18	0·5	B8H (2)	A, E, M, R
7601A	19	110	12·6	0·3	17	0·5	B8H (1)	A, E, M, R
7701A	23	110	12·6	0·3	17	0·5	B8H (1)	A, E, M, R
A47–13W	19	110	6·3	0·3	18	0·55	B8H (1)	A, E, M, R
A47–14W	19	110	6·3	0·3	20	0·4	B8H (1)	A, E, M, R
A59–11W	23	110	6·3	0·3	18	0·7	B8H (1)	A, E, M, R
A59–15W	23	110	6·3	0·3	20	0·7	B8H (1)	A, E, M, R
A59–16W	23	110	6·3	0·3	18	0·4	B8H (1)	A, E, M, R
AW43–80	17	85	6·3	0·3	16	0·5	B12A (7)	A, E, IT, M, R
AW43–88	17	110	6·3	0·3	16	0·5	B8H (1)	A, E, M, R
AW53–88	21	110	6·3	0·3	16	0·4	B8H (1)	A, E, M, R
MW36–44	14	65	6·3	0·3	14	0·41	B12A (5)	IT, M, R
MW43–69	17	65	6·3	0·3	16	0·41	B12A (5)	A, IT, M, R
MW53–80	21	85	6·3	0·3	18	0·5	B12A (5)	A, IT, M, R
SE17–70	17	70	6·3	0·3	18	0·5	B12A (6)	A, E, IT, M, R
MAZDA								
A47–13W	19	110	6·3	0·3	20	0·7	B8H (1)	A, E, M, R
A47–14W	19	110	6·3	0·3	20	0·7	B8H (1)	A, E, M, R

Table 11.1   *Continued*

Type	Size (in)	Deflection Angle (deg)	Heater Volts	Heater Amps	Final Anode kV (max)	First Anode kV (max)	Base (see end of Table)	Notes (see end of Table)
A47–17W	19	110	6·3	0·3	20	0·7	B8H (1)	A, E, M, R, RG
A59–12W	23	110	6·3	0·3	20	0·7	B8H (1)	A, E, M, R, RG
A59–13W	23	110	6·3	0·3	20	0·7	B8H (1)	A, E, M, R
A59–15W	23	110	6·3	0·3	20	0·7	B8H (1)	A, E, M, R
AW47–90	19	110	6·3	0·3	17	0·5	B8H (1)	A, E, M, R
AW47–91	19	110	6·3	0·3	20	0·7	B8H (1)	A, E, M, R
AW59–90	23	110	6·3	0·3	17	0·5	B8H (1)	A, E, M, R
CME1101	11	110	6·3	0·3	15	0·55	B8H (1)	A, E, M, R, RG
CME1201	12	110	6·3	0·3	13·5	0·55	B8H (1)	A, E, M, R, RG
CME1601	16	110	6·3	0·3	17	0·55	B8H (1)	A, E, M, R
CME1602	16	110	6·3	0·3	17	0·55	B8H (1)	A, E, M, R, RG
CME1702	17	90	12·6	0·3	16	0·4	B12A (2)	A, E, M, R
CME1703	17	110	12·6	0·3	16	0·4	B8H (1)	A, E, M, R
CME1705	17	110	12·6	0·3	16	0·5	B8H (2)	A, E, M, R
CME1713	17	110	6·3	0·3	18	0·7	B8H (1)	A, E, M, R, RG
CME1902	19	110	6·3	0·3	17	0·5	B8H (1)	A, E, M, R
CME1903	19	110	6·3	0·3	20	0·7	B8H (1)	A, E, M, R
CME1905	19	110	6·3	0·3	20	0·7	B8H (1)	A, E, M, R, RG
CME1906	19	110	6·3	0·3	20	0·7	B8H (1)	A, E, M, R
CME1907	19	110	6·3	0·3	20	0·7	B8H (1)	A, E, M, R, RG
CME1908	19	110	6·3	0·3	20	0·7	B8H (1)	A, E, M, R
CME1913	19	110	6·3	0·3	20	0·7	B8H (1)	A, E, M, R, RG
CME2013	20	110	6·3	0·3	20	0·7	B8H (1)	A, E, M, R, RG
CME2101	21	110	12·6	0·3	16	0·4	B8H (1)	A, E, M, R
CME2104	21	110	12·6	0·3	18	0·5	B8H (2)	A, E, M, R
CME2301	23	110	12·6	0·3	17	0·5	B8H (1)	A, E, M, R
CME2302	23	110	6·3	0·3	17	0·5	B8H (1)	A, E, M, R
CME2305	23	110	6·3	0·3	20	0·7	B8H (1)	A, E, M, R, RG
CME2306	23	110	6·3	0·3	20	0·7	B8H (1)	A, E, M, R
CME2308	23	110	6·3	0·3	20	0·7	B8H (1)	A, E, M, R
CME2312	23	110	6·3	0·3	20	0·7	B8H (1)	A, E, M, R, RG
CME2313	23	110	6·3	0·3	20	0·7	B8H (1)	A, E, M, R, RG
CME2413	24	110	6·3	0·3	20	0·7	B8H (1)	A, E, M, R, RG
CME2501	25	110	6·3	0·3	20	0·7	B8H (1)	A, E, M, R, RG
CRM141	13·5	67	12·6	0·3	14	0·4	B12A (1)	A, IT
CRM142	13·5	67	12·6	0·3	14	0·4	B12A (1)	A, IT
CRM171	17	70	12·6	0·3	16	0·4	B12A (1)	A, IT, R
CRM172	17	70	12·6	0·3	16	0·4	B12A (1)	A, IT, M, R
CRM173	17	90	12·6	0·3	16	0·4	B12A (1)	A, IT, M, R
CRM174	17	70	12·6	0·3	16	0·4	B12A (1)	A, IT, M, R
CRM211	21	70	12·6	0·3	16	0·4	B12A (1)	A, IT, M, R
CRM212	21	90	12·6	0·3	20	0·4	B12A (1)	A, IT, M, R
MULLARD								
A28–14W	11	90	11·0	0·07	12	0·35	7-pin	A, E, M, R
A44–120W/R	17	110	6·3	0·3	18	0·7	B8H (1)	A, E, M, R
A47–11W	19	110	6·3	0·3	18	0·7	B8H (1)	A, E, M, R
A47–13W	19	110	6·3	0·3	18	0·55	B8H (1)	A, E, M, R
A47–14W	19	110	6·3	0·3	20	0·4	B8H (1)	A, E, M, R

Table 11.1 *Continued*

Type	Size (in)	Deflection Angle (deg)	Heater Volts	Heater Amps	Final Anode kV (max)	First Anode kV (max)	Base (see end of Table)	Notes (see end of Table)
A47–26W	19	110	6·3	0·3	20	0·4	B8H (1)	A, E, M, R
A47–26W/R	19	110	6·3	0·3	20	0·7	B8H (1)	A, E, M, R
A50–120W/R	20	110	6·3	0·3	20	0·7	B8H (1)	A, E, M. R
A59–11W	23	110	6·3	0·3	18	0·7	B8H (1)	A, E, M, R
A59–15W	23	110	6·3	0·3	20	0·7	B8H (1)	A, E, M, R
A59–16W	23	110	6·3	0·3	18	0·4	B8H (1)	A, E, M, R
A59–23W	23	110	6·3	0·3	20	0·4	B8H (1)	A, E, M, R
A59–23W/R	23	110	6·3	0·3	20	0·7	B8H (1)	A, E, M, R
A65–11W	25	110	6·3	0·3	20	0·4	B8H (1)	A, E, M, R
AW21–11	8·5	90	11·5	0·06	16	0·8	B8H (1)	A, E, M, R
AW43–80Z	17	90	6·3	0·3	16	0·3	B12A (3)	A, M, R
AW43–88	17	110	6·3	0·3	16	0·5	B8H (1)	A, E, M, R
AW43–89	17	110	6·3	0·3	16	0·7	B8H (2)	A, E, M, R
AW53–80	21	85	6·3	0·3	16	0·5	B12A (7)	A, E, IT, M, R
AW53–88	21	100	6·3	0·3	16	0·5	B8H (1)	A, E, M, R
MW36–44	14	65	6·3	0·3	14	0·41	B12A (5)	IT, M, R
MW43–69Z	17	70	6·3	0·3	16	0·41	B12A (5)	M, R
MW43–80Z	17	90	6·3	0·3	16	0·41	B12A (5)	A, M, R
MW53–80	21	85	6·3	0·3	18	0·5	B12A (5)	A, IT, M, R

Notes on table: A = Aluminised screen
E = Electrostatic focusing
IT = Ion trap

M = External conductive coating
R = Rectangular screen
RG = Ringuard protected tube

BASES: (viewed from beneath, number refers to Table).

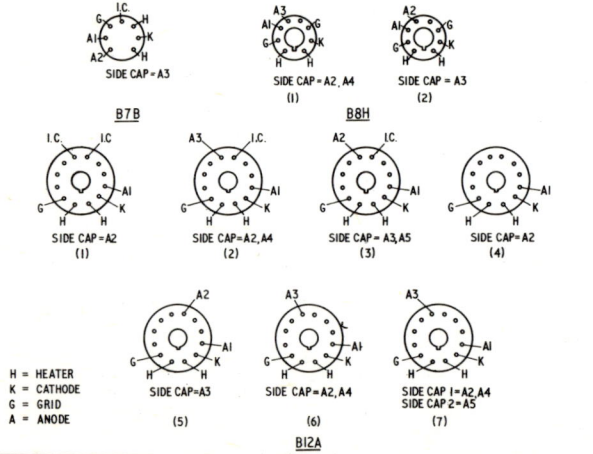

H = HEATER
K = CATHODE
G = GRID
A = ANODE

**Table 11.2   COLOUR PICTURE TUBE DATA
(CURRENT TYPES ONLY)**

Type	Size (in)	Angle (deg)	Heater Volts	Heater Amps	Final Anode kV (max)	First Anode kV (max)	(see end of Table)	Notes (see end of Table)
MAZDA								
A49–11X	19	90	6·3	0·9	27·5	1	B14G	RG
A49–200X	19	90	6·3	0·9	27·5	1	B14G	RG
A55–141X	22	90	6·3	0·9	27·5	1	B14G	RG
A56–120X	22	90	6·3	0·9	27·5	1	B14G	RG
A63–200X	25	90	6·3	0·9	27·5	1	B14G	RG
A63–11X	25	90	6·3	0·9	27·5	1	B14G	RG
MULLARD								
A49–11X	19	90	6·3	0·9	25	0·500	B12–244	—
A49–120X	19	90	6·3	0·9	25	0·500	B12–244	—
A56–120X	22	92	6·3	0·9	25	0·500	B12–244	—
A63–11X	25	90	6·3	0·9	25	0·500	B12–244	—
A63–120X	25	90	6·3	0·9	25	0·500	B12–244	—
A66–120X	26	92	6·3	0·9	25	0·500	B12–244	—
R.C.A.								
A49–15X	19	90	6·3	0·9	27·5	0·685	B14G	—
A49–191X	19	90	6·3	0·9	27·5	0·685	B14G	—
A55–14X	22	90	6·3	0·9	27·5	0·685	B14G	—
A63–17X	25	90	6·3	0·9	27·5	0·685	B14G	—
A63–200X	25	90	6·3	0·9	27·5	0·685	B14G	—

Notes on table: All above are shadowmask tubes with rectangular screen, electrostatic focusing and
external conductive coating.
    RG = Rimguard protected tube.
BASES: (pin connections viewed from beneath—connections similar on both B14G and B12–244
    bases but not interchangeable safely).

H = HEATER
G = GRID
K = CATHODE
A = ANODE
(R) = RED
(G) = GREEN
(B) = BLUE

SIDE CAP = A3, A4

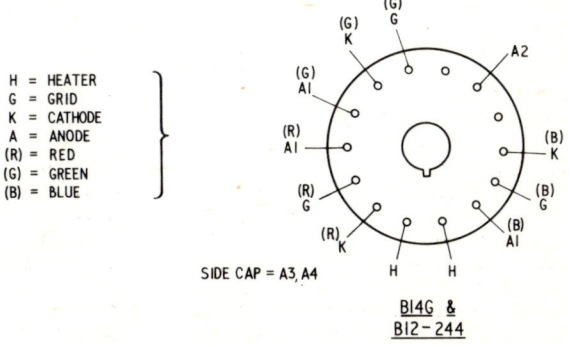

B14G &
B12–244

Table 11.3 CATHODE-RAY TUBE EQUIVALENTS

Type	Equivalents*	Type	Equivalents*
A31–18W	CME1201	AW47–91	CME1903, C19/10A, A47–14W
A40–11W	CME1601		
A47–11W	A47–26W	AW47–97	CME1901, 7601A, A47–14W
A47–13W	CME1906, C19/10AP, A47–26W	AW53–80	21CLP4
A47–14W	CME1908	AW53–88	C21AA, C21/7A
A47–15W	A47–14W	AW59–90	C23AK, CME2302, A59–15W
A47–17W	CME1905, A47–26W		
A47–17W	A47–26W	AW59–91	CME2303, C23/10A, A59–15W
A47–18W	A47–26W		
A47–27W	A47–26W	AW59–95	CME2301, A59–15W
A47–28W	A47–26W	C9A	CRM92, CRM92A
A47–28W/R	A47–26W/R	C12/1	C12FM, 121K, 12XP4A, MW31–74
A49–11X	A49–120X		
A49–15X	A49–120X		
A49–18X	A49–120X	C12A	CRM121, CRM121A, CRM121B
A49–191X	A49–120X		
A49–200X	A49–120X	C12B	12MW3A
A59–11W	A59–23W	C12D	12MW3
A59–12W	CME2305, A59–23W	C12FM	121K, 12XP4, MW31–16, MW31–74, P12, T12/100, C12/1
A59–13W	CME2306, C23/10AP, A59–23W		
A59–14W	CME2307, C23AKT, A59–23W	C14/3A	C14PM, SE14/70, AW36–20
A59–15W	CME2308	C14PM	C14/3A, SE14/70, AW36–20
A59–16W	A59–23W		
A59–25W	A59–23W	C15B	15MW3A
A63–11X	A63–120X	C17/1	171K, 17ASP4, 17AXP4, TR17/21, MW43–64
A63–200X	A63–120X		
A65–11W	CME2501		
AW36–20	C14PM, SE14/70, C14/3A	C17–1A	MW43–69, TR17/22
		C17–2	MW43–69
AW36–21	P14ES	C17/4A	MW43–80
AW43–80	P17EN, 17BTP4, 174K, C17/5A	C17/5A	AW43–80
		C17/7A	AW43–88, C17AA
AW43–80Z	C17SM	C17AA	AW43–88, C17/7A
AW43–88	C17AA, P17EH, C17/7A	C17AF	AW43–89
		C17FM	CRM174
AW47–90	C19AK, CME1902, C19/7A, A47–14W	C175M	AW43–80Z

Table 11.3 *Continued*

Type	Equivalents*	Type	Equivalents*
C19AK	AW47–90, CME1902, C19/7A	CME2101	7503A
		CME2301	7701A, AW59–95
C19/7A	CME1902, AW47–90, C19AK	CME2302	AW59–90, C23AK, C23
C19/10A	CME1903, AW47–91	CME2303	AW59–91, C23/10A
C19/10AP	CME1906, A47–13W	CME2305	A59–12W, A59–11W
C21/1A	MW53–80	CME2306	A59–13W, C23/10AP, A59–16W
C21/7A	AW53–88, C21AA		
C21AA	AW53–88, C21/7A	CME2307	A59–14W, C23AKT
C21KM	MW53–80	CME2308	A59–15W
C21NM	MW53–20, 21CJP4	CME2312	A59–25W
C21SM	AW53–80	CME2313	A59–23W
C21TM	CRM212, 7502A, MW53–80	CME2413	A61–120W
		CME2501	A65–11W
C23/7A	CME2302, AW59–90, C23AK	CRM71	MW18–2
		CRM91	MW22–3, C9A
C23/10A	CME2303, AW59–91	CRM92, A	MW22–3, C9A
C23/10AP	CME2306, A59–13W	CRM121, A, B	C12A, CRM121
C23AK	CME2302, AW59–90, C23/7A	CRM144	7204A
C23AKT	CME2307, A59–14W	CRM172	7404A
CME1201	A31–18W	CRM212	7502A
CME1402	7205A	MW6–2	3NP4
CME1601	A40–11W	MW18–2	CRM71
CME1602	A40–12W	MW22–3	CRM92
CME1703	7405A, AW43–88	MW31–16	C12FM, 121K, 12XP4, MW31–74, C12/1
CME1705	7406A, AW43–89		
CME1706	AW43–88	MW31–18	112K
CME1713	A44–121W	MW31–74	C12FM, 121K, 12XP4A, C12/1
CME1713R	A44–120W/R		
CME1901	7601A, AW47–97	MW36–24	141K, 14KP4A, 14LP4, C36–24
CME1902	AW47–90, C19AK, C19/7A		
		MW41–1	T901, T901A
CME1903	AW47–91, C19/10A	MW43–64	172K, MW43–69, C17/1, TR17/21
CME1905	A47–17W		
CME1906	A47–13W, C19/10AP	MW43–69	C17–2, 172K, 173K, P17A, TR17/22
CME1907	A47–25W		
CME1908	A47–14W	MW53–80	C21KM, C21/1A, 212K, TR21/22
CME1913	A47–28W		
CME2013	A50–120W	MW53–88	C21KM

Table 11.3   *Continued*

Type	Equivalents*	Type	Equivalents*
P12	MW31–16, MW31–74, T12/100, C12FM, 12XP4A, 121K	14KP4A	141K, MW36–24, C36–24
		15MW3A	C15B
P14	C14BM, TR14/13, 7201A	17ASP4	171K, C17/1
		17AXP4	171K (near)
P14ES	AW36–21	112K	MW31–18, C12FM, 12XP4
P17	MW43–64, TR17/21, 17ASP4, 172K	121K	MW31–74, etc.
P17A	MW43–69, C17–1A, 172K, 173K	141K	MW36–24, P141, 14KP4A, C36–24
P17AM	CRM172, 7404A	171K	17ASP4, C17/1
P17EH	AW43–88, C17AA	172K	MW43–64, MW43–69, C17–2, P17A
P17EN	AW43–80, 174K		
P17MN	MW43–80	173K	MW43–69, C17–2, P17A
P141	141K, 14KP4, 14KP4A		
SE14/70	C14PM, C14/3A	212K	MW53–80, C21/1A
SE17/70	C17PM	7204A	CRM144
T12/2	C12D	7205A	CME1402
T12/100	MW31–74	7404A	CRM172
T901, A	MW41–1	7405A	CME1703, AW43–88
12MW3	C12D, T12/46	7406A	CME1705, AW43–89
12MW3A	C12B	7502A	CRM212, MW53–80
12XP4	C12FM, MW31–16, MW31–74	7503A	CME2101, AW53–88
		7504A	CME2104
12XP4A	MW31–74, 121K, C12/1	7601A	CME1901, A47–14W
		7701A	CME2301, A59–15W

*Note. Not all types shown are direct equivalents and many require a tube manufacturer's modification kit or instructions.

12

COLOUR CODES AND USEFUL ADDRESSES

Resistors

The information on resistors given by current colour coding systems includes value, tolerance and grade. These characteristics are indicated either (1) by a series of three or more colour rings which are read from the end of the resistor towards its centre (*Figure 12.1*); or, alternatively, (2) by reading first the body colour; secondly, the tip colour; thirdly, the spot or band colour (*Figure 12.2*). In system (2) the fourth colour (tolerance) is indicated by marking the second tip but, since the colours normally differ from those used to designate value, no confusion is likely to arise.

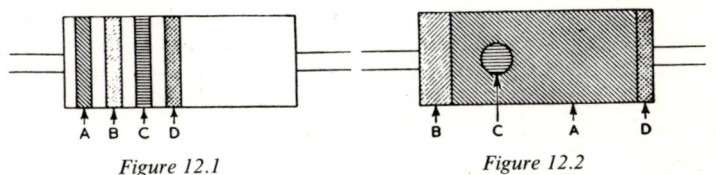

Figure 12.1   Figure 12.2

In each system, the first colour to be read indicates the first figure of the value; the second colour gives the second figure of the value; and the third colour gives the number by which the first two figures should be multiplied in order to arrive at the true value of the resistor. The fourth colour shows the tolerance; the accepted tolerances being $\pm$ 1 per cent, $\pm$ 2 per cent, $\pm$ 5 per cent, $\pm$ 10 per cent and $\pm$ 20 per cent. Where no tolerance is indicated, it may be assumed that the tolerance is $\pm$ 20 per cent.

Grade 1, high-stability, composition resistors are coded as (1) above, the grade being denoted by either a fifth band of salmon pink, or the body being of that colour.

Table 12.1 RESISTOR COLOUR CODE

Colour	1st Figure (A)	2nd Figure (B)	Multiplier (T)	Tolerance (D)
Black . .	—	0	1	—
Brown . .	1	1	10	$\pm 1\%$
Red . .	2	2	100	$\pm 2\%$
Orange .	3	3	1 000	—
Yellow .	4	4	10 000	—
Green . .	5	5	100 000	—
Blue . .	6	6	1 000 000	—
Violet . .	7	7	10 000 000	—
Grey . .	8	8	100 000 000	—
White . .	9	9	1 000 000 000	—
Gold . .	—	—	0·1	$\pm 5\%$
Silver . .	—	—	0·01	$\pm 10\%$
No colour .	—	—	—	$\pm 20\%$

Examples.   A resistor with a blue body, a grey tip and an orange spot would have a value of 68 000 Ω with a tolerance of ± 20 per cent. The addition of a silver band or tip would indicate a tolerance of ± 10 per cent.

A resistor with four bands of colour, the end one being orange, the next orange, followed by brown and gold would have a value of 330 Ω with a tolerance of ± 5 per cent. In this case the body colour would have no significance, unless salmon pink, which would indicate a Grade 1 resistor.

Capacitors

Although many capacitors continue to be marked directly with their value and rating, several systems of colour coding are also in use. These differ according to the type of capacitor and the extent of the information to be conveyed, though in all cases the same basic code to that used for resistors is adopted, except for the 0·1 and 0·01 multipliers. Information that may be shown by colour coding includes: value, temperature coefficient, tolerance and voltage rating. In addition, the connection to the outer foil of tubular paper capacitors may be indicated by a band of colour, usually black, being placed on the casing close to the appropriate connection. All values are colour coded in picofarads (to convert to microfarads divide by 1 000 000).

Ceramic Dielectric. There are two preferred methods in use, the five-band or dot colour code (Table 12.2) and the six-band or dot colour code (Table 12.3).

Tubular, Metallised-paper The values may be colour coded in picofarads, indicated by three dots, having the same significance as in the third, fourth and fifth columns in Table 12.2 for ceramic dielectric capacitors.

Alternative Methods While the above systems are those recommended for current usage, several other methods may be met in

Table 12.2 CERAMIC CAPACITOR FIVE-BAND COLOUR CODE

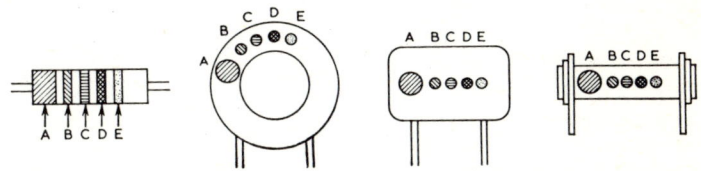

Colour	Temperature Coefficient		Capacitance				
	Rated Value Tolerance (A)		First Figure (B)	Second Figure (C)	Multi-plier (D)	Tolerance (E)	
						> 10 pF	⩽ 10 pF
Black .	0	±30	—	0	1	±20%	—
Brown .	−33	±30	1	1	10	±1%	±0·1 pF
Red . .	−75	±30	2	2	10^2	±2%	±0·25 pF
Orange .	−150	±30	3	3	10^3	—	—
Yellow .	−220	±30	4	4	10^4	—	—
Green .	−330	±60	5	5	—	±5%	—
Blue .	−470	±90	6	6	—	—	—
Violet .	−750	±120	7	7	—	—	—
Grey .	—	—	8	8	10^{-2}	—	—
White	−330	±500	9	9	10^{-1}	±10%	±1 pF

Note. If it is inconvenient to apply the colour code, the temperature coefficient should be indicated by a single body colour in accordance with this code, and the capacitance and tolerance given in figures.

practice. For example, one colour only may be used to denote tolerance, two colours to denote tolerance and voltage rating, three colours to denote capacitance in picofarads, five colours to denote capacitance in picofarads (first three colours), tolerance (fourth dot) and voltage rating (fifth dot). The order in which the dots are to be read is sometimes indicated by an arrow, but in all cases is from left to right, the first dot being that nearest to one end.

In such instances, tolerance and voltage rating are coded as Table 12.4.

Coding of American capacitors also differs slightly from that described above: the RMA three-dot code is used for capacitors

Table 12.3 CERAMIC CAPACITOR SIX-BAND COLOUR CODE

Colour	Temperature Coefficient			Capacitance				
	Significant Figure		Multiplier (B)	First Figure (C)	Second Figure (D)	Multiplier (E)	Tolerance (F)	
	Value Tolerance* (A)						> 10 pF	≤ 10 pF
Black .	—	—	− 1	—	0	1	±20%	—
Brown .	—	—	− 10	1	1	10	±1%	±0·1 pF
Red .	1·0	±0·15	− 10²	2	2	10²	±2%	±0·25 pF
Orange .	1·5	±0·25	− 10³	3	3	10³	—	—
Yellow .	2·2	±0·35	− 10⁴	4	4	10⁴	—	—
Green .	3·3	±0·6	− 1	5	5	—	±5%	—
Blue .	4·7	±0·9	—	6	6	—	—	—
Violet .	7·5	±1·2	10²	7	7	—	—	—
Grey .	—	—	10³	8	8	10⁻²	—	—
White .	—	—	10⁴	9	9	10⁻¹	±10%	±1 pF

*Or ± 40 parts per million per deg. C, whichever is the greater.

Table 12.4

Colour	Tolerance, %	Voltage Rating
Black	—	—
Brown	1	100
Red	2	200
Orange	3	300
Yellow	4	400
Green	5	500
Blue	6	600
Violet	7	700
Grey	8	800
White	9	1 000
Silver	10	—
Gold	5	—

having a tolerance of 20 per cent, the dots indicating the capacitance in picofarads; the RMA six-dot code gives (*top row*) first, second and third significant figures; (*bottom row*) voltage rating, tolerance and decimal multiplier. American fixed ceramic capacitors have a broad band followed by four narrow bands or dots giving temperature co-efficient, first significant figure, second significant figure, decimal multiplier and tolerance, this system being similar to that described for British capacitors of this type.

USEFUL ADDRESSES

Broadcasting Authorities

British Broadcasting Corporation
Broadcasting House
Portland Place
LONDON, W1A 1AA   Tel: 01-580 4468

Independent Broadcasting Authority (formerly the ITA)
70 Brompton Road
LONDON, SW3   Tel: 01-584 7011

Radio Telefís Éireann
Donnybrook
DUBLIN, 4   Tel: Dublin 69311

Service Depots – U.K. Receiver Manufacturers

Rank-Bush-Murphy   (Bush, Murphy)
Drayton Road
BOREHAMWOOD, Hertfordshire   Tel: 01-953 6151

British Radio Corporation   (Thorn, Ferguson, Ultra, etc.)
PO Box 121
Lea Valley Trading Estate
Angel Road
Edmonton
LONDON, N18 3SP   Tel: 01-807 3060

Decca Radio & Television Ltd   (Decca)
Ingate Place
Queenstown Place
Queenstown Road
LONDON, SW8   Tel: 01-622 6677

Combined Electronic Services Ltd   (Philips, Pye, etc.)
604 Purley Way
Waddon
CROYDON, CR9 4DR   Tel: 01-686 0505

GEC Radio and Television Group   (GEC, Sobell, etc.)
East Lane
WEMBLEY, Middlesex   Tel: 01-904 4388

ITT Consumer Products Services Ltd   (KB, STC)
PADDOCK WOOD, Kent   Tel: 089-283 4422

Some other Organisations

British Radio Equipment Manufacturers' Association (BREMA)
49 Russell Square
LONDON, WC1   Tel: 01-580 3586

British Radio Valve Manufacturer' Association (BVA)
156-162 Oxford Street
LONDON, W1   Tel: 01-580 8562

City and Guilds of London Institute
76 Portland Place
LONDON, W1   Tel: 01-580 3050

Electronic Engineering Association
Berkeley Square House
Berkeley Square
LONDON, W1   Tel: 01-499 4501

Electronic Valve and Semiconductor Manufacturers' Associations
(VASCA)
156-162 Oxford Street
LONDON, W1   Tel: 01-580 8562

Institution of Electrical and Electronics Technician Engineers
26 Bloomsbury Square
LONDON, WC1   Tel: 01-580 5927

Radio & Television Retailers' Association
19 Conway Street
Fitzroy Square
LONDON, W1   Tel: 01-387 6046

Radio Trades Examination Board
33 Bedford Street
LONDON, WC2   Tel: 01-240 0926

Royal Television Society
166 Shaftesbury Avenue
LONDON, WC2   Tel: 01-836 3330

INDEX

365